T0276823

Mobile Robotics

Mobile Robotics

Edited by **Jared Kroff**

New Jersey

Published by Clanrye International,
55 Van Reypen Street,
Jersey City, NJ 07306, USA
www.clanryeinternational.com

Mobile Robotics
Edited by Jared Kroff

International Standard Book Number: 978-1-63240-352-0 (Hardback)

Printed in the United States of America.

Contents

Preface

Mobile robotics is gradually becoming a stimulating area because of the efforts of various researchers. As numerous researches are being conducted in the field of robotics; and mobile robots have always formed an essential part of such researches. Mobile robotics is a growing integrative field associated with several other fields like mechanical, electronic and electrical engineering, cognitive, social and computer sciences. It is associated with the computational vision, robotics and artificial intelligence because of being involved in the design of automated systems. This book covers numerous topics by integrating contributions from many researchers around the globe. It lays stress on the computational techniques of programming mobile robots. It consists of two sections: "Mobile Robots Navigation" and "Methods for Control". It aims to serve as a guide to students and researchers studying mobile robotics.

After months of intensive research and writing, this book is the end result of all who devoted their time and efforts in the initiation and progress of this book. It will surely be a source of reference in enhancing the required knowledge of the new developments in the area. During the course of developing this book, certain measures such as accuracy, authenticity and research focused analytical studies were given preference in order to produce a comprehensive book in the area of study.

This book would not have been possible without the efforts of the authors and the publisher. I extend my sincere thanks to them. Secondly, I express my gratitude to my family and well-wishers. And most importantly, I thank my students for constantly expressing their willingness and curiosity in enhancing their knowledge in the field, which encourages me to take up further research projects for the advancement of the area.

Editor

Part 1

Mobile Robots Navigation

1

Fuzzy Logic Based Navigation of Mobile Robots

Amur S. Al Yahmedi and Muhammed A. Fatmi
Sultan Qaboos University,
Oman

1. Introduction

Robots are no longer confined to engineered, well protected sealed corners, but they are currently "employed" in places closer and closer to "us". Robots are getting out of factories and are finding their way into our homes and to populated places such as, museum halls, office buildings, schools, airports, shopping malls and hospitals.

The gained benefit of the potential service and personal robots comes along with the necessity to design the robot in a way that makes it safe for it to interact with humans and in a way that makes it able to respond to a list of complex situations. This includes at least the possibility to have the robot situated in an unknown, unstructured and dynamic environment and to navigate its way in such an environment. One of the fundamental issues to be addressed in autonomous robotic system is the ability to move without collision. An "intelligent" robot should avoid undesirable and potentially dangerous impact with objects in its environment. This simple capability has been the subject of interest in robotic research.

Behavior based navigation systems (Arkin, 1987, 1989; Arkin & Balch, 1997; AlYahmedi et al., 2009; Brooks, 1986, 1989; Fatmi et al. 2006 and Ching-Chih et al. 2010) have been developed as an alternative to the more traditional strategy of constructing representation of the world and then reasoning prior to acting. The main idea of behavior based navigation is to identify different responses (behaviors) to sensory inputs. For example, a behavior could be "avoiding obstacles" in which sonar information about a close obstacle should result in a movement away from the obstacle. A given set of behaviors is then blended in a certain way to produce either a trade off behavior or a more complex behavior. However, a number of issues with regard to behavior based navigation are still under investigation. These issues range from questions concerning the design of individual behaviors to behavior coordination issues, to intelligently improve "behaviors" through learning.

An important problem in autonomous navigation is the need to deal with the large amount of uncertainties of the sensory information received by the robot which is incomplete and approximate as well as with the fact that the environment in which such robots operate contains dynamics and variability elements.

A fuzzy logic behavior based navigation approach is introduced in this chapter in order to deal with the uncertainty and ambiguity of the information the system receives. Issues of individual behavior design and action coordination of the behaviors will be addressed using fuzzy logic.

The approach described herein, consists of the following four tasks,

- The use of fuzzy sets to represent the approximate positions and possibly shapes of objects in the environment.
- The design of simple fuzzy behaviors (avoiding obstacles, goal reaching, wall following...etc.).
- The blending of the different fuzzy behaviors.

2. Behavior based navigation

One of the long standing challenging aspect in mobile robotics is the ability to navigate autonomously, avoiding modeled and unmodeled obstacles especially in crowded and unpredictably changing environment. A successful way of structuring the navigation task in order to deal with the problem is within behavior based navigation approaches (Arkin, 1987, 1989; Arkin & Balch, 1997; AlYahmedi et al., 2009; Brooks, 1986, 1989; Fatmi et al. 2006; Ching-Chih et al. 2010; Maes, 1990; Mataric, 1997; Rosenblatt et al. 1989, 1994, 1995; Saffiotti, 1997 and Seraji & Howard, 2002).

2.1 Introduction

The basic idea in behavior based navigation is to subdivide the navigation task into small easy to manage, program and debug behaviors (simpler well defined actions) that focus on execution of specific subtasks. For example, basic behaviors could be "avoid obstacles" or "moving to a predefined position". This divide-and-conquer approach has turned out to be a successful approach, for it makes the system modular, which both simplifies the navigation solution as well as offers a possibility to add new behaviors to the system without causing any major increase in complexity. The suggested outputs from each concurrently active behaviors are then "blended" together according to some action coordination rule. The task then reduces to that of coupling actuators to sensory inputs, with desired robot behaviors. Each behavior can take inputs from the robot's sensors (e.g., camera, ultrasound, infrared, tactile) and/or from other behaviors in the system, and send outputs to the robot's actuators(effectors) (e.g., wheels, grippers, arm, and speech) and/or to other behaviors.

A variety of behavior-based control schemes have been inspired by the success of (Brooks, 1986, 1989), with his architecture which is known by the subsumption architecture. In this architecture behaviors are arranged in levels of priority where triggering a higher level behavior suppresses all lower level behaviors. (Arkin, 1987, 1989; Arkin & Balch, 1997), has described the use of reactive behaviors called motor schemas. In this method, potential field is used to define the output of each schema. Then, all the outputs are combined by weighted summation. Rosenblatt et al. (Rosenblatt et al. 1989, 1994, 1995), presented DAMN architecture in which a centralized arbitration of votes provided by independent behaviors combines into a "voted" output. Others (Saffiotti, 1997), (Seraji et al., 2001, 2002), (Yang et al. 2004, 2005; Selekwa et al., 2005 and Aguirre & Gonzales, 2006) used fuzzy logic system to represent and coordinate behaviors.

2.2 Fuzzy behavior based navigation

An important problem in autonomous navigation is the need to deal with the large amount of *uncertainties* that has to do with the sensory information received by the robot as well as with the fact that the environment in which such robots operate contains elements of *dynamics* and *variability* that limit the utility of prior knowledge. Fuzzy theory has the features that enable it to cope with uncertain, incomplete and approximate information. Thus, fuzzy logic stirs more and more interest amongst researchers in the field of robot navigation. Further, in the majority of fuzzy logic applications in navigation, a mathematical model of the dynamics of the robot nor the environment is needed in the design process of the motion controller.

The theory of fuzzy logic systems is inspired by the remarkable human capacity to reason with perception-based information. Rule based fuzzy logic provides a formal methodology for linguistic rules resulting from reasoning and decision making with uncertain and imprecise information.

In the fuzzy logic control inputs are processed in three steps (Fuzzification, Inference and Deffuzification) as seen in Fig. 1.

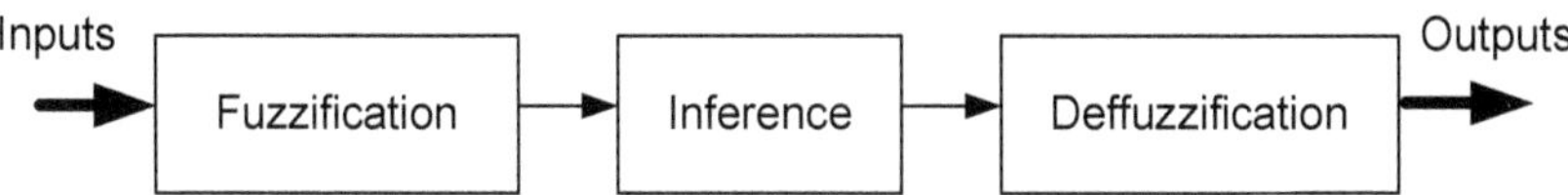

Fig. 1. Fuzzy logic control steps.

In the fuzzification block one defines for example fuzzy set A in a universe of discourse X defined by its membership function $\mu_A(x)$ for each x representing the degree of membership of x in X. In fuzzy logic control, membership functions assigned with linguistic variables are used to fuzzify physical quantities. Next, in the inference block, fuzzified inputs are inferred to a fuzzy rules base. This rules base is used to characterize the relationship between fuzzy inputs and fuzzy outputs. For example, a simple fuzzy control rule relating input v to output u might be expressed in the condition-action form as follows,

$$IF\ v\ is\ W\ then\ u\ is\ Y \tag{1}$$

Where W and Y are fuzzy values defined on the universes of v and u, respectively.

The response of each fuzzy rule is weighted according to the degree of membership of its input conditions. The inference engine provides a set of control actions according to fuzzified inputs. Since the control actions are in fuzzy sense. Hence, a deffuzification method is required to transform fuzzy control actions into a crisp value of the fuzzy logic controller.

In behavior based navigation the problem is decomposed into simpler tasks(independent behaviors). In fuzzy logic behavior based navigation systems each behavior is composed of a set of fuzzy logic rule statements aimed at achieving a well defined set of objectives, for example a rule could be:

If goal is near and to the left then turn left and move forward with a low speed

In general the actions recommended by different behaviors are compiled to yield the most appropriate action according to certain criteria.

2.3 Behavior coordination

The main problem in robot behavior based navigation is how to coordinate the activity of several behaviors, which may be active concurrently with the possibility of having behavior conflict. For example, one may have "goal reaching" behavior and " obstacle avoidance" behavior active at the same time as seen in Fig 2.

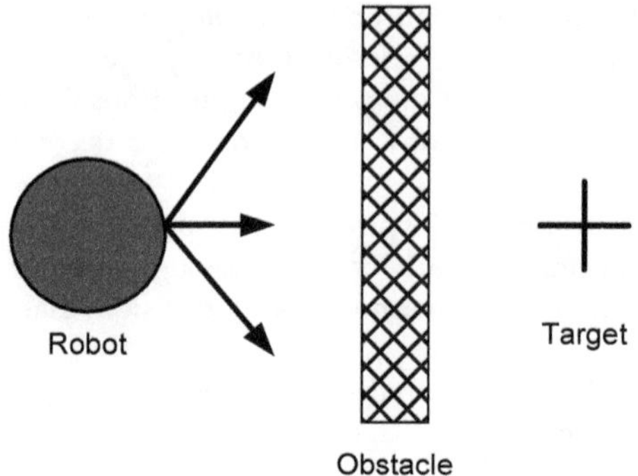

Fig. 2. Conflicts in bahviors.

The coordination task will be to reach a trade-off conclusion that provides the suitable command to the robot actuators which can result in choosing one behavior or a combination of all activated behaviors as shown in Fig.3. Behavior coordination is the point at which most strategies differ. Some of the earlier strategies are based on Brooks subsumption architecture (Brooks, 1986, 1989) uses a switching type of behavior coordination. In the Subsumption approach a prioritization scheme is used in which recommendation of only one behavior with the highest priority is selected, while recommendations of the remaining competing behaviors are ignored. This approach however, leads to inefficient results or poor performance in certain situations. For example if a robot is to encounter an obstacle right in front of it the action that will be selected is "avoid obstacle", the robot then decides to turn left to avoid the obstacle while the goal is to the right of the robot, so the "seek goal" behavior is affected in a negative way.

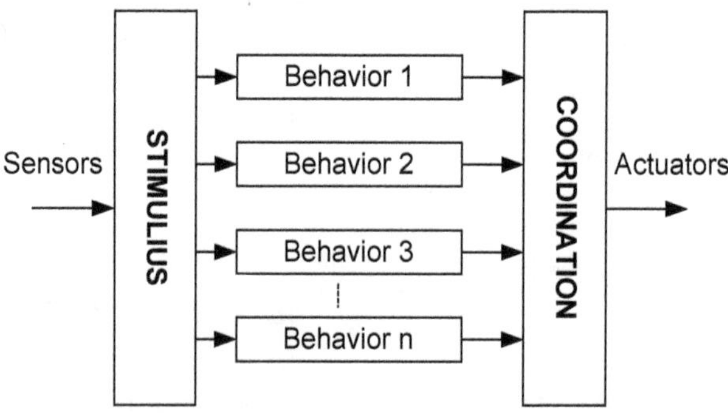

Fig. 3. Structure of behavior based navigation.

Other techniques combine the output of each behavior based on predetermined weighting factors, for example Arkin's motor schema approach(Arkin, 1987, 1989; Arkin & Balch, 1997), or Philipp A. and H.I. Christensen (ALthaus & Christensen, 2002) and the work of Rosenblatt(Rosenblatt et al. 1989, 1994, 1995), who developed the distributed architecture for mobile robot navigation, in which a centralized arbitration of votes provided by independent behaviors. In this method each behavior is allowed to vote for or against certain vehicle actions. The action that win the vote is carried out. These techniques may as

well lead to poor performance in certain situations, for example if the robot is to encounter an obstacle right in front of it the "avoid obstacle" behavior may recommend the robot to turn left, while the "seek goal" behavior may request the robot to turn right since the goal is to the right of the robot, this may lead to trade off command that directs the robot forward resulting in a collision with the obstacle.
To deal with these limitations other schemes were recommended that achieve the coordination via considering the situation in which the robot is found, i.e each behavior is allowed to affect the robot motion based on the situational context. (Saffiott, 1997) uses the process of context-dependent-blending in which the current situation is used to decide the action taken using fuzzy logic. Independently (Tunstel et al., 1997) developed an approach similar to context-dependent-blending, in which adaptive hierarchy of multiple fuzzy behaviors are combined using the concept of degree of applicability. In this case certain behaviors are allowed to affect the overall behavior as required by the current situation and goal. The behavior fusion methodology in this chapter is motivated by the approaches used by Saffiot and Tunstel et al.

2.4 Context-dependent behavior coordination

The robot navigation tasks are divided into small independent behaviors that focus on execution of a specific subtask. For example, a behavior focuses on reaching the global goal, while another focuses on avoiding obstacles. Each behavior is composed of a set of fuzzy logic rules aimed at achieving a given desired objective. The navigation rules consist of a set of fuzzy logic rules for robot *velocity* (linear velocity m/s) and *steering* (angular velocity rad/s) of the form

$$IF\ C\ then\ A \tag{2}$$

Where the condition C is composed of fuzzy input variables and fuzzy connectives (And) and the action A is a fuzzy output variable. Equation (2) represents the typical form of natural *linguistic* rules .This rules reflect the human expert and reason to ensure logic, reliable and safe navigation. For example, obstacle avoidance behavior has inputs sensory data which can be represented by fuzzy sets with linguistic labels, such as {Near, Medium, Far}, corresponding to distance between robot and obstacle. Typical examples of fuzzy rules are as follow,

If *Front left* is Near And *Front right* is Far, Then *Steering* is Right

If *Front left* is Far And *Front right* is Near, Then *Velocity* is Zero

Where *Front left* and *Front right* are the distances acquired from sensors located in different locations on the robot.
Many behaviors can be active simultaneously in a specific situation or *context*. Therefore, a coordination technique, solving the problem of activation of several behaviors is needed. We call the method *context dependent behavior coordination*. The coordination technique employed herein is motivated by the approaches used by Saffiotti (Saffiotti, 1997). The supervision layer based on the context makes a decision as to which behavior(s) to process (activate) rather than processing all behavior(s) and then blending the appropriate ones, as a result time and computational resources are saved. Fig.4 and Fig.5 represent the architecture of the compared method and our approach, respectively.

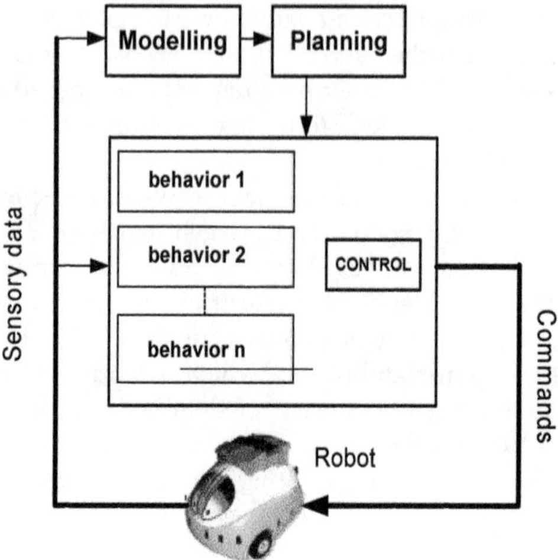

Fig. 4. Architecture of compared method.

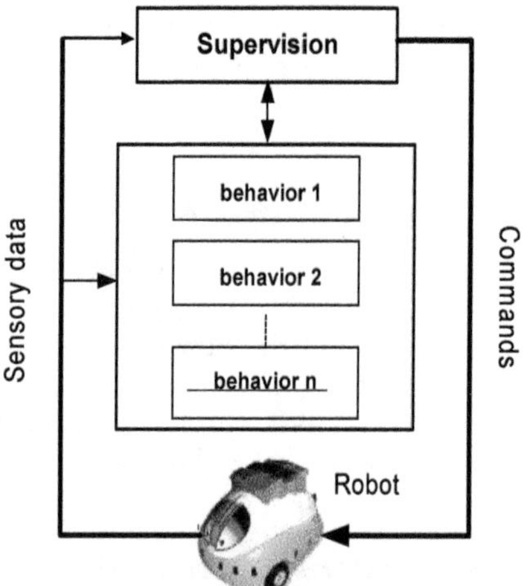

Fig. 5. Architecture of Context Dependent Behavior Coordination.

Our approach consists of the following characteristics.

- The robot navigation is comprised of four behaviors as follows,
  1. Goal reaching behavior
  2. Wall following behavior
  3. Emergency situation behavior
  4. Obstacle avoidance behavior.
- Each behavior is composed of a set of fuzzy logic rules achieving a precise goal.
- The output of each behavior represents the Steering angle and the Velocity.
- The supervision layer defines the priority of each behavior. It selects or activates and blends behaviors depending on situation or context.

3. Simulation & experimental results

3.1 Introduction

The navigation task can be broken down to a set of simple behaviors. The behaviors can be represented using fuzzy *if-then* rules. A context dependent coordination method can be used to blend behaviors.

To validate the applicability of the method simulation and experimental studies were performed.

3.2 Design of individual behaviors

To provide the robot with the ability to navigate autonomously avoiding modeled and unmodeled obstacles especially in crowded and unpredictably dynamic environment the following behaviors were designed: Goal reaching , Emergency situation, Obstacle avoidance, Wall following. Each behavior was represented using a fuzzy *if- then* rule base. The fuzzy rule base comprises the following *if-then* rules:

$$R^{(l)}: IF\ x_1 \text{ is } A_1^l \text{ and ... and } x_n \text{ is } A_n^l, THEN\ y \text{ is } B^l \tag{3}$$

Where $l = 1...m$, and m is the number of rules in a given fuzzy rule base, x_1 ... x_n are the input variables which are the sensor data of the mobile robot, $A^l_1...A^l_2$ are the input fuzzy sets, B^l is the output fuzzy set and y is the output variable.

3.2.1 Goal reaching behavior

The goal reaching behavior tends to drive the robot from a given initial position to a stationary or moving target position. This behavior drives the robot to the left to the right or forward depending on θ_{error}, the difference between the desired heading (the heading required to reach the goal) and the actual current heading.

Fig.6 gives a schematic block diagram of the goal reaching architecture. From this figure we can notice that the inputs of the goal reaching controller are the distance robot to goal (D_{rg}) and θ_{error} which are given by Equation (4) and Equation (5), respectively.

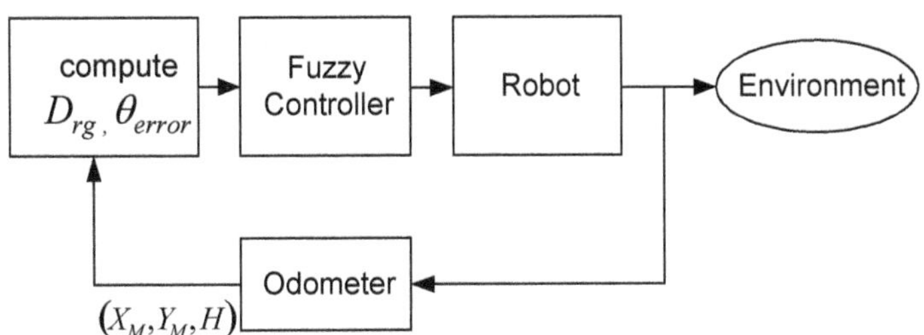

Fig. 6. Control architecture for goal reaching Behavior.

$$D_{rg} = \sqrt{(X_{Goal} - X_M)^2 + (Y_{Goal} - Y_M)^2} \tag{4}$$

$$\theta_{error} = \tan^{-1}\left(\frac{Y_{Goal} - Y_M}{X_{Goal} - X_M}\right) - H \tag{5}$$

Where, (X_M, Y_M, H) are the robot position and the heading measured by the robot odometer. (X_{Gaol}, Y_{Gaol}) is the target position

Although, there is no restriction on the form of membership functions, the appropriate membership functions for D_{rg} (in mm) and θ_{error} (in degrees) shown in Fig.7 were chosen.

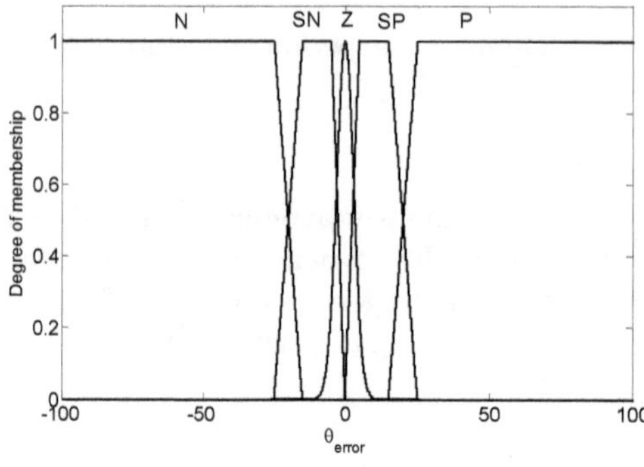

N: Negative, SN: Small Negative, Z: Zero, SP: Small Positive, P: Positive.

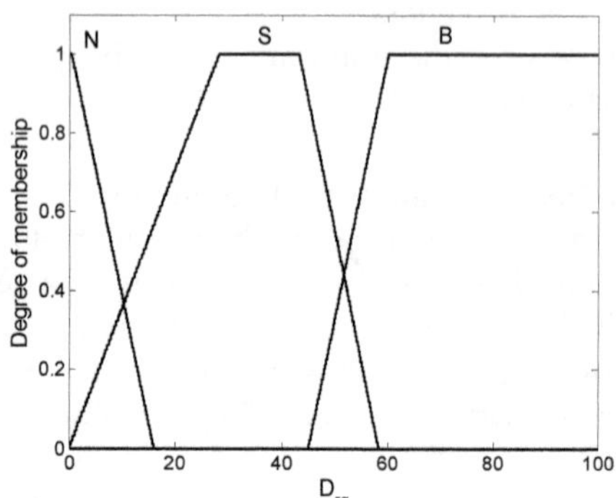

N for Near, S for Small, B for Big.

Fig. 7. Input membership functions for D_{rg} and θ_{error}.

The Goal reaching is expected to align the robot's heading with the direction of the goal so when θ_{error} is positive, the Steering is left or when θ_{error} is negative, the Steering is Right in a way that minimizes θ_{error}. For The Velocity it is proportional to the distance to goal D_{rg}. Example of goal reaching fuzzy rules.

If θ_{error} is *Positive* And D_{rg} *is Big then Velocity is Small Positive*
If θ_{error} is *Positive* And D_{rg} *is Big then Steering is Left*
If θ_{error} is *Negative* And D_{rg} *is Small then Velocity is Small Positive*
If θ_{error} is *Negative* And D_{rg} *is Small then Steering is Right*
If θ_{error} is *Small Negative* And D_{rg} *is Big then Velocity is Positive*
If θ_{error} is *Small Negative* And D_{rg} *is Big then Steering is Right Front*

Velocity

D_{rg} \ θ_{error}	Z	SN	N	SP	P
Near	Z	Z	Z	Z	Z
Small	P	P	SP	SP	SP
Big	P	P	SP	P	SP

Steering

D_{rg} \ θ_{error}	Z	SN	N	SP	P
Near	F	RF	R	LF	L
Small	F	RF	R	LF	L
Big	F	RF	R	LF	L

Table 1. Fuzzy table rules for goal reaching behavior.

For other behaviors the robots needs to acquire information about the environment. The Pekee Robot (Pekee is the robotic platform that will be used to validate the functionality of the proposed scheme experimentally) is endowed with 14 infrared sensors (See Fig.8). These sensors are used to detect obstacles in short-range distances and in a cone of 10 degrees. These sensors are clustered into 6 groups in the simulated robot in such a way as to be similar to Pekee(the simulated robot is very similar kinematically to Pekee).

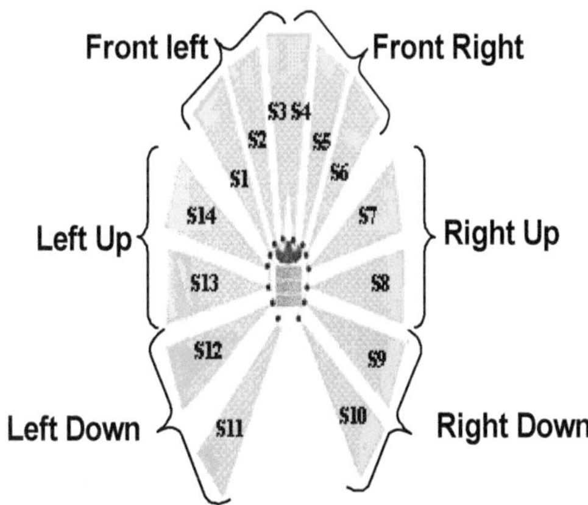

Fig. 8. Clustered sensors arrangement.

For each group, we refer to the minimum of distance measurement by group of sensors as:

Right Down= *min* (D_9, D_{10}) *Right Up*= *min* (D_7, D_8) *Front Right*= *min* (D_4, D_5, D_6)

Left Down=*min* (D_{11}, D_{12}) *Left Up*=*min* (D_{13}, D_{14}). *Front Left*=*min* (D_1, D_2, D_3),

D_i is the distance acquired by the sensor S_i, i=1….14.

These distances represent the inputs of the fuzzy controller for behaviors like Emergency situation, Obstacle avoidance, Wall following behaviors. Each distance is fuzzified using the following membership function described in Fig.9.

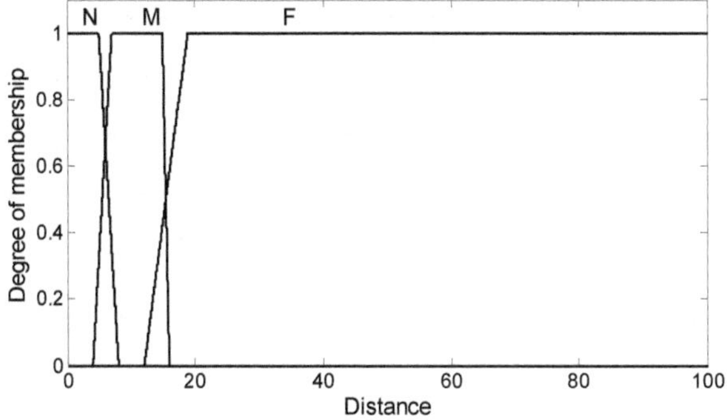

Fig. 9. Input membership function for distance to obstacle, N:Near, M:Medium, F:Far

The output variable are the Steering and the Velocity. Fig.10 illustrates the membership functions for these outputs.

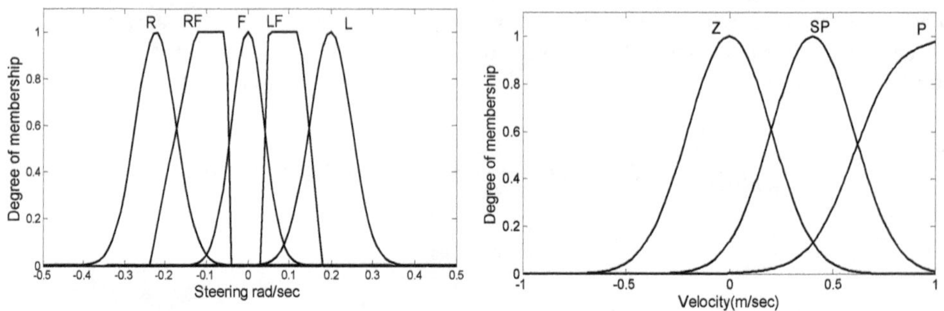

R: Right, RF: Right Front, FL: Front Left, F: Front, L: Left

Z: Zero, SP: Small Positive, P: Positive

Fig. 10. Output membership functions for steering & velocity.

3.2.1 Obstacle avoidance behavior

The obstacle avoidance behavior tends to avoid collisions with obstacles that are in the vicinity of robot. We take into account different cases as shown in table2.

	Inputs						Outputs	
Cases	Right Down	Right Up	Front Right	Front Left	Left Up	left Down	Steering	Velocity
1			F	N			R	Z
			F	M			R	SP
2			N	F			L	Z
			M	F			L	SP
3		N					L	Z
		M					L	SP
4					N		R	Z
					M		FR	P
5						N	R	Z
6	N						L	Z
7		N	N	F			L	Z
		M	M	F			L	SP

Cases	Inputs						Outputs	
	Right Down	Right Up	Front Right	Front Left	Left Up	left Down	Steering	Velocity
8			F	N	N		R	Z
			F	M	M		R	SP
9			N	N	F		L	Z
			M	M	F		L	SP
			F	F			F	P
10		N	F	F	N		F	SP
		N		M	N		F	SP
		N	M		N		F	SP

Table 2. Fuzzy table rules for obstacle avoidance behavior.

Example of obstacle avoidance rules
If Front left is Near And Front right is Far, Then Steering is Right
If Front left is Near And Front right is Far, Then Velocity is Zero
If Front left is Far And Front right is Near, Then Steering is left
If Front left is Far And Front right is Near, Then Velocity is Zero

3.2.2 Wall following behavior

The objective of the control of the wall following behavior is to keep the robot at a safe close distance to the wall and to keep it in line with it. Example of wall following fuzzy rule:
If Right down is Medium and Front Right is Medium then Steering is Front
If Right down is Medium and Front Right is Medium then Velocity is Positive

Cases	Inputs						Outputs	
	Right Down	Right Up	Front Right	Front Left	Left Up	left Down	Steering	Velocity
	M	M	F	F			F	P
	M	M	M	F			F	P
			F	F	M	M	F	P
			F	M	M	M	F	P

Table 3. Fuzzy table rules for wall following behavior.

3.2.3 Emergency situation behavior

The emergency situation behavior drives the robot to the left or to the right when it is surrounded by obstacles in away depicted in Table 4.

	Inputs				Outputs	
Cases	Right Up	Front Right	Front Left	Left Up	Steering	Velocity
	Far	M	M	M	R	SP
	M	M	M	F	L	SP
	M		F	M	L	Z
	M	F		M	L	Z
	M	M	M	M	R	Z

Table 4. Fuzzy table rules for "emergency" behavior.

3.3 Blending of behaviors

The question to answer once the behaviors are designed is how best decide what the actuators shall receive(in terms of steering and velocity)taking into account the context in which robot happens to be in and relative importance of each behavior. To achieve that the work herein proposes the following architecture with details.

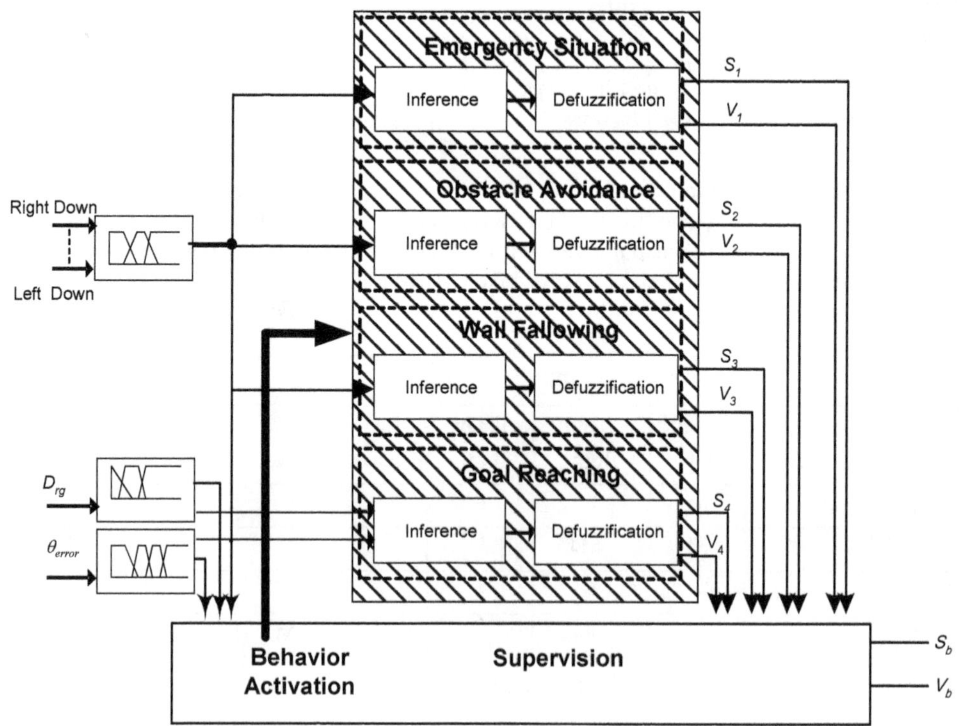

Fig. 11. Supervision Architecture.

S_i and V_i are the output Steering and Velocity of each behavior, i=1...4
S_b and V_b are the output steering and velocity to motor.
The inputs of supervision layer are the degree of membership of each group of sensors in all membership function, D_{rg} and θ_{error},in addition to the Steering and Velocity outputs of all behaviors (S_1, V_1, S_2, V_2, S_3, V_3, S_4 *and* V_4).
The supervision layer program is based on fuzzy rules such as,

$$IF\ context\ then\ behavior \tag{6}$$

Cases	RD	RU	FR	FL	LU	LD	θ_{error}	D_{rg}	Behavior
		F	F	F	F				Goal Reaching
								N	
				F	F	F	SP		Goal Reaching
				F	F	F	P		
				F	F	F	Z		
			N	N			SN		Fusion
			N	N			N		
			M	M			SN		
			M	M			N		
	F	F	F				SP		Goal Reaching
	F	F	F				P		Goal Reaching
	M	M		F					Wall Following
			F		M	M			Wall Following

Table 5. Context priority.

S_b and V_b are the appropriate velocity and steering control commands sent to the motors in a given robot situation as a result of the decision to activate a certain behavior (avoid obstacle, wall following...). However, S_b and V_b can be the fusion result of many behaviors. As seen at Table 5 the direction of robot to avoid obstacles is taken as default to the left, but, the supervision layer takes into account the position of goal by θ_{error} in such away as to minimize the distance traveled towards the goal as shown in Fig.12.
The advantage of this method is the fact that the number of rules are reduced by reducing the number of input variables to the fuzzy rules of each behavior, for example in the obstacle avoidance behavior there was no need to add θ_{error} as an input.

3.4 Simulation results

To verify the validity of the proposed scheme, some typical cases are simulated in which a robot is to move from a given current position to a desired goal position in various unknown environment. In all cases the robot is able to navigate its way toward the goal while avoiding obstacles successfully.

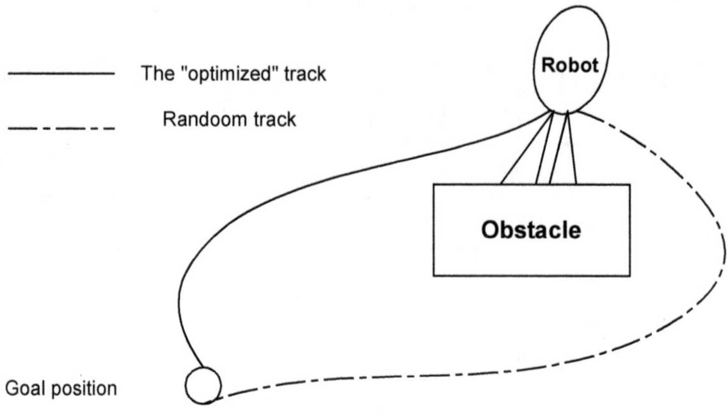

Fig. 12. Fusion of obstacle avoidance and goal reaching behaviors.

In the experimental work, the functionality of the proposed scheme was tested in an environment mimicking a crowded dynamic environment. The environment was designed taking into account several situations such as: simple corridor, narrow space and an area with many different shapes as obstacles (round, rectangle, trapezoidal)in away that mimics an image of office or equipped room (indoor environment).

In experiment (1) Fig.13 the robot has to reach Goal 1 for the start point placed between two walls. The robot begins to execute the behavior according to the rule base of the supervision layer depending on the current context. First, the robot follows wall 1 with maximum velocity until it senses obstacle 1, then it changes its behavior to obstacle avoidance at point A up to point B during which the robot crosses a narrow space between obstacle 1 and wall 2. The goal reaching behavior and the obstacle avoidance behavior are activated separately or fused to leave the narrow space, until point C. The robot then was encountered by obstacle 2, wall 2 and wall 3, so the emergency situation behavior was active. Next, the presence of obstacle 2 in front of the robot makes the obstacle avoidance behavior active until point D. For the route between of point D to E the robot just follows wall 2. From point E, three behaviors are activated (wall following, obstacle avoidance and goal reaching). The wall follow behavior is activated when a medium distance between the robot and obstacle 4 and 5 is established(corresponding to the last two cases in Table 5). The goal reaching behavior is activated to guide the robot to the goal. Between point F and G the robot is situated far from obstacles which in turn makes the goal reaching behavior active up to the presence of obstacle 6 at which point both goal reaching and obstacle avoidance are active. Finally, the goal reaching behavior is activated to reach the goal when it is near to the robot.

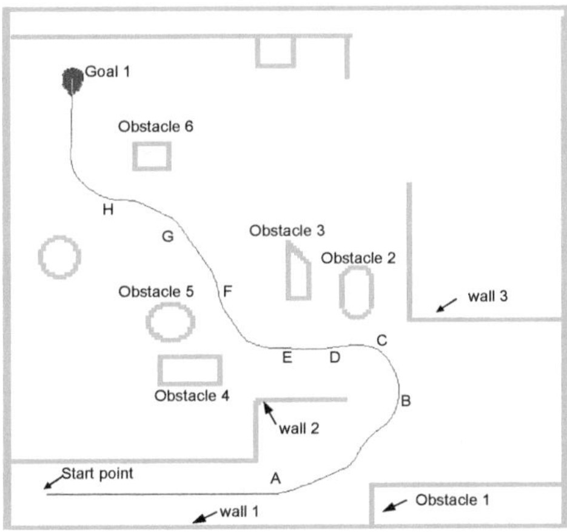

Fig. 13. Experiment 1.

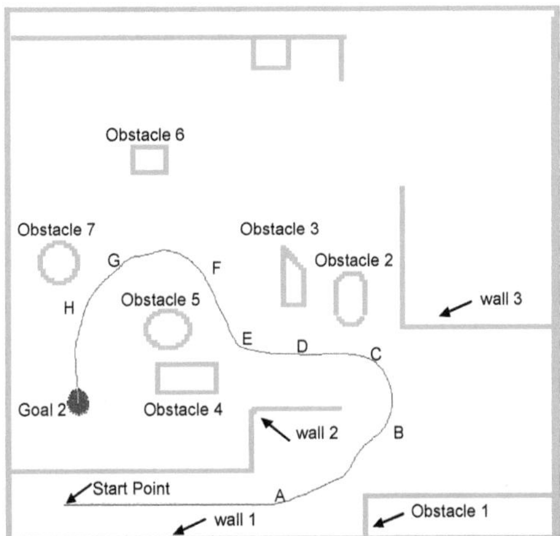

Fig. 14. Experiment 2.

For the experiment (2) Fig.14 the robot had to reach goal 2. It uses the same path as experiment 1 up to the point C. Then, avoiding obstacle 2 and goal reaching behavior are activated up to point D. From this point the robot follows obstacle 3 as a wall up to point E. The presence of obstacle 4 changes the behavior of the robot to obstacle avoidance. The robot avoids obstacle 4 then it follows obstacle 5 as wall by keeping medium distance to it. After that, the robot is situated in a free space and goal reaching behavior is activated up to point G. The obstacle avoidance behavior is activated when the robot senses obstacle 7 the robot avoids obstacle 7 and the behavior is changing to wall following when a medium distance to obstacle is measured by the robot. Finally the robot reaches the goal.

Figs.15-16 illustrate the levels of activation of each behavior during experiment 1 and experiment 2, respectively.

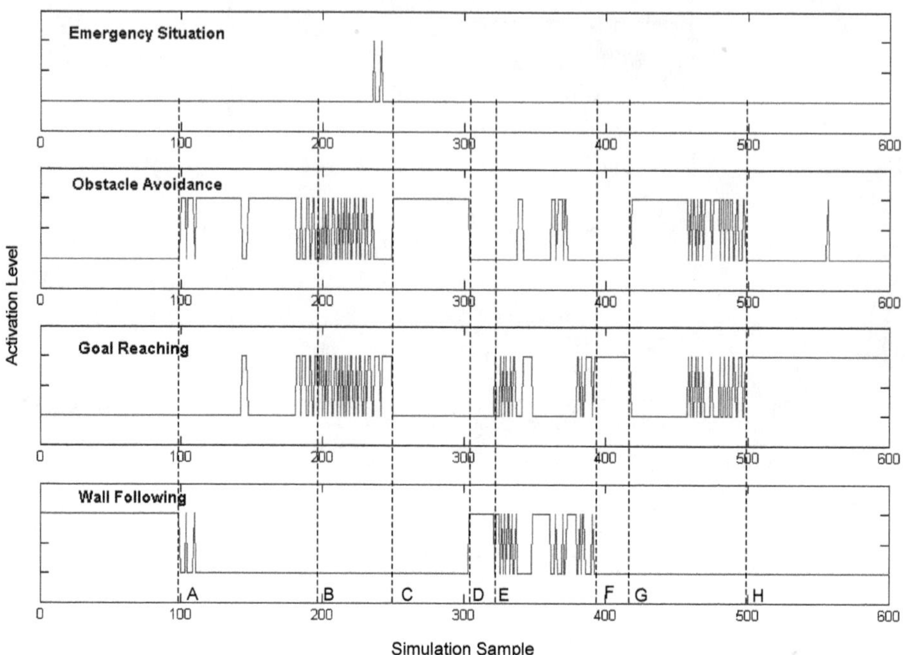

Fig. 15. Levels of activation of each behavior during Experiment 1.

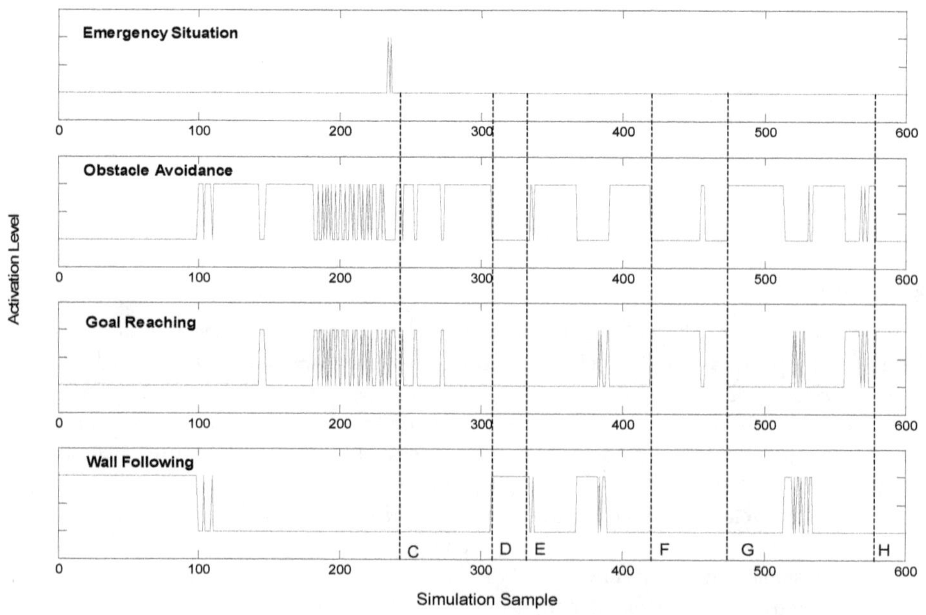

Fig. 16. Levels of activation of each behavior during Expriment 2.

The aim of experiment (3) (see Fig.17) is to show the ability of the robot to manage to escape from U shape obstacle and reaching a goal between two walls. From the start point the robot is situated in a U shape. In this context three behaviors are activated (the goal reaching behavior, obstacle avoidance and wall following). The avoid obstacle behavior always guided by the activation of goal reaching behavior especially at point A and B as shown in Fig.18.

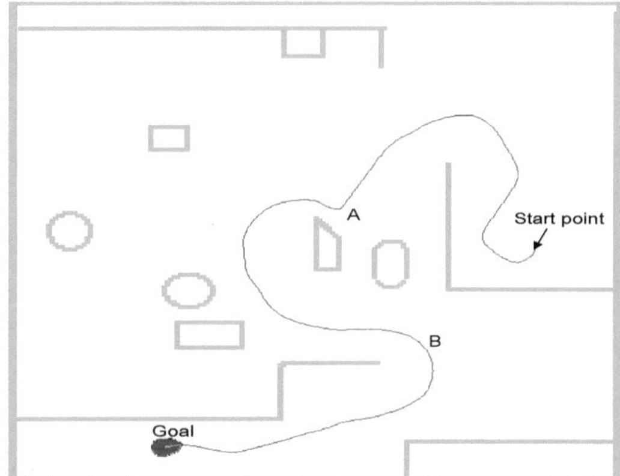

Fig. 17. Expriment 3.

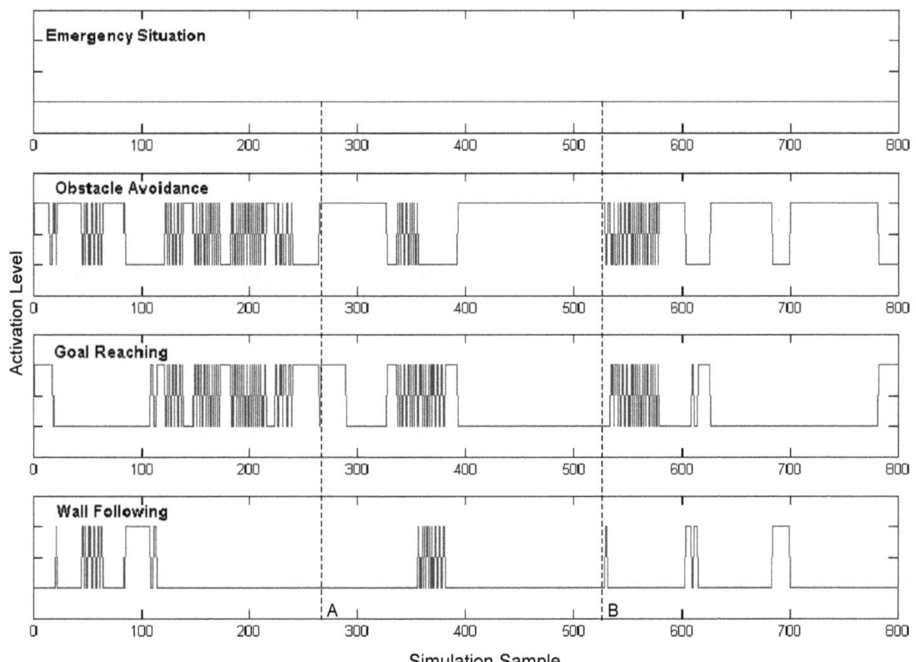

Fig. 18. Levels of activation of each behavior during Experiment 3.

Experiment (4) (Fig.19) shows the ability of the robot to escape from a trap situation and searching another path for reaching the goal.

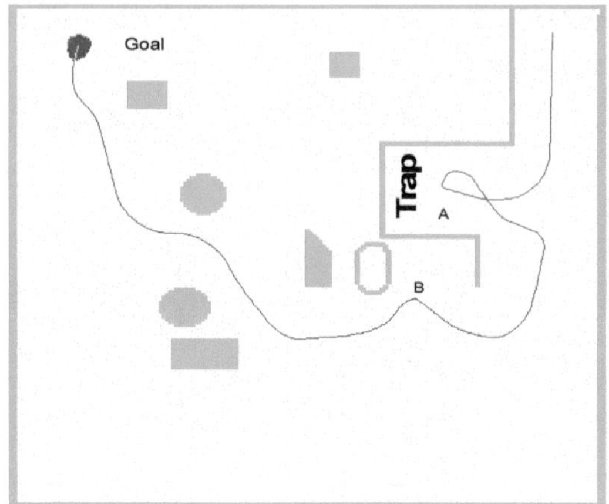

Fig. 19. Expriment 4.

After crossing the corridor, no obstacles are in front of the robot. Thus, the robot goes to the direction of the goal. At point A the robot senses the existence of obstacles around it, in the front, left and right, and then it made a round to right to escape from this trap and continues its navigation looking for another path.
In point B, the activation of goal reaching behavior and obstacle avoidance behavior is observed to take place concurrently. Here, the supervision layer fused these 2 behaviors to get the appropriate action as shown in section 3.3 Fig.12. The orientation of the robot depends on the goal position.
The Fig.20 shows the level of activation of emergency situation in the point A and the activation of goal reaching behavior and obstacle avoidance at the point B.

3.5 Experimental work

The effectiveness of the suggested navigation approach was experimentally demonstrated on a robotic platform named Pekee (Pekee™ robot is an open robotic development toolkit of Wany Robotics).

3.5.1 Pekee mobile robot

Pekee is equipped with two driving wheels with an additional supporting wheel. Its length is 40 cm and width is 25.5 cm, max speed 1 meter/second rotation 360 degree in a circle of 70 cm. The velocities of driven wheels are independently controlled by a motor drive unit. In addition the robot is endowed by,

- 2 odometers (180 impulses/wheel-turn).
- 2 gyro meters (pan and tilt)
- 2 temperature sensors.
- 1 variable frequency buzzer.
- Infrared link for communication between robots and peripherals.
- Serial infrared link for data transfer between Pekee and docking station or PC.

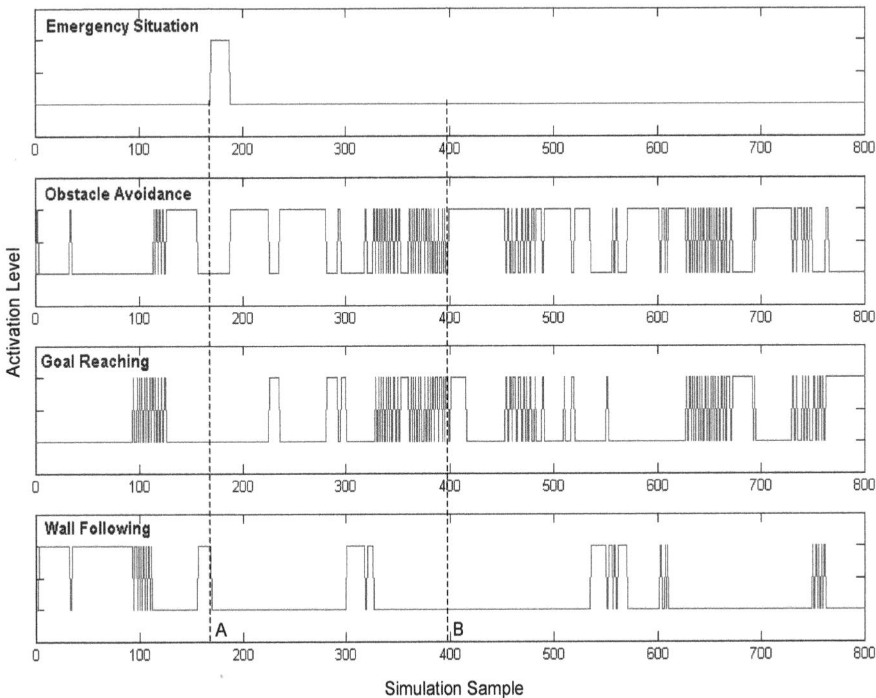

Fig. 20. Levels of activation of each behavior during expriment 4.

- Embedded Pc X86
- Camera
- 16-Mhz Mitsubishi micro-controller (16-bit), with 256 KB Flash-ROM and 20 KB RAM.
- 15 infrared sensors infrared telemeters (up to 1 measurement) arranged as shown in Fig21

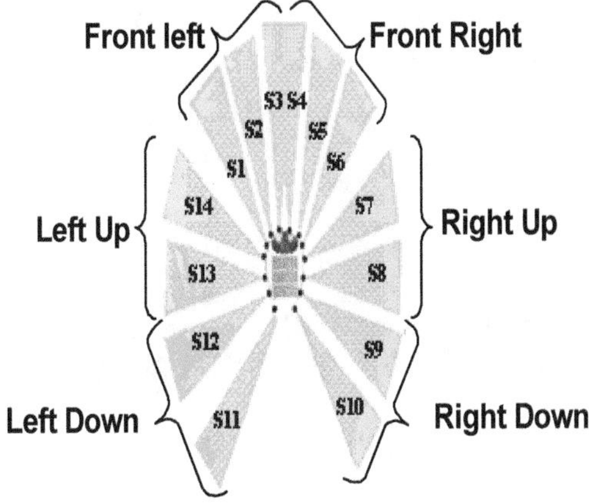

Fig. 21. IR sensors arrangement.

3.5.2 Pekee to PC communication

In the experimental work, the robot and the PC are connected via TCP/IP protocol using 2 network wireless cards one on the robot and the other on the computer or using a RJ45 cable.

The PC is equipped by a 54G network card compiles with the IEEE 802.11b standard in order to communicate with other 802.11b compliant wireless devices at 11 Mbps (the robot wireless network card). The card runs at speed of up to 54Mbps and operates on the same 2.4 GHz frequency band as 802.11b WI-FI products. This frequency band is suitable in industrial, science, and medical band operation. The work space of this card is as maximum 300 m.

The PC compiles, links the source code and executes the program. Then, it transmits the frames to the robot embedded PC via TCP\IP. These frames will be transmitted to the micro controller via the OPP bus. The micro controller is responsible to execute frames and transmits order to actuators (motors, buzzer…) also it sends data about the robot sensors status, robot position, measured distances ….

3.5.3 Experimental results

In order to validate the results of simulated experiments, a navigation task has been tested in a real world as an environment similar to a small equipped room. Fig. 22-26 shows that Pekee was able to navigate from a given starting point to a target point while avoiding obstacles.

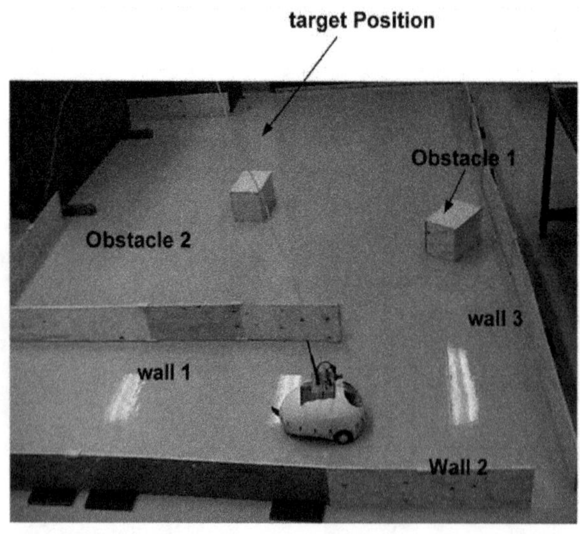

Fig. 22. Pekee Navigates in a crowded environment.

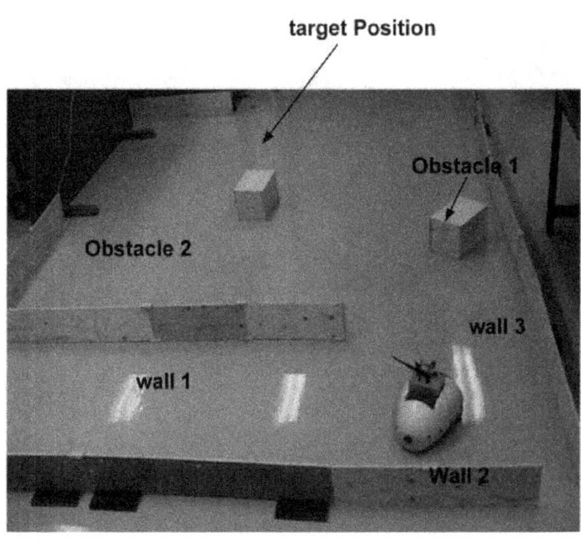

Fig. 23. Pekee Navigates in a crowded environment.

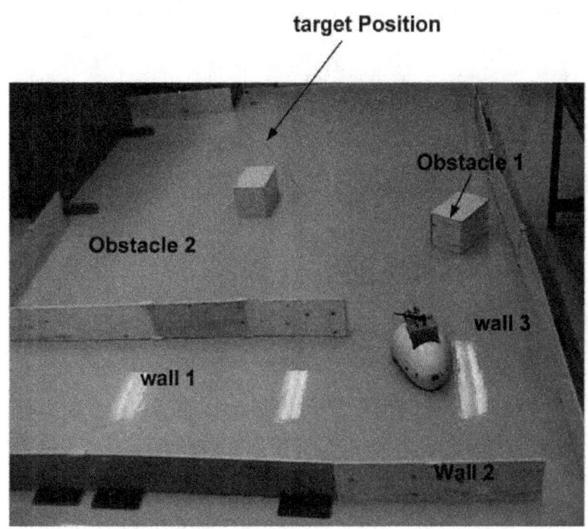

Fig. 24. Pekee Navigates in a crowded environment.

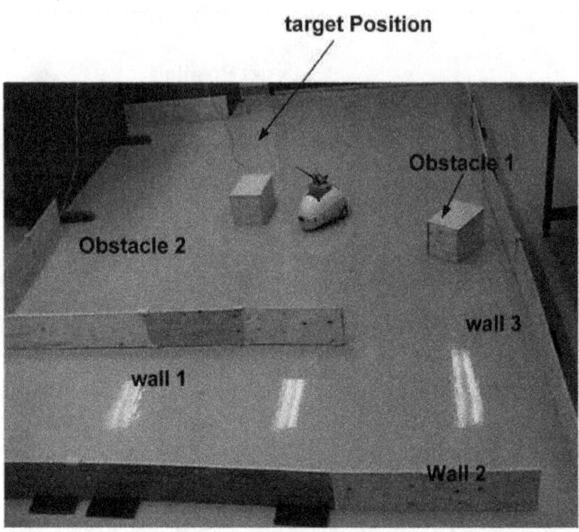

Fig. 25. Pekee Navigates in a crowded environment.

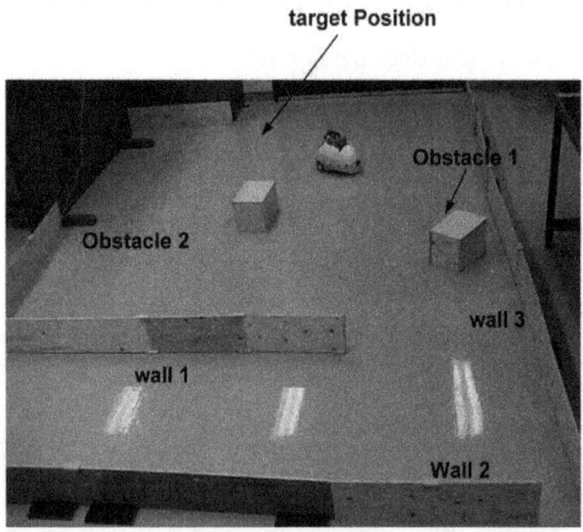

Fig. 26. Pekee Navigates in a crowded environment.

3.6 Conclusions

A successful way of structuring the navigation task in order to deal with the problem of mobile robot navigation is demonstrated. Issues of individual behavior design and action coordination of the behaviors were addressed using fuzzy logic. The coordination technique employed in this work consists of two layers. A Layer of primitive basic behaviors, and the supervision layer which based on the context makes a decision as to which behavior(s) to

process (activate) rather than processing all behavior(s) and then blending the appropriate ones, as a result time and computational resources are saved. Simulation and experimental studies were done to validate the applicability of the proposed strategy.
As an improvement to our implemented system we are planning to incorporate learning to improve the effectiveness of the navigation approach.
As a further work the algorithm is to be tested on a "robotic cane", which is a device to help the blind or visually impaired users navigate safely and quickly among obstacles and other hazards. During operation, the user will be able to push the lightweight "Robotic Cane" forward. When the "Robotic Cane's ultrasonic sensors detect an obstacle, the behavior based navigation that will be developed determines a suitable direction of motion that steers the user around it. The steering action results in a very noticeable force felt in the handle, which easily guides the user without any conscious effort on his/her part. The navigation methodology that will be followed in the robotic cane will be based on the behavior based navigation developed herein, in particular the work will use fuzzy logic based navigation scheme to steer the "Robotic Cane". Further the user is continually interacting with the robotic cane, this calls for the need to address the issue of cooperation and/or conflict resolution between the user and the robotic cane.

4. References

Aguirre E. & Gonzales A. (2000). Fuzzy behaviors for mobile robot navigation:design, coordination and fusion, *Int. J. of Approximate Reasoning*, Vol. 25, pp. 255-289.

Althaus P. & Christensen H. I.(2002). Behavior coordination for navigation in office environment, *Proceedings of 2002 IEEE/RSJ Int. Conference on Intelligent Robots and Systems*, pp. 2298-2304, Switzerland, 2002

AlYahmedi, A. S., El-Tahir, E., Pervez, T. (2009). Behavior based control of a robotic based navigation aid for the blind, *Control & Applications Conference, CA2009*, July 13-July 15, 2009, Cambridge, UK.

Arkin, R. C. (1987). Towards Cosmopolitan Robots: Intelligent Navigation in Extended Man-made Environments, *PhD Thesis*, University of Massachusetts, Department of Computer and Information Science.

Arkin, R. C. (1989). Motor schema-based mobile robot navigation, *Int. J. of Robotic Research*, Vol 8, pp. 92-112.

Arkin, R. C. & Balch, T.(1997) AuRA: Principles and Practice in Review, *Journal of Experimental and Theoretical Artificial Intelligence(JETAI)*, Vol. 9, No. 2/3, pp. 175-188.

Brooks R. A.(1986). A Robust Layered Control System for a Mobile Robot, *IEEE Journal of Robotics and Automation*, Vol. 2, No. 1, (March 1986), pp. 14–23.

Brooks R. A.(1989). A Robot that Walks; Emergent Behavior from a Carefully Evolved Network, *IEEE International Conference on Robotics and Automation*, Scottsdale, AZ, pp. 292–296, May 1989

Ching-Chih, T., Chin-Cheng, C., Cheng-Kain, C, Yi Yu, L. (2010). Behavior-based navigation using heuristic fuzzy kohonen clustering network for mobile service robots, *International Journal of Fuzzy Systems, Vol. 12, No. 1, March, 2010* , pp. 25-32.

Fatmi, A., ALYahmedi, A. S., Khriji, L., Masmoudi, N(2006). A fuzzy logic based navigation of a mobile robot, *World academy of science, Engineering and Technology*, issue 22, 2006, pp. 169-174.

Huq, R., Mann, G. K. I., Gosine, R. G.(2008). Mobile robot navigation using motor schema and fuzzy context dependent behavior modulation, *Applied soft computing,* 8, 2008, pp. 422-436.

Langer D., Rosenblatt J.K. & Hebert M. (1994). A Behavior-Based System For Off-Road Navigation, *IEEE Journal of Robotics and Automation* , Vol. 10, No. 6, pp. 776-782.

Maes P. (1990). How to do the Right Thing, *Connection Science Journal, Special Issue on Hybrid Systems,* Vol. 1.

Mataric´ M. J.(1997). Behavior-Based Control: Examples from Navigation, Learning, and Group Behavior, *Journal of Experimental and Theoretical Artificial Intelligence,* special issue on Software Architectures for Physical Agents, Vol. 9, No.2/3, pp. 323-336

Rosenblatt J. & Payton D. W.(1989). A Fine-Grained Alternative to the Subsumption Architecture for Mobile Robot Control, *Proceedings of the IEEE/INNS International Joint Conference on Neural Networks,* Washington DC, June 1989, vol. 2, pp. 317-324.

Rosenblatt J.(1995). DAMN: A Distributed Architecture for Mobile Navigation, *Ph.D. dissertation, Carnegie Mellon University Robotics Institute Technical Report CMU-RI-TR-97-01,* Pittsburgh, PA, 1995.

Saffiotti A.(1997). The uses of fuzzy logic for autonomous robot navigation: a catalogue raisonn'e, *Soft Computing Research journal,* Vol. 1, No. 4, pp. 180-197.

Selekwa M. F., Damion D., & Collins, Jr. E. G. (2005). Implementation of Multi-valued Fuzzy Behavior Control for Robot Navigation in Cluttered Environments, *Proceedings of the 2005 IEEE International Conference on Robotics and Automation,* Barcelona, Spain, pp., 3699-3706, April 2005

Seraji H. & Howard A.(2002). Behavior-based robot navigation on challenging terrain: A fuzzy logic approach, *IEEE Trans. Rob. Autom.* Vol. 18, No. 3, pp. 308-321

Seraji H., Howard A. & Tunstell E.(2001). Terrain-Based Navigation of Planetary Rovers: A Fuzzy Logic Approach, *Proceeding of the 6th International Symposium on Artificial Intelligence and Robotics & Automation in Space,* Canada, June 18-22, 2001.

Tunstel E., Lippincott T. & Jamshidi M.(1997). Behavior Hierarchy for Autonomous Mobile Robots: Fuzzy-behavior modulation and evolution, *International Journal of Intelligent Automation and Soft Computing, Special Issue: Autonomous Control Engineering at NASA ACE Center,* Vol. 3,No. 1,pp. 37--49.

Yang S. X., Li H., Meng M. Q.-H , & Liu P. X. (2004). An Embedded Fuzzy Controller for a Behavior-Based Mobile Robot with Guaranteed Performance, *IEEE Transactions on Fuzzy Systems,* Vol. 12, No. 4, pp.436-446.

Yang S. X., Moallem M., & Patel R. V. (2005). A Layered Goal-Oriented Fuzzy Motion Planning Strategy for Mobile Robot Navigation, *IEEE transactions on systems, man, and cybernetics – part b: cybernetics,* Vol. 35, no. 6, 1214-1224.

2

Fictitious Fuzzy-Magnet Concept in Solving Mobile–Robot Target Navigation, Obstacle Avoidance and Garaging Problems

Srđan T. Mitrović[1] and Željko M. Đurović[2]
[1]*Defense University, Military Academy*
[2]*University of Belgrade, School of Electrical Engineering*
Serbia

1. Introduction

A human driver can successfully execute different vehicle navigation tasks, such as: forward garaging, reverse garaging, parallel parking, diagonal parking, following a given trajectory, stopping at a given point, avoiding one or more obstacles, and the like. These processes are either too difficult to model accurately or too complex to model at all, while experienced drivers are able to successfully execute the tasks without any knowledge of the vehicle's mathematical model, even when they switch vehicles. Existing knowledge about how such problems are solved and the ability to express this knowledge, are benefits which can be put to good use when designing controllers for wheeled mobile robot (WMR) navigation tasks. It is well known that fuzzy logic is the most effective tool for solving a problem for which there is a human solution (Zadeh, 2001), because it offers a mathematical apparatus in the background of an inaccurate, qualitative description, which projects input variables into output variables through several stages. However, describing the tasks executed by a driver in spoken language is not a simple process. When we drive a car, we rely on our driving skills and prefer to use our hands and feet, rather than our brain. It is safe to assume that compiling rules based on driver skills will not be an easy task.

In view of the above, this chapter presents a new approach to the modeling of driver skills, based on a fuzzy model and an original virtual fuzzy-magnet concept. Driver experience is used to solve simple tasks (turn left or right, slow down or accelerate), while the entire problem is solved by specifying an appropriate number of fictitious fuzzy magnets and defining their tasks. Each fictitious fuzzy magnet is defined by its position and a set (or subset) of fuzzy rules which determine its action. For example, to solve a target navigation problem in open space requires only one fuzzy magnet which attracts the mobile robot to the target, while a more complex problem, such as bidirectional garaging of a mobile robot, requires two fuzzy magnets: one immediately in front of the garage and the other inside the garage. The first point is used to approach the garage and while the vehicle is entering the garage, it serves as a reference point for proper orientation, similar to human driving skills. The second point is both a target and a reference point. Compared to other algorithms which address this type of problem, the proposed algorithms are very simple; they are not based on the WMR model. The proposed fuzzy controllers are of the Takagi-Sugeno (T-S) type; they are manually generated (in a manner similar to human driving skills), and have clearly-defined physical parameters.

The first section of this chapter describes the fictitious fuzzy-magnet concept and its application in solving target navigation problems in an obstacle-free environment. It further analyzes the application of a pair of fuzzy magnets in solving a garaging problem involving a differential-drive mobile robot. The fictitious fuzzy magnet concept allows navigation to the target in a single maneuver, without changing the direction of WMR travel. The symmetry of the differential-drive WMR is utilized fully, such that the algorithm provides a bidirectional solution to the WMR garaging problem. The robot is automatically parked from the end of the robot that is closer to the garage entrance. The algorithm can be applied when the control variable is of the discrete type and where there are relatively few quantization levels. The efficiency and shortfalls of the proposed algorithm are analyzed by means of both detailed simulations and multiple re-runs of a real experiment. Special attention is devoted to the analysis of different initial robot configurations and the effect of an error in the estimation of the current position of the robot on garaging efficiency.

The third section of this chapter analyzes the possibility of applying the fictitious fuzzy-magnet concept to navigate a mobile robot to a target in a workspace which contains one or more obstacles. At the preprocessing stage, the left and right side of the obstacle in the robot's line of sight are determined based on the mutual positions of the robot and the obstacle. Since controller inputs are based on the relative dimensions of the obstacle, the controller can be applied to obstacles of different shapes and sizes. Following the design of the above-described controllers, and the successful testing of navigation efficiency to the goal in a workspace which includes a single stationary obstacle, the algorithm was applied, without modification, to avoid an obstacle moving along a straight line at a constant speed equal to one-third of the maximum speed of the robot. Here the proposed algorithm demonstrated limited efficiency. The final section of the chapter proposes a method for a simple but efficient generalization of the algorithm to support a group of obstacles, using parallel data processing, illustrated by several computer simulations of WMR navigation to the target point through a group of obstacles. The disadvantages of the proposed algorithm when applied in complex environments are discussed at the end of the chapter.

2. Fictitious fuzzy magnets concept

A given point $A(x_A, y_A)$, is assumed to be in the Cartesian coordinate system and to represent the position of the fictitious fuzzy magnet FM. The FM is defined as an arranged pair comprised of its position A and an added sub-set of fuzzy rules, $FRsS$ (Fuzzy Rules subSet):

$$FM = (A, FRsS) \tag{1}$$

The $FRsS$ enables the determination of the zone of influence of the fuzzy magnet, as well as the definition of the action it causes. To establish the structure of the fuzzy rules, it is necessary to adopt input and output variables (i.e., premise variables and conclusion variables). Figure 1 shows the position of the WMR relative to the point A, which represents the position of the fictitious fuzzy magnet, FM.

Point (x_R, y_R) is the center of the WMR, and its orientation relative to the x axis is denoted by θ_R, such that the configuration of the WMR q_R is unequivocally defined by three coordinates: $q_R(x_R, y_R, \theta_R)$. The bold line denotes the front end of the robot. The angle formed by the

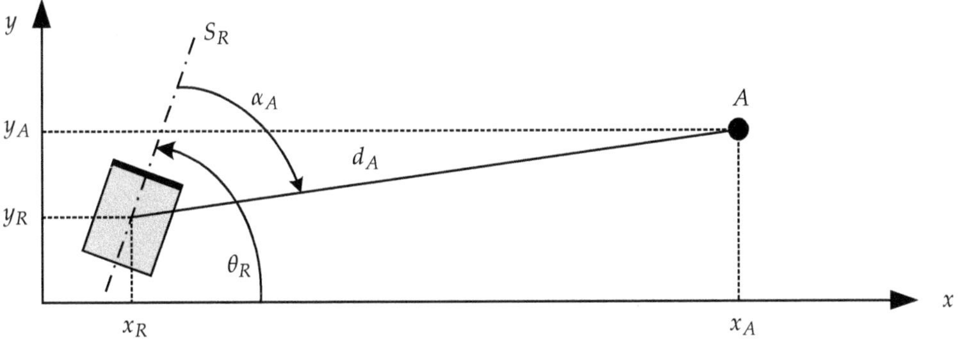

Fig. 1. Mutual spatial positions of wheeled mobile robot and fictitious fuzzy magnet.

longitudinal line of symmetry of the WMR, S_R, and the segment which connects the points (x_R, y_R) and A is denoted by α_A. If the coordinates of the WMR center are denoted by $(x_R(k), y_R(k))$ at the sample time k, the distance between the WMR and the fictitious fuzzy magnet, $d_A(k)$, is:

$$d_A(k) = \sqrt{(x_A - x_R(k))^2 + (y_A - y_R(k))^2}. \tag{2}$$

The position of the WMR at the sample time k, relative to the point A, is unequivocally determined by its orientation relative to the point A – angle $\alpha_A(k)$ and the distance $d_A(k)$, which will be adopted as input variables of the *FRsS*. In the general case, we will state that the output velocity variables are v_1 and v_2. The *FRsS* of r rules for the discrete T-S fuzzy system is defined as:

$$\begin{array}{ll} \textit{Control Rule } i: \text{If } d_A(k) \text{ is } \mu_{i1} \text{ and } \alpha_A(k) \text{ is } \mu_{i2} \\ \qquad\qquad\qquad\quad \text{then } v_1(k) = C_{i1} \text{ and } v_2(k) = C_{i2} \end{array} \quad i = 1, 2, \ldots, r \tag{3}$$

where $d_A(k)$ and $\alpha_A(k)$ are premise variables. The membership function which corresponds to the ith control rule and the jth premise variable is denoted by μ_{ij}, and C_{ij} are constants. System outputs are $v_1(k)$ and $v_2(k)$, obtained from:

$$\begin{bmatrix} v_1(k) \\ v_2(k) \end{bmatrix} = \frac{\sum\limits_{i=1}^{r} \mu_{i,1}(d_A(k)) \cdot \mu_{i,2}(\alpha_A(k)) \cdot [C_{i,1}\ C_{i,2}]^T}{\sum\limits_{i=1}^{r} \mu_{i,1}(d_A(k)) \cdot \mu_{i,2}(\alpha_A(k))} \tag{4}$$

where $\mu_{ij}(x_j(k))$ is the degree of membership of $x_j(k)$ in μ_{ij}. As is well known, the control of a differential drive mobile robot can be designed in two ways: using the speed of the left and right wheels, or using the linear and angular velocities of the robot. Consequently, the general definition of the *FRsS* can be written more precisely in two ways. The navigation of a differential drive mobile robot was designed in (Mitrović & Đurović, 2010a) based on wheel speed control, such that the output variables of the *FRsS* are the speed of the left wheel – v_L and the speed of the right wheel – v_R, such that (3-4) become:

$$\begin{array}{ll} \textit{Control Rule } i: \text{If } d_A(k) \text{ is } \mu_{i1} \text{ and } \alpha_A(k) \text{ is } \mu_{i2} \\ \qquad\qquad\qquad\quad \text{then } v_L(k) = C_{i1} \text{ and } v_R(k) = C_{i2} \end{array} \quad i = 1, 2, \ldots, r \tag{5}$$

$$\begin{bmatrix} v_L(k) \\ v_R(k) \end{bmatrix} = \frac{\sum_{i=1}^{r} \mu_{i,1}(d_A(k)) \cdot \mu_{i,2}(\alpha_A(k)) \cdot [C_{i,1}\ C_{i,2}]^T}{\sum_{i=1}^{r} \mu_{i,1}(d_A(k)) \cdot \mu_{i,2}(\alpha_A(k))} \tag{6}$$

while in (Mitrović & Đurović, 2010b) the output variables of the *FRsS* are linear and angular velocities of the robot v_R and ω_R, and in this case (3-4), can be written as:

$$\begin{array}{l} \textit{Control Rule } i: \text{If } d_A(k) \text{ is } \mu_{i1} \text{ and } \alpha_A(k) \text{ is } \mu_{i2} \\ \qquad \text{then } v_R(k) = C_{i1} \text{ and } \omega_R(k) = C_{i2} \end{array} \quad i = 1, 2, \ldots, r \tag{7}$$

$$\begin{bmatrix} v_R(k) \\ \omega_R(k) \end{bmatrix} = \frac{\sum_{i=1}^{r} \mu_{i,1}(d_A(k)) \cdot \mu_{i,2}(\alpha_A(k)) \cdot [C_{i,1}\ C_{i,2}]^T}{\sum_{i=1}^{r} \mu_{i,1}(d_A(k)) \cdot \mu_{i,2}(\alpha_A(k))} \tag{8}$$

2.1 Navigation in an obstacle-free environment

The application of the proposed concept will be illustrated by an example involving WMR navigation to a target in an obstacle-free environment. The WMR configuration (x_R, y_R, θ_R) and the target position (x_A, y_A) are shown in Fig. 1. We will assume the robot control inputs (v_R, ω_R) to be the outputs of the fuzzy logic controller for navigation in an obstacle free environment (FLC_{OFE}). The WMR angle relative to the target α_A and the distance to the target d_A are fuzzy controller inputs, and the corresponding membership functions are shown in Fig. 2.

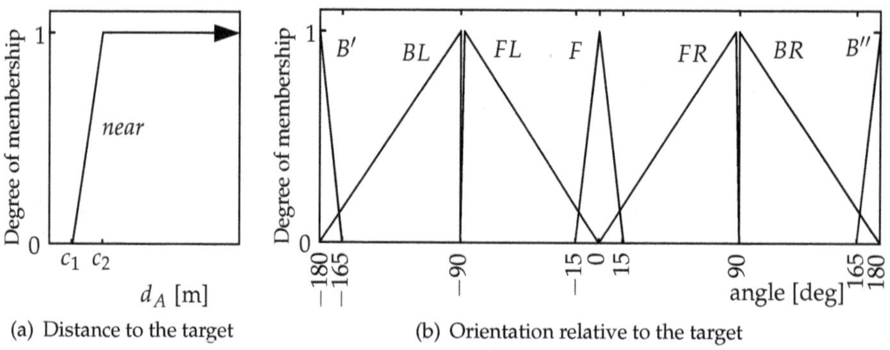

(a) Distance to the target (b) Orientation relative to the target

Fig. 2. Membership functions of input linguistic variables: (a) Orientation relative to the target – α_A, (b) Distance relative to the target d_A.

The linguistic variable α_A is defined through the following membership functions:

$$\alpha_a \{Front, FrontLeft, FrontRight, Back, BackLeft, Back, Right\} \tag{9}$$

or abbreviated as:

$$\alpha_A \{F, FL, FR, B, BL, BR\} \tag{10}$$

as shown in Figure 2(b). In order to enable bidirectional navigation, the variable α_A is strictly divided into two groups – the angles related to orientation at the front $\{F, FL, FR\}$, and the angles related to orientation at the rear $\{B, BL, BR\}$. The method proposed in this chapter analyzes mobile objects with an equal ability of maneuvering by front and rear pace; therefore, the objective of defined membership functions and fuzzy rules is to provide identical performance in both cases. For this reason, the membership function B (*Back*) is divided into two sub-functions B' and B'', in the following manner:

$$B = S(B', B'') \tag{11}$$

where S represents the operator of $S - norm$, which corresponds to the fact that the union of sets B' and B'' produces set B. Since the sets B' and B'' are disjunctive, the calculation of $S - norm$ is not important (Mitrović & Đurović, 2010a).

The distance-to-target linguistic variable d_A is described by means of a single membership function – *near* (Fig. 2(a)), and its task is to stop the WMR in close proximity to the target. This is ensured by adjustment of free parameters c_1 and c_2. The designed FLC is of the T-S type, with two inputs: the distance between the robot and the target – d_A and the orientation of the robot relative to the target – α_A, and two outputs: linear velocity v_R and angular velocity ω_R of the robot. The fuzzy controller rules for robot navigation to the target in an obstacle-free environment are shown in Table 1.

	B	BL	FL	F	FR	BR
v_R	$-V_m$	$-V_m/2$	$V_m/2$	V_m	$V_m/2$	$-V_m/2$
ω_R	0	ω_m	$-\omega_m$	0	ω_m	$-\omega_m$

Table 1. Fuzzy Logic Controller for navigation in an Obstacle–Free Environment (FLC_{OFE}) – fuzzy rules base.

The maximum linear velocity of the analyzed robot is $V_m = 0.5\,\text{m/s}$ and its maximum angular velocity is $\omega_m = 7.14\,\text{rad/s}$. The approximate dynamic model of a mobile robot (Mitrović, 2006) was used for the simulations, and the distance between robot wheels was 7 cm. Figure 3 shows WMR trajectories to the target (coordinate origin), from three initial configurations. Figure 4 shows controller outputs for the navigation scenarios depicted in Fig. 3. It is apparent from Fig. 4 that during navigation, one of the controller outputs assumes maximum values and this reduces the duration of the navigation process.

3. Bidirectional garage parking

The WMR garage parking (garaging) problem can also be viewed as a navigation problem in an environment with an obstacle, where the garage is the obstacle of a specific shape, but is also the target. When navigation problems with obstacles are solved, the goal is to achieve the ultimate configuration without colliding with the obstacle, which involves obstacle avoidance. Numerous obstacle avoidance and mobile robot navigation methods are discussed in literature. The potential field method, originally proposed by Khatib (1986) for obstacle avoidance in real-time by manipulators and mobile robots, plays a major role in solving this type of problem. A robot is navigated in a potential field which represents the sum of attractive forces originating from the target, and repulsive forces originating from the obstacles, while the analytical solution is the negative gradient of a potential function. Additionally, the resulting potential field may contain local minima where navigation ends without reaching the target, which is the greatest disadvantage of this method (Koren & Borenstein, 1991). The most frequent example of a local minimum is the case where obstacles

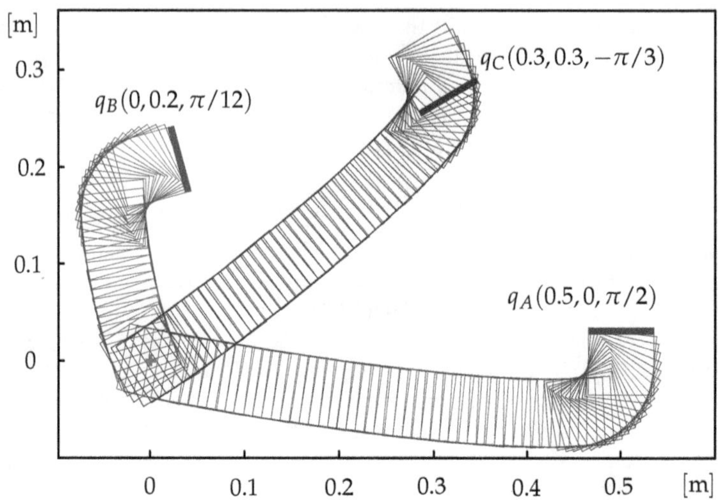

Fig. 3. Robot navigation trajectories from three initial postures (x_R, y_R, θ_R) to the target $(0,0)$.

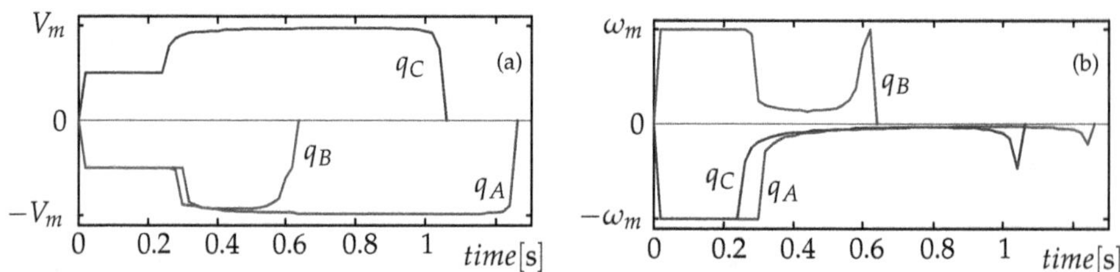

Fig. 4. Graphs for scenarios depicted in Fig. 3: (a) WMR linear velocity, (b) WMR angular velocity.

are deployed in such a way that they form a concave **U** barrier, such that the algorithm enters into an infinite loop or halts at the local minimum (Krishna & Kalra, 2001; Motlagh et al., 2009). Since the garage is a concave obstacle, the main shortfall of the potential field method could also be a solution to the garaging problem; however, analyzed literature does not provide an example of the use of the potential field method to solve mobile robot garaging problems. A probable reason for this is that the goal of garaging is to reach the target configuration in a controlled manner, with limited deviations, while the completion of navigation at the local minimum of the potential field is accompanied by oscillatory behavior of the object being parked, or the entry into an infinite loop (Koren & Borenstein, 1991), which is certainly not a desirable conclusion of the process.

The garaging/parking problem implies that a mobile robot is guided from an initial configuration $(x_{R(0)}, y_{R(0)}, \theta_{R(0)})$ to a desired one (x_G, y_G, θ_G), such that it does not collide with the garage. Figure 5 shows the garage and robot parameters; the position of the garage is defined by the coordinates of its center – C_m, while its orientation is defined by the angle of the axis of symmetry of the garage – S_G relative to the x axis, identified as β in the figure. The width of the garage is denoted by W_G and the length by L_G, and it is understood that garage dimensions enable the garaging of the observed object. The implication is that the goal

(x_G, y_G) is inside the garage and is set to coincide with the center of the garage C_m, such that $C_m(x_G, y_G)$. In the case of bidirectional garaging, the targeted configuration is not uniquely defined because the objective is for the longitudinal axis of symmetry of the robot to coincide with that of the garage, such that $\delta = 0$ or $\delta = \pi$, meaning that there are two solutions for the angle $\theta_G : \theta_{G_1} = \beta$, $\theta_{G_2} = \beta + \pi$. For reasons of efficiency, the choice between these two possibilities should provide the shortest travel distance of the mobile robot. The controller proposed in this section does not require *a priori* setting of the angle θ_G, because it has been designed in such a way that the mobile robot initiates the garaging process from the end closer to the garage.

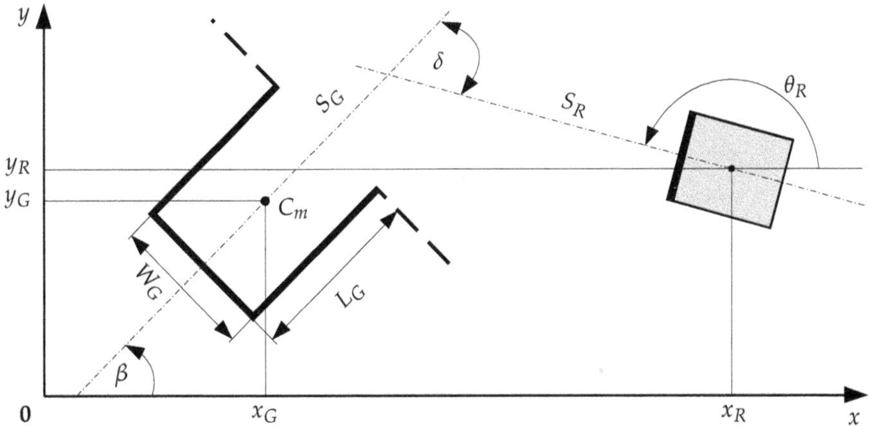

Fig. 5. Mutual spatial positions of wheeled mobile robot and garage.

The bidirectional garaging problem is similar to the stabilization problem, also known as the "parking problem", formulated by Oriolo et al. (2002): "the robot must reach a desired configuration $q_G(x_G, y_G, \theta_G)$ starting from a given initial configuration $q_{(0)}(x_{R(0)}, y_{R(0)}, \theta_{R(0)})$". The differences are that there are two solutions for the desired configuration in bidirectional garaging problems and that the stabilization problem need not necessarily involve constrains imposed by the presence of the garage.
Although the bidirectional garaging problem is addressed in this chapter as a stabilization problem, the proposed solution can also be used as a garaging trajectory generator because it includes non-holonomic constraints of the mobile robot.

3.1 System of two fictitious fuzzy magnets

From a human control skill perspective, the garaging of a vehicle is comprised of at least two stages: approach to the garage and entry into the garage. The adjustment of the vehicle's position inside the garage might be the third stage, but the need for this stage depends on the success of the second stage (i.e., driver skills). Since there is much less experience in differential drive mobile vehicle control than in car and car-like mobile robot control, and based on recommendations in (Sugeno & Nishida, 1985), the proposed WMR garaging system is a model of operator control actions, designed by means of two fictitious fuzzy magnets, one of which is located immediately in front of the garage and the other inside the garage. The first point is used to approach the garage and while the vehicle is entering the garage, it serves as a reference point for proper orientation, similar to human driving skills. The second point is both the target and a reference point for the WMR when it enters the garage and when it completes the garaging process. Contrary to the well-known point-to-point navigation approach, navigation aided by a system of two fuzzy magnets is executed in a single stage since the garaging process requires the robot to approach the area in front of the garage, but not to also reach the point which is the center of this area. The fictitious fuzzy magnets are denoted by FM_{Fm} (*Forward magnet*) and FM_{Cm} (*Central magnet*), and created according to (1):

$$FM_{Fm} = (F_m, FRsS_{Fm}), \tag{12}$$

$$FM_{C_m} = (C_m, FRsS_{Cm}). \tag{13}$$

In Fig. 6, their positions are denoted by C_m and F_m. The point F_m lies on the garage axis of symmetry S_G at a distance d_F from the front line. Let us imagine that these points are fictitious fuzzy magnets with attraction regions around them and if the mobile object/vehicle finds itself in that region, the attraction force will act on it. The activity of these fictitious fuzzy magnets will be neutralized in point C_m, thus finishing the garaging process.

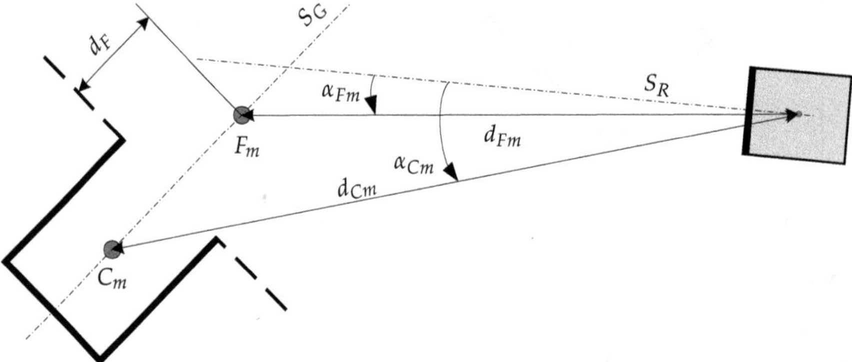

Fig. 6. Mutual spatial positions of wheeled mobile robot and fictitious fuzzy magnet.

The implementation of this concept requires the definition of the interaction between the fictitious fuzzy magnets and the vehicle, consistent with the definition of the fictitious fuzzy magnets. Figure 6 also shows the parameters which define the relationship between the fictitious fuzzy magnets and the vehicle. The orientation of the WMR relative to the point F_m is denoted by α_{Fm}, and relative to the point C_m by α_{Cm}. The distance between the WMR and the fictitious fuzzy magnet C_m is d_{Cm}, while the distance to the fictitious magnet F_m is d_{Fm}. The key step in the design of such an approach is the observation of these two distances as fuzzy sets which need to be attributed appropriate membership functions. The first task of the FM_{Fm} is to guide the WMR to the area immediately in front of the garage, such that the linguistic variable d_{Fm} is assigned a single membership function, denoted by far in Fig. 7(a). The membership function was not assigned to the linguistic variable $\mu_{far}(d_{Fm})$ in the standard manner because the entire universe of discourse is not covered (since $F_1 > 0$). This practically means that FM_{Fm} will have no effect on the robot when $d_{Fm} < F_1$, but FM_{Cm} will; the membership function for the distance to its position C_m is shown in Fig. 7(b). The membership function $\mu_{near}(d_{Cm})$ allows the action of fuzzy magnet C_m to affect the WMR only when the WMR is near FM_{Cm}, or when the distance of d_{Cm} is less than C_4. Even though no linguistic variable related to distance covers the entire set of possible values, the WMR is in the region

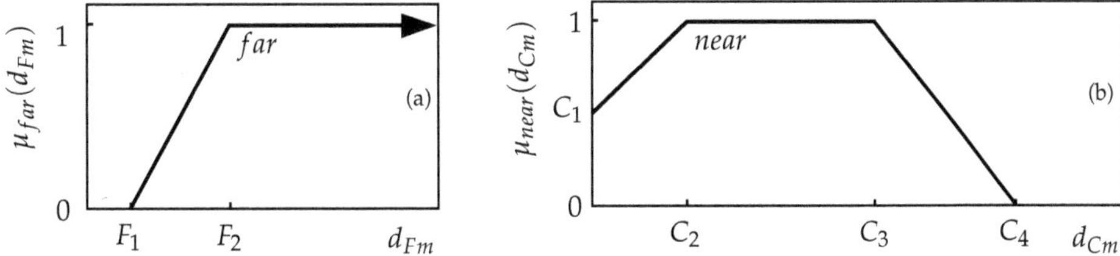

Fig. 7. Definition of membership functions for distances: (a) d_{Fm}, (b) d_{Cm}.

of attraction of at least one fictitious fuzzy magnet at all times. A more detailed procedure for the selection of parameters and fuzzy rules will be presented later in the text.
The next step in the definition of fuzzy rules subsets is the definition of the second input variable, associated with the WMR orientation relative to the fictitious fuzzy magnets. The angles are denoted by α_{Fm} and α_{Cm} in Fig. 6. If angle α_{Fm} is close to zero, then the vehicle's front end is oriented toward the fictitious fuzzy magnet F_m, and if the value of angle α_{Fm} is close to π, the vehicle's rear end is oriented toward F_m. This fact is important, as will be shown later in the text, from the perspective that the objective of vehicle control will be to reduce the angles α_{Fm} and α_{Cm} to a zero or π level, depending on initial positioning conditions. Linguistic variables α_{Fm} and α_{Cm} are together identified as *Direction* and are described by membership functions which should point to the orientation of the WMR relative to the fictitious fuzzy magnets according to (9) and (10), whose form is shown in Fig. 8.

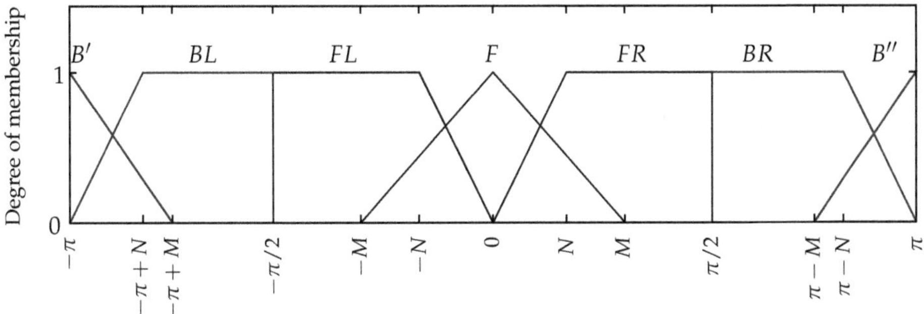

Fig. 8. Membership functions of linguistic variable *Direction*.

3.2 Definition of fuzzy rules

Since the WMR bidirectional garaging system is comprised of two fictitious fuzzy magnets, FM_{F_m} and FM_{C_m}, the fuzzy rules base is created from the rules subsets of each fuzzy magnet. It follows from the definition of the concept (3) that a fuzzy rules subset is determined by two inputs (WMR distance and orientation relative to fuzzy magnet position), meaning that the number of fuzzy controller inputs is equal to $K \times 2$, where K is the number of fuzzy magnets which make up the system, such that the number of inputs into a fuzzy system comprised of two fuzzy magnets is four. This concept is convenient because inherence rules for each fuzzy magnet are generated from inputs related to that magnet, such that the addition of a fuzzy magnet increases the number of rules in the controller base only in its subset of fuzzy rules. We will now specifically define two fuzzy rules subsets, $FRsS_{F_m}$ and $FRsS_{C_m}$. Since the

fuzzy rules subsets are created for the same object, the output variables are defined equally, regardless of whether concept (5-6) or (7-8) is applied.
The definition of output variables of the FLC requires a clear definition of the object of control. The algorithm described in this chapter is primarily intended for the control of two-wheeled, differential drive mobile robots. The wheels of this type of vehicle are excited independently and their speeds are defined autonomously. Guidance is controlled by the wheel speed difference, and the fuzzy rules subsets will be generated in the form of (5-6). We will assume that the physical characteristics of the actuators of both wheels are identical, and that the object has identical front- and back-drive capabilities. Since the inherence rules will be defined in general, we will assume certain general characteristics of the object of control, which will not affect the practical application of the concept. The maximum wheel speed will be denoted by the positive constant V_m, where $-V_m$ means that the wheel is rotating at maximum velocity which induces the object to move in reverse. We will also assume a nonnegative constant V_1, which satisfies the condition:

$$V_1 \leq V_m. \tag{14}$$

Fuzzy rules are defined based on the following principles (Mitrović & Đurović, 2010a):

- As the distance from the vehicle to the garage grows, the total speed of the vehicle should increase;
- As the distance from the vehicle to the garage shrinks, the total speed of the vehicle should drop;
- The difference between the wheel speeds causes turning, which must depend on the robot's orientation toward the garage;
- In the case of good orientation, the robot speed may be maximal.

We will identify the fuzzy logic controller for bidirectional garaging using the fictitious fuzzy magnet concept as FLC_{BG}. The fuzzy rules base of the FLC_{BG} is comprised of two subsets: the $FRsS_{Fm}$ subset, which corresponds to the fictitious fuzzy magnet FM_{Fm}, shown in Table 2(a), and the $FRsS_{C_m}$ subset shown in Table 2(b).

Rule	d_{Fm}	α_{Fm}	v_L	v_R
1.	far	B	$-V_m$	$-V_m$
2.	far	BL	0	$-V_m$
3.	far	FL	0	V_m
4.	far	F	V_m	V_m
5.	far	FR	V_m	0
6.	far	BR	$-V_m$	0

(a) Fuzzy rules subset $FRsS_{Fm}$

Rule	d_{Cm}	α_{Cm}	v_L	v_R
7.	$near$	B	$-V_m$	$-V_m$
8.	$near$	BL	0	$-V_1$
9.	$near$	FL	0	V_1
10.	$near$	F	V_m	V_m
11.	$near$	FR	V_1	0
12.	$near$	BR	$-V_1$	0

(b) Fuzzy rules subset $FRsS_{Cm}$

Table 2. Fuzzy rules base for bidirectional garaging fuzzy controller – FLC_{BG}.

Membership functions of input variables d_{Fm} and d_{Cm} independently activate particular rules (Table 2), an action which results in a considerable reduction in the number of rules. The rule which produces zero commands on both wheels does not exist; this might lead to the wrong conclusion that the vehicle never stops and does not take a final position. Since the points C_m and F_m are on the garage axis of symmetry, when the vehicle finds itself near its final position the difference between orientation angles α_{Fm} and α_{Cm} is close to $\pm\pi$. In this case, at least two rules generating opposite commands are activated, and their influence becomes annulled (Mitrović & Đurović, 2010a).

3.3 Selection of fuzzy controller parameters

However, with regard to the endless set of possible combinations of input variables, the selection of fuzzy rules does not guarantee that the vehicle will stop at the desired position. It is therefore, necessary to pay close attention to the selection of parameters d_F, M, N, F_1, $F2$, C_1, C_2, C_3, and C_4.

3.3.1 Parameter d_F

FM_{Fm} is described by (12) and, before the parameters which define the membership functions of the linguistic variables of $FRsS_{Fm}$ are determined, the position F_m of FM_{Fm} needs to be established. In Fig. 6, the point F_m is located on the longitudinal axis of symmetry of the garage and its distance from the garage door is d_F. The parameter d_F should be adopted in the first iteration, observing the following limitation:

$$d_F > \sqrt{(W_R/2)^2 + (L_R/2)^2}, \tag{15}$$

where W_R is the robot width, and L_R is the robot length, with the recommendation $d_F \approx W_R$.

3.3.2 Parameters M and N

The selection of parameters M and N, in their geometrical sense as presented in Fig. 8, influences the nature of the maneuver (curve and speed of turning) performed by the vehicle during garaging, especially at the beginning of the garaging process when the robot is at some distance from the garage. When the values of M and N are low, the vehicle rotates with a very small curve diameter; as these values increase, the arches circumscribed by the vehicle also increase (Mitrović & Đurović, 2010a). During the selection of these values, a compromise must be made to maintain the maneuvering capabilities of the vehicle, as well as the constraints imposed by geometry, namely, the ratio between the vehicle dimensions and the width of the garage. Generally, these parameters are adjusted in such a way that the vehicle circumscribes larger arches when distant from the garage, whereas more vivid maneuvers are needed in the vicinity of the garage. As such, in the second iteration, parameters M and N related to input α_{Fm} are adjusted, followed by those related to the input α_{Cm}, with the following constraints:

$$\pi/2 > M \geq N > 0. \tag{16}$$

3.3.3 Coefficients C_3, C_4, F_1, and F_2

The trapezoidal membership function *far* of the linguistic variable d_{Fm} (Fig. 7(a)) is determined by the coefficients F_1 and F_2, while the membership function *near* of the linguistic variable d_{Cm} is determined by the coefficients C_1, C_2, C_3 and C_4 (Fig. 7(a)). The coefficients C_3, C_4, F_1 and F_2 enable gradual activation of fictitious magnet C_m and the deactivation of fictitious magnet F_m, as the robot approaches its destination. The selection of these parameters has a critical impact on the performance of the entire fuzzy controller. Inequalities (17) and (18) guarantee the form of the membership functions shown in Fig. 7(a) and Fig. 7(b), respectively.

$$C_4 > C_3 > 0 \tag{17}$$

$$F_2 > F_1 \geq 0 \tag{18}$$

Figure 9(a) illustrates the action of FM_{F_m}. Outside the circle of radius F_2 , with its center at F_m (dark gray in the figure), this fuzzy magnet exhibits maximum activity. When the WMR is within this region, the distance between the robot and the point F_m, which was earlier denoted by d_{Fm} is greater than F_2 and $\mu_{far}(d_{F_m} > F_2) = 1$, as shown in Fig. 7(a).

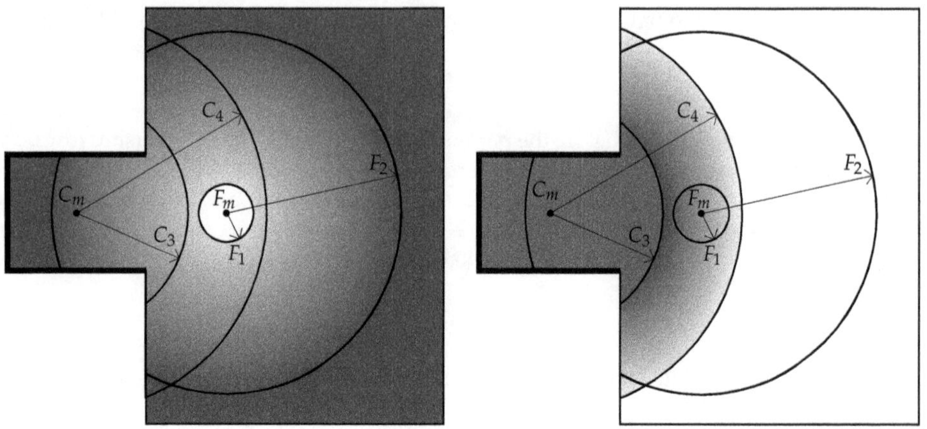

(a) Activity of fictitious fuzzy magnet FM_{Fm}. (b) Activity of fictitious fuzzy magnet FM_{Cm}.

Fig. 9. Illustration of fictitious fuzzy magnet activity.

The activity of FM_{Fm} declines as the position of the robot approaches the point F_m, through to the edge of the circle of radius F_1, inside which the activity of this fuzzy magnet equals zero (i.e., $\mu_{far}(d_{Fm} \leq F_1) = 0$). To facilitate adjustment of the overlap region of the two fuzzy magnets, it is convenient for the point C_m, which determines both the position of FM_{Cm} and the position of the center of the garage, to lie within the circle of radius F_2, which can be expressed, based on Fig. 9(a), by the following inequality:

$$F_2 \geq L_G/2 + d_F. \tag{19}$$

The coefficients C_1 and C_2 (Fig. 7(b)) should initially be set at $C_1 = 1$ and $C_2 = 0$. Figure 9(b) illustrates the activity of FM_{Cm}. Outside the circle of radius C_4, whose center is at C_m, the activity of this fuzzy magnet is equal to zero (i.e., $\mu_{near}(d_{Cm} > C_4) = 0$). The activity of FM_{Cm} increases as the robot approaches the point C_m, through to the edge of the circle of radius C_3, within which the activity of this fuzzy magnet is maximal (i.e., $\mu_{near}(d_{Cm} \leq C_3) = 1$), shown in dark gray in the figure). To ensure that at least one fuzzy magnet is active at any point in time, the following constraint must strictly be observed:

$$C_4 > L_G/2 + d_F + F_1. \tag{20}$$

This constraint ensures that the circle of radius F_1 is included in the circle of radius C_4, which practically means that the vehicle is within the region of attraction of at least one fuzzy magnet at all times. The dominant impact on vehicle garaging is obtained by overlapping fictitious fuzzy magnet attraction regions, and proper vehicle garaging is achieved through the adjustment of the region in which both fictitious fuzzy magnets are active.

3.3.4 Coefficients C_1 and C_2

The selection of the above parameters might enable vehicle stopping near point C_m. The simultaneous adjustment of the garaging path and stopping at the target position, due to a large number of free parameters, requires a compromise which may have a considerable impact on the quality of the controller. The introduction of parameters C_1 and C_2 enables unharnessed adjustment of vehicle stopping. The selection of parameters C_1 and C_2, enabled by theoretically exact stopping of vehicles at point C_m, will cause oscillations in movement around point C_m due to the dynamics of the vehicle, a factor which was neglected in the process of controller design. Accordingly, the selected desired final point of garaging is in the immediate vicinity of the fictitious magnet C_m location. Coefficients C_1 and C_2 take into account the neglected dynamics of the vehicle, and their adjustment is performed experimentally, observing the following constraints:

$$d_{Cm} < L_G/2 \Rightarrow d_{Fm} \leq L_G/2 + d_F. \tag{21}$$

The above relation ensures that at least two opposed rule-generating commands are always activated around the point C_m. If the variable d_{Fm} assumes a value larger than $L_G/2 + d_F$, meaning that the vehicle has practically surpassed the target point, the angles α_{Cm} and α_{Fm} will become nearly equal, and results in the activation of rules that generate commands of the same sign, which in turn causes oscillations around the point C_m.

3.4 Simulation results

An approximate dynamic model of the *Hemisson* robot (Mitrović, 2006) was used for the simulations. This is a mobile vehicle of symmetrical shape, with two wheels and a differential drive. Wheels are independently excited, and their speeds may be defined independently from each other. Guidance is controlled by the difference in wheel speeds. Each of these speeds can be set as one of the integer values in the interval $[-9, 9]$, (Table 3), where a negative value means a change in the direction of wheel rotation. The dependence of the command on wheel speed is not exactly linear causing certain additional problems in the process of vehicle control (Mišković et al., 2002). Specifically, for the subject case, the selected output variables of the FLC are the speed commands of the left and right wheels: VL and VR, respectively. Accordingly, this discrete set of integer values from the interval $[-9, 9]$ is selected as the domain of membership functions attributed to the output variables.

Wheel Speed Command	0	1	2	3	4	5	6	7	8	9
Wheel Speed $[\mathrm{mm/s}]$	0	3	18	42	64	86	109	134	156	184

Table 3. Wheel speed of the Hemisson robot as a function of wheel speed command.

The sampling period of the FLC was $T_s = 0.2\,\mathrm{s}$. The size of the garage was 16 cm × 20 cm, and the size of the robot was 10 cm × 12 cm. The FLC_{BG} was designed to conform to the constraints defined in the previous section, with the objective of ensuring successful garaging. The experimentally-determined values of these parameters, for the examined case of the Hemisson robot, are presented in Table 4.

Figure 10 shows WMR garaging trajectories for a set of 17 different initial conditions. Depending on the initial configuration, the WMR is guided to the target by forward or back drive, which illustrates the bidirectional feature of the proposed fictitious fuzzy magnets concept.

Coeff.	Value	Coeff.	Value
$M(\alpha 1)$	$7\pi/36$	C_1	$20/63$
$N(\alpha 1)$	$\pi/12$	C_2	$4.3\,\text{cm}$
$M(\alpha 2)$	$\pi/4$	C_3	$16\,\text{cm}$
$N(\alpha 2)$	$\pi/18$	C_4	$40.0\,\text{cm}$
F_1	$4.0\,\text{cm}$	d_F	$11.0\,\text{cm}$
F_2	$44.5\,\text{cm}$	V_1	Command 3

Table 4. FLC_{BG} – coefficient values.

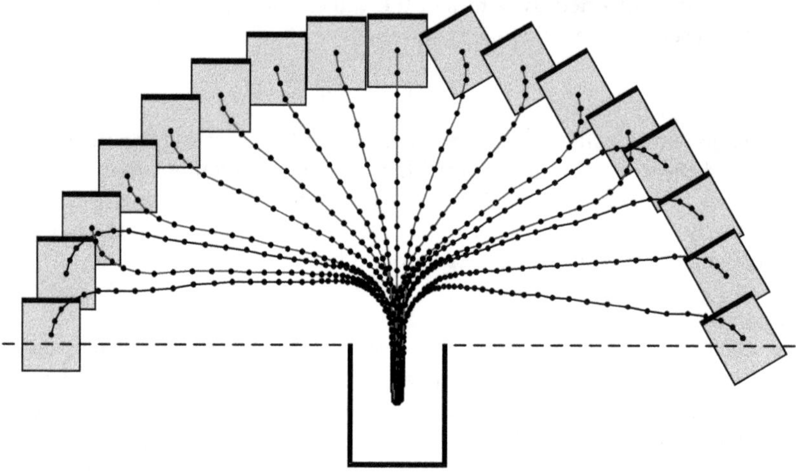

Fig. 10. Illustrative trajectories for front- and back-drive garaging

To test the efficiency and applicability of the proposed method using an extended set of initial configurations, a new test set of N initial conditions was created. This set of initial configurations ($x_{Ri,(t=0)}, y_{Ri,(t=0)}$, and $\theta_{Ri,(t=0)}$), $i = 1 : N$, was formed such that the robot is placed at equidistant points on the x and y axes spaced 1 cm apart, while the angle $\theta_{Ri,(t=0)}$ was a uniform-distribution random variable in the interval $[-\pi/2, \pi/2]$. The test configuration set was divided into two sub-sets: the first satisfies the condition that the robot in its initial configuration is at an adequate distance from the garage and contains N_d elements, while the other includes initial configurations near the garage and contains $N_c = N - N_d$ elements. Garaging performance was measured by the distance of the robot from the target position d_{Cm} (Fig. 6), and the angle δ (Fig. 5), which showed that the robot and garage axes of symmetry did not coincide:

$$\delta = |(\beta - \theta_R) \bmod \pi| \tag{22}$$

The results of simulations with the FLC_{BG}, for the set of initial conditions N_d, are shown in Fig. 11. In all N_d cases, the garaging process was completed with no collision occurring between the robot and the garage. In view of the non-linear setting of the speeds and the long sampling period, the conclusion is that the results of garaging were satisfactory. The average deviation in N_d cases was $\overline{d}_{Cm} = 1.1\,\text{cm}$, while the average angle error was $\overline{\delta} = 1.4°$.

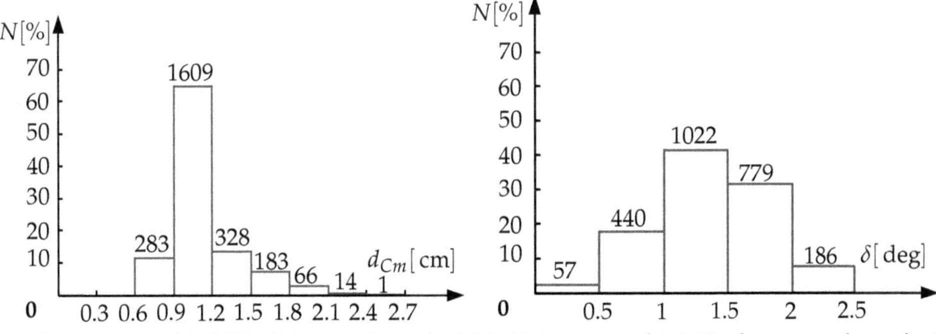

(a) Histograms of WMR distances from final position

(b) Histograms of WMR deviations from final orientation

Fig. 11. Results of simulations with FLC_{BG}, for the set of initial conditions N_d.

3.5 Robustness of the system

Since the system was designed for use with a real vehicle, its efficiency needed to be examined under less-than-ideal conditions. To operate the system, the positions and orientations of the robot for each selection period must be determined, whereby the robot position and orientation need not necessarily coincide with the real position and orientation. Since inputs into the system consist of two distances and two angles whose accuracy depends directly on the accuracy of the determination of the position of the robot during the garaging process, the effect of inaccuracies of the vehicle coordinates on the efficiency of the garaging process were analyzed. The set of experiments was repeated for the FLC_{BG} with N_d initial conditions, and the WMR coordinate determination error was modeled by noise with uniform distribution within the range $[-1\,\text{cm}, 1\,\text{cm}]$. Figure 12 shows histograms of distance and orientation deviations from the targeted configuration, under the conditions of simulated sensor noise. It was found that the system retained its functionality but that the deviations were greater than those seen in the experiment illustrated in Fig. 11.

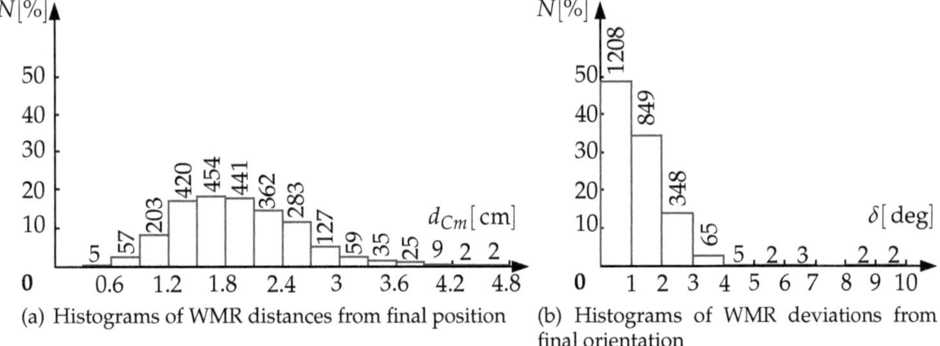

(a) Histograms of WMR distances from final position

(b) Histograms of WMR deviations from final orientation

Fig. 12. Results of simulations with FLC_{BG} with simulated sensor noise, for the N_d initial conditions.

Table 5 shows extreme and mean values derived from the garaging experiments. The subscript N denotes that the simulations were conducted under simulated sensor noise conditions. The sensor noise mostly affected the maximum distances from the target position $d_{Cm\,\max}$; the mean values of angle error $\overline{\delta}$ were unexpectedly lower under simulated sensor noise conditions.

	$\overline{\delta}\,[°]$	$\delta_{\max}\,[°]$	$\overline{d}_{Cm}\,[\mathrm{cm}]$	$d_{Cm\max}\,[\mathrm{cm}]$
FLC_{BG}	1.37	2.35	1.10	2.41
FLC_{BG_N}	1.19	9.52	1.93	4.80

Table 5. Extreme and mean values derived from garaging experiments.

3.6 Limitations of the proposed FLC

Simulations were conducted for the sub-set N_c of initial conditions, under which the robot was not at a sufficient distance from the garage and was placed at the points of an equidistant grid, spaced 0.25 cm apart, where the initial orientation of the robot $\theta_{Ri,(t=0)}$ was a uniform-distribution random variable in the interval $[-\pi/2, \pi/2]$. Figure 13 shows the regions of initial conditions where the probability of collision of the robot with the garage is greater than zero. Full lines identify the limits of the region under ideal conditions, while dotted lines denote simulated sensor noise conditions. Figure 13 shows that sensor noise has little effect on the restricted area of initial conditions, which is an indicator of the robustness of the system to sensor noise.

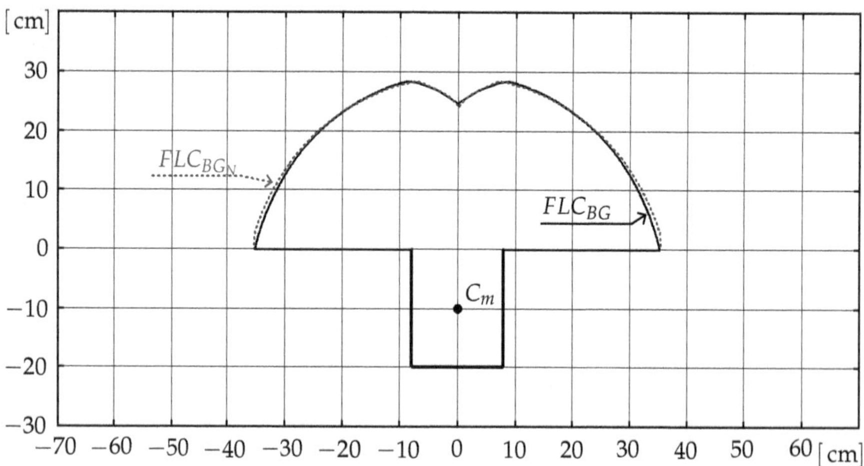

Fig. 13. Limits of the system – region of initial configurations with probable collision of the WMR with the garage: full line – ideal conditions (FLC_{BG}); dotted line – with sensor noise conditions (FLC_{BG_N}).

3.7 Experimental results

An analogous experiment was performed with a real mobile vehicle (Hemisson mobile robot) and a real garage whose dimensions are 16 cm × 20 cm. A block diagram of the garaging experiment is shown in Fig. 14.

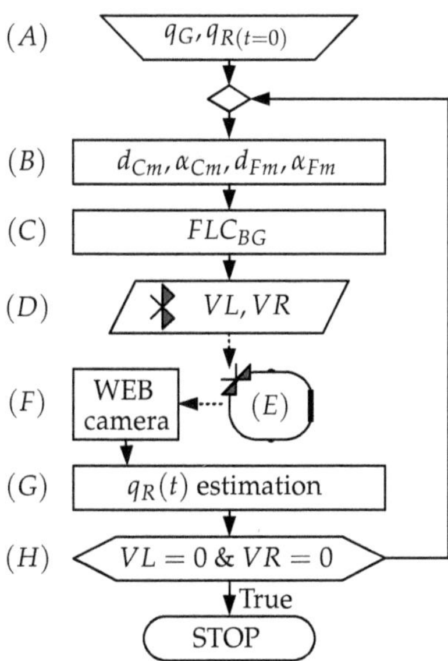

Fig. 14. Block diagram of real-time experiment: (A) Determining of initial conditions; (B) Calculation of: d_{Cm}, d_{Fm}, α_{Cm} and α_{Fm}; (C) Fuzzy logic controller for bidirectional garaging; (D) Bluetooth interface; (E) Hemisson mobile robot; (F) Web camera; (G) Estimation of the WMR posture; (H) Examination of VL and VR values.

A personal computer with a Bluetooth interface (D) and web camera (F) – resolution 640×480 pixels, were used during the experiment. The initial conditions were determined prior to the initiation of the garaging process, namely (A): the position and orientation of the garage q_G, and the posture (position and orientation) of the robot $q_{R(t=0)}$. Based on the posture of the robot and the position and orientation of the garage, in block (B), input variables d_{Cm}, d_{Fm}, α_{Cm} and α_{Fm} were calculated for the FLC_{BG} (C). FLC outputs were wheel speed commands for the Hemisson robot (VL and VR), which were issued to the robot via the Bluetooth interface. The web camera performed a successive acquisition of frames in real time, with a 0.2 s repetition time, which dictated the sampling period for the entire fuzzy controller. In block (G), web camera frames were used to estimate the position of the robot and compute its orientation (Đurović & Kovačević, 1995; Kovačević et al., 1992). During the garaging process, wheel speed commands were different from zero. When both commands became equal to zero, garaging was completed and block (H) halted the execution of the algorithm.

Figures 15(a) and 15(b) show 12 typical trajectories obtained during the process of garaging of a Hemisson robot in a real experiment. The illustrative trajectories shown in Fig. 15(a) deviate from expected trajectories and are mildly oscillatory in nature, which is a result of robot wheel eccentricity and the ultimate resolution of the camera used to estimate the position and orientation of the robot. The system which tracks the WMR to the target can be upgraded by a non-linear algorithm (Đurović & Kovačević, 2008) or instantaneous acquisition by two sensors (Đurović et al., 2009). Figure 15(b) shows trajectories obtained during the course of garaging from "difficult" initial positions (initial angle between the axis of symmetry of the robot and the axis of symmetry of the garage near $\pi/2$). These trajectories were obtained equally under front- and back-drive conditions (the front of the robot is identified by a thicker line). All of the illustrated trajectories are indicative of good performance of the proposed garaging algorithm.

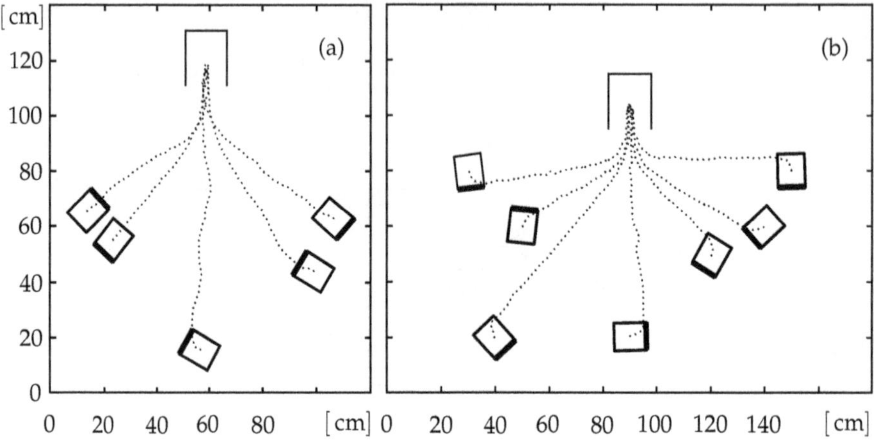

Fig. 15. Real-time garaging experiment results.

4. Obstacle avoidance

The navigation of a mobile robot in a workspace with obstacles is a current problem which has been extensively researched over the past few decades. Depending on the definition of the problem, the literature offers several approaches based on different methodologies. Relative to the environment, these approaches can be classified depending on whether the vehicle moves in a known (Kermiche et al., 2006), partially known (Rusu et al., 2003), or totally unknown environment (Pradhan et al., 2009). Further, the environment can be stationary or dynamic – with moving obstacles (Hui & Pratihar, 2009), or a moving target (Glasius et al., 1995). With regard to the ultimate target, the mobile robot is navigated to a given point (Mitrović et al., 2009), or a given configuration – a point and a given orientation at that point (Mitrović & Đurović, 2010a), or navigated to reach and follow a given trajectory (Tanaka & Sano, 1995). The problems are solved both in real-time and off-line.

Numerous obstacle avoidance methods are discussed in literature. The potential field method (Khatib, 1986) plays a major role in solving this type of problem. The potential function has to be derived for each individual obstacle and this can be a drawback from a real-time perspective if there are a large number of obstacles. Additionally, the resulting potential field may contain local minima where navigation ends without reaching the goal (Koren & Borenstein, 1991).

Fuzzy-logic-based controllers are widely applied to address obstacle avoidance and mobile robot navigation problems, and they range from reactive fuzzy controllers (which do not include a planning stage and generate outputs based on real-time sensor signals) to controllers which resolve the trajectory planning problem of a WMR through the synthesis of fuzzy logic with genetic algorithms, neural networks, or the potential field approach. Reignier (1994) proposed a reactive FLC for WMR navigation through a workspace with obstacles, based on data from 24 ultrasonic sensors. Abdessemed et al. (2004) proposed a reactive controller inspired by the potential field method, with a rule base optimized by evolutionary techniques. Hui & Pratihar (2009) compared the potential field method with soft computing methods used to solve moving obstacle avoidance problems.

This section proposes a new methodology for robot navigation in a workspace which contains one or more obstacles. The approach is based on the fictitious fuzzy magnet concept but, contrary to the system described in the previous section, the rules are generated such that the action of the fuzzy magnet is to repel the vehicle. It should be noted that the inputs into the controller based on the fuzzy magnet concept are the distance and orientation relative to the fuzzy magnet position, which is a point, while the obstacles have dimensions which must be taken into account at the controller design stage. The controller proposed in this section uses information about the relative dimensions of the obstacle. At the preprocessing stage, the left and right side of the obstacle in the robot's line of sight are determined based on the mutual positions of the robot and the obstacle. Since controller inputs are based on the relative dimensions of the obstacle, the controller can be applied to obstacles of different shapes and sizes.

4.1 Navigation in a single-obstacle environment

In general, WMR navigation to a target in a workspace with obstacles is comprised of two tasks: avoidance of obstacles and navigation to the target. The controller may be designed to integrate these two tasks, or the tasks can be performed independently but in parallel. Parallel processing of target and obstacle data, contrary to integrated approaches, offers a number of advantages, ranging from fewer FLC inputs to parallel distribute compensation discussed by Wang et al. (1996). To allow for the algorithm to be used for navigation in an environment which contains an unknown number of obstacles, the parallel fuzzy controller data processing approach was selected. The FLC_{OFE} controller discussed in Section 2.1 was used to execute a navigation task to a target in an obstacle-free environment. Instead of v_R and ω_R, in this section its outputs will be denoted by v_{OFE} and ω_{OFE}, respectively, since notations with R are reserved for outputs from the entire navigation system.

Variables which will reflect the mutual positions of the robot and the obstacle need to be defined and efficient navigation in real-time enabled, regardless of the position and size of the obstacle relative to the target and the current configuration of the robot. Figure 16 shows the WMR and obstacle parameters. Since we address the WMR navigation problem in planar form, we will assume that the obstacle in the robot's workspace is a circle defined by its center coordinates and radius: $O(x_O, y_O, r_O)$. Figure 16 shows a real obstacle whose radius is r_r, in light gray. Because of the finite dimensions of the robot, it is extended by a circular ring of width r_{cl} (dark gray). Hereafter we will use r_O for the radius of the obstacle, computed from the equation:

$$r_O = r_r + r_{cl}, \qquad r_{cl} > W_G/2. \tag{23}$$

In Fig. 16, the distance between the WMR center and the obstacle is denoted by d_{RO}. Point P_C lies at the intersection of the WMR course line and the obstacle's radius which form a right angle and represent a virtual collision point. When this intersection is an empty set, the obstacle does not lie on the WMR trajectory and point P_C is not defined. The portion of the radius which is located to the left of the point P_C relative to the WMR is denoted by l_L, while the portion of the radius to the right of P_C is denoted by l_R. We will use variables l_L and l_R to define the obstacle avoidance algorithm. Since l_L and l_R are dimensional lengths and since

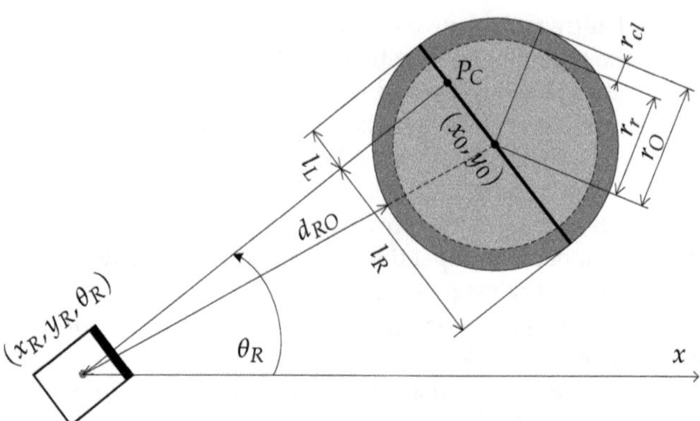

Fig. 16. Relative positions of obstacle and mobile robot.

their values depend on the WMR position and obstacle dimensions, we will normalize them based on the equation:

$$\begin{aligned} L_L &= l_L/(2 \cdot r_O), \\ L_R &= l_R/(2 \cdot r_O), \end{aligned} \tag{24}$$

where L_L and L_R are the left and right side of the obstacle, respectively, and can assume values from the interval $[0, 1]$, regardless of obstacle dimensions, since $L_L + L_R = 1$. If L_L or L_R is equal to 1, the robot will follow a course which will bypass the obstacle. If point P_C is not defined, we will assume that L_L and L_R are equal to zero, meaning that the obstacle does not lie on the robot's course. Based on the position of the virtual collision point P_C and the linear velocity of the robot v_R, we will generate the obstacle avoidance controller input using:

$$Side = \begin{cases} [-1, -0.5) & \text{if } L_L \in (0.5, 1] \text{ and } v_R \le 0 \\ [-0.5, 0) & \text{if } L_R \in [0.5, 1] \text{ and } v_R \le 0 \\ (0, 0.5] & \text{if } L_R \in [0.5, 1] \text{ and } v_R > 0 \\ (0.5, 1] & \text{if } L_L \in (0.5, 1] \text{ and } v_R > 0 \\ 0 & \text{if } L_R = 0 \quad \text{and } L_L = 0\,. \end{cases} \tag{25}$$

The variable *Side* provides information about the position of the obstacle relative to the robot and is equal to zero if the obstacle does not lie on the robot's course. The corresponding membership functions are shown in Fig. 17(a). In the name of the membership function, subscript B stands for *Backward* and F for *Forward*. To ensure efficient bypassing of the obstacle, in addition to the position of the obstacle relative to the WMR, the controller has to be provided with the distance between the WMR and the obstacle, denoted by d_{RO} in Fig. 16, whose membership functions are shown in Fig. 17(b).

The outputs of the obstacle avoidance controller are the linear velocity v_{OA} and the angular velocity ω_{OA} of the robot. The respective fuzzy rules are shown in Table 6.

It is important to note that the outputs of the obstacle avoidance controller are equal to zero when the obstacle does not lie on the robot's course. For this reason the algorithm for WMR navigation to the target point in a single-obstacle environment must integrate FLC_{OFE} and FLC_{OA} solutions. The linear velocity v_R and the angular velocity ω_R of the robot are computed

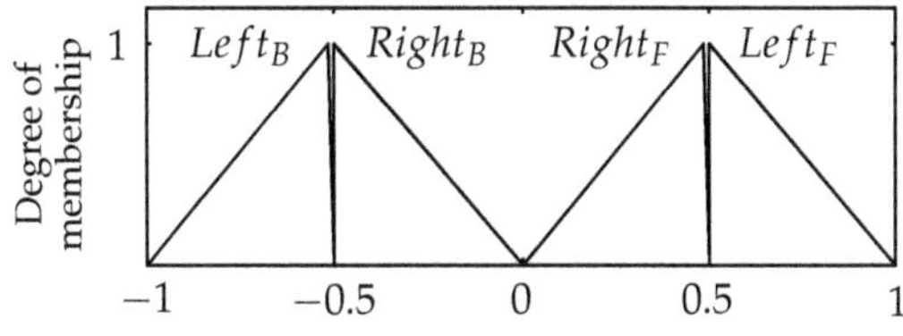

(a) Membership functions of linguistic variable *Side*.

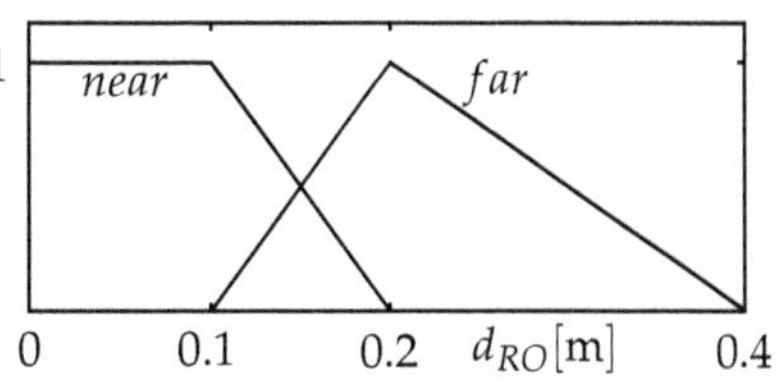

(b) Membership functions of linguistic variable distance to obstacle – d_{RO}.

Fig. 17. Membership functions of obstacle avoidance fuzzy controller FLC_{OA}.

Rule	*Side*	d_{OR}	v_{OA}	ω_{OA}
1.	$Left_F$	*near*	$V_m/2$	$-\omega_m$
2.	$Left_B$	*near*	$-V_m/2$	ω_m
3.	$Right_F$	*near*	$V_m/2$	ω_m
4.	$Right_B$	*near*	$-V_m/2$	$-\omega_m$
5.	$Left_F$	*far*	V_m	$-\omega_m/2$
6.	$Left_B$	*far*	$-V_m$	$\omega_m/2$
7.	$Right_F$	*far*	V_m	$\omega_m/2$
8.	$Right_B$	*far*	$-V_m$	$-\omega_m/2$

Table 6. FLC_{OA} – fuzzy rules base.

from the following relation:

$$\omega_R = \begin{cases} \omega_{OFE} & \text{if } \omega_{OA} = 0 \\ \omega_{OA} & \text{if } \omega_{OA} \neq 0 \end{cases}; \quad v_R = \begin{cases} v_{OFE} & \text{if } v_{OA} = 0 \\ v_{OA} & \text{if } v_{OA} \neq 0 \end{cases}. \tag{26}$$

To decrease the duration of the navigation process, the controller output v_R was post-processed to ensure maximum linear velocity v_{Rmax} of the robot based on (Mitrović & Đurović, 2010b):

$$v_{Rmax}(\omega_R, v_R) = sgn(v_R) \cdot (V_m - |\omega_R| \cdot D/2), \tag{27}$$

where D is the distance between the robot wheels. The maximum linear velocity of the robot is a function of the angular velocity ω_R and the physical constraints of the robot. Figure 18 illustrates the trajectories followed by the WMR to avoid two obstacles: $O_1(-0.4, 0, 0.15)$ and $O_2(0.3, 0.1, 0.05)$, from four initial positions: $q_A(-0.8, 0, \pi)$, $q_B(-0.7, 0.2, \pi/4)$, $q_C(0.45, 0.25, 0)$, and $q_D(0.6, 0.2, -2\pi/3)$.

4.1.1 Moving obstacle avoidance

The proposed algorithm, without modification, was applied in a single moving obstacle avoidance scenario. It was assumed that the obstacle travels along a straight line at a constant speed v_{obs}, which is equal to one-third of the maximum speed of the robot. The outcome was successful when the robot overtook or directly passed by the obstacle. However, if it attempted to pass by at an angle, the robot collided with the obstacle. This was as expected, since the controller does not take the speed of the obstacle into account. Figure 19 illustrates the scenarios in which the robot overtakes, directly passes by, and meets with the moving obstacle at an angle.

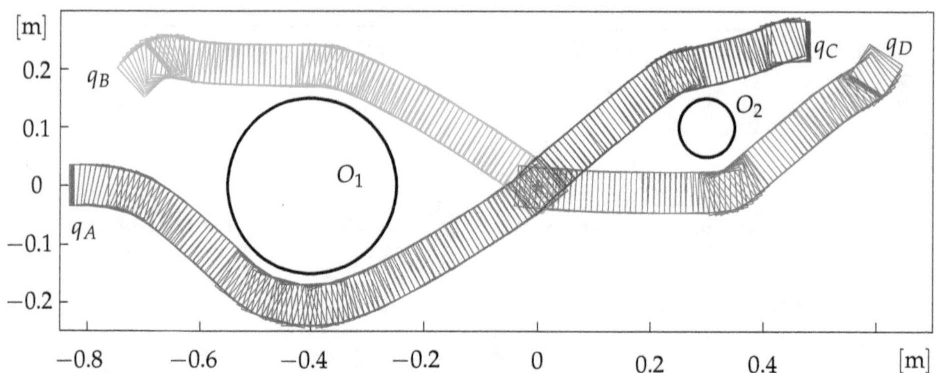

Fig. 18. Bypass trajectories of the robot from four initial positions around two obstacles of different dimensions.

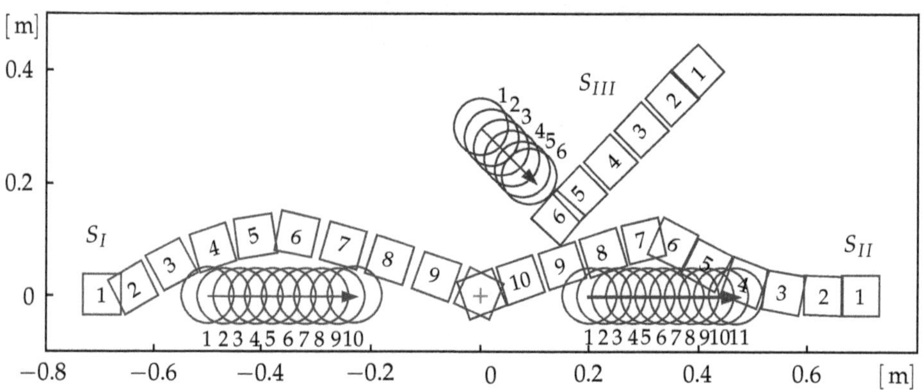

Fig. 19. Avoidance of moving obstacles: S_I – overtaking, S_{II} – directly passing, S_{III} – meeting at an angle.

4.2 Navigation trough a group of obstacles

We will now address the scenario in which there is a group of N obstacles in the workspace of the robot, where $N > 1$. To benefit from the approach described in the previous section, we will use the single-obstacle FLC_{OA} for each individual obstacle and thus obtain N values for the robot's linear and angular velocities. We will then compute the final angular velocity from the following equation:

$$\omega_{OA} = \left(\sum_{k=1}^{N} C_k \cdot \omega_{OAk} \cdot d_{ROk}^{-1}\right) \cdot \left(\sum_{k=1}^{N} C_k \cdot d_{ROk}^{-1}\right)^{-1}, \tag{28}$$

where ω_{OAk} is the FLC_{OA} output for the kth obstacle and d_{ROk} is the distance between the robot and the kth obstacle. As the distance from the vehicle to the kth obstacle shrinks, its influence to the controller outputs should increase. Using the coefficient C_k, we will take into account only those obstacles which lie on the robot's course. It is computed using (25), as follows:

$$C_k = \begin{cases} 0 & \text{if } Side_k = 0 \\ 1 & \text{if } Side_k \neq 0 \end{cases} . \tag{29}$$

We will derive the speed for the system of N obstacle avoidance controllers in a manner similar to (28), from:

$$v_{OA} = \left(\sum_{k=1}^{N} C_k \cdot v_{OAk} \cdot d_{ROk}^{-1} \right) \cdot \left(\sum_{k=1}^{N} C_k \cdot d_{ROk}^{-1} \right)^{-1} , \tag{30}$$

where v_{OAk} is the FLC_{OA} output for the kth obstacle. Outputs from the system of N FLC_{OA} and FLC_{OFE} controllers are integrated by means of (26), and the speed of the robot is maximized based on (27). Figure 20 shows a scenario with several randomly-distributed obstacles. Initial positions and the target (coordinate origin) are beyond the obstacle area. It is apparent from the figure that the proposed methodology efficiently solves the problem of navigation through an area which includes obstacles.

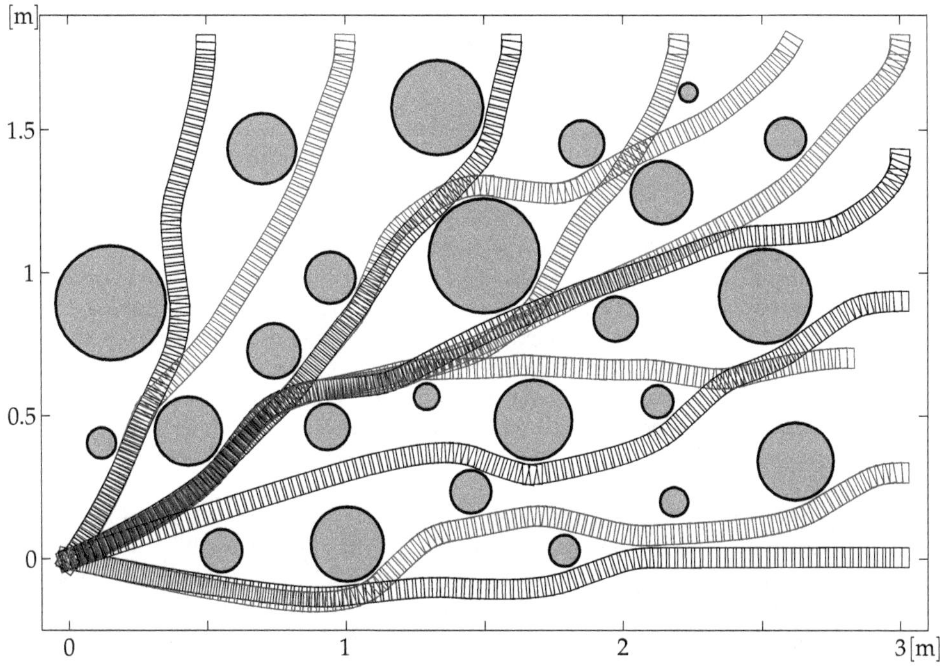

Fig. 20. Navigation to the goal (coordinate origin) through a group of obstacles.

4.2.1 Constraints of the system

Although the positions of the obstacles and the target are known, a closer analysis of the proposed approach leads to the conclusion that navigation to the target takes place in an unknown environment. The position of the target is known in all of the illustrated scenarios, but obstacles are taken into account only after they have triggered appropriate FLC rules, that is, when the distance between the robot and the obstacle is not greater than 40 cm (universe of

discourse of the linguistic variable d_{RO}, Fig. 17(b)). This basically means that the robot takes into account obstacles within a 40 cm radius.
It is well-known that a basic FLC cannot solve navigation through a workspace with arbitrarily distributed obstacles, if prior information about them is not available. A special problem is encountered when the obstacles are deployed such that they create a concave, **U**, barrier, in which case the robot enters an infinite loop or local minima (Krishna & Kalra, 2001; Motlagh et al., 2009).

5. Conclusion

A fictitious fuzzy magnet concept was introduced and its application analyzed in this chapter, using different navigation problems of differential drive mobile robots. The concept allows for navigation to the target in a single maneuver, without changing the direction of travel of the mobile robot. The symmetricity of the differential drive mobile robot is fully utilized, such that approaches based on the fictitious fuzzy magnet concept provide bidirectional solutions to mobile robot navigation problems. For example, during garaging, the mobile robot is automatically parked from the end of the robot which is closer to the garage entrance. The algorithm can be applied when the control variable is of the discrete type and where there are relatively few quantization levels.
When applied to a mobile robot garaging problem, a detailed analysis of simulation and experimental results illustrates the efficiency of the proposed algorithm and its robustness in the case of a random or systematic WMR position estimation error, as well as its limitations. The most significant shortfall of the proposed algorithm is that it does not provide a solution which will ensure that regardless of initial conditions the garage parking process is completed with no collision with the garage. The geometrical position of the initial conditions which lead to a collision of the robot with the garage is a compact, finite area which is discussed in the chapter. Some of the constraints mentioned here could be overcome in further research aimed at improving the proposed algorithm through a higher level of FLC complexity. Additionally, if a larger number of fictitious fuzzy magnets are introduced, the concept could be used to perform more intricate garage parking tasks. Re-configuration of controller outputs would render the proposed algorithm applicable for garage parking of a broader class of car-like mobile robots.
Additionally, it was shown that a larger number of fictitious fuzzy magnets can be introduced to execute tasks more complex than garage parking. Specifically, the feasibility was explored of applying the fictitious fuzzy magnet concept to execute the task of mobile robot navigation to a target in an environment which contains one or more obstacles. It was shown that the algorithm designed to avoid a single stationary obstacle can be successfully generalized to a multiple obstacles avoidance problem, preserving the bidirectional capability. The proposed algorithm can be extended by fuzzification of the coefficients C_k, taking into account the probability of collision with obstacles which lie on the robot's course.

6. References

Abdessemed, F., Benmahammed, K. & Monacelli, E. (2004). A fuzzy-based reactive controller for a non-holonomic mobile robot, *Robotics and Autonomous Systems* 47(1): 31–46.

Đurović, Ž. & Kovačević, B. (1995). QQ–plot approach to robust Kalman filtering, *International Journal of Control* 61(4): 837–857.

Đurović, Ž. & Kovačević, B. (2008). A sequential LQG approach to nonlinear tracking problem, *International Journal of Systems Science* 39(4): 371–382.

Đurović, Ž., Kovačević, B. & Dikić, G. (2009). Target tracking with two passive infrared non-imaging sensors, *IET Signal Processing* 3(3): 177–188.

Glasius, R., Komoda, A. & Gielen, S. (1995). Neural network dynamics for path planning and obstacle avoidance, *Neural Networks* 8(1): 125–133.

Hui, N. & Pratihar, D. (2009). A comparative study on some navigation schemes of a real robot tackling moving obstacles, *Robotics and Computer Integrated Manufacturing* 25(4-5): 810–828.

Kermiche, S., Saidi, M., Abbassi, H. & Ghodbane, H. (2006). Takagi–Sugeno based controller for mobile robot navigation, *Journal of Applied Sciences* 6(8): 1838–1844.

Khatib, O. (1986). Real-time obstacle avoidance for manipulators and mobile robots, *The International Journal of Robotics Research* 5(1): 90–98.

Koren, Y. & Borenstein, J. (1991). Potential field methods and their inherent limitations for mobile robot navigation, *Proceedings of the IEEE Conference on Robotics and Automation*, Sacramento, pp. 1398–404.

Kovačević, B., Đurović, Ž. & Glavaški, S. (1992). On robust Kalman filtering, *International Journal of Control* 56(3): 547–562.

Krishna, K. & Kalra, P. (2001). Perception and remembrance of the environment during real-time navigation of a mobile robot, *Robotics and Autonomous Systems* 37: 25–51.

Mišković, Lj., Đurović, Ž. & Kovačević, B. (2002). Application of the minimum state error variance approach to nonlinear system control, *International Journal of Systems Science* 33(5): 359–368.

Mitrović, S. (2006). Design of fuzzy logic controller for autonomous garaging of mobile robot, *Journal of Automatic Control* 16: 13–16.

Mitrović, S. & Đurović, Ž. (2010a). Fuzzy logic controller for bidirectional garaging of a differential drive mobile robot, *Advanced Robotics* 24(8–9): 1291–1311.

Mitrović, S. & Đurović, Ž. (2010b). Fuzzy–based controller for differential drive mobile robot obstacle avoidance, *in* G. Indiveri & A. M. Pascoal (eds), *7th IFAC Symposium on Intelligent Autonomous Vehicle*, Vol. 7, IFAC–PapersOnLine, Lecce, Italy.

Mitrović, S., Đurović, Ž. & Aleksić, M. (2009). One approach for mobile robot navigation throw area with obstacles, *eProc. of LIII ETRAN Conference*, Vol. 53, Vrnjačka Banja, Serbia, pp. AU3.5 1–4.

Motlagh, O., Hong, T. & Ismail, N. (2009). Development of a new minimum avoidance system for a behavior-based mobile robot, *Fuzzy Sets and Systems* 160: 1929–1945.

Oriolo, G., Luca, A. & Vendittell, M. (2002). WMR control via dynamic feedback linearization: Design, implementation, and experimental validation, *IEEE Trans. Control Systems Technology* 10: 835–852.

Pradhan, S., Parhi, D. & Panda, A. (2009). Fuzzy logic techniques for navigation of several mobile robots, *Applied Soft Computing Journal* 9(1): 290–304.

Reignier, P. (1994). Fuzzy logic techniques for mobile robot obstacle avoidance, *Robotics and Autonomous Systems* 12(3-4): 143–153.

Rusu, P., Petriu, E., Whalen, T., Cornell, A. & Spoelder, H. (2003). Behavior–based neuro–fuzzy controller for mobile robot navigation, *IEEE Transactions on Instrumentation and Measurement* 52(4): 1335–1340.

Sugeno, M. & Nishida, M. (1985). Fuzzy control of model car, *Fuzzy Sets and Systems* 16(2): 103–113.

Tanaka, K. & Sano, M. (1995). Trajectory stabilization of a model car via fuzzy control, *Fuzzy Sets and Systems* 70(2-3): 155–170.

Wang, H., Tanaka, K. & Griffin, M. (1996). An approach to fuzzy control of nonlinear systems: Stability and design issues, *IEEE Transactions on Fuzzy Systems* 4: 14 – 23.

Zadeh, L. (2001). Outline of a computational theory of perceptions based on computing with words, *in* N. Sinha, M. Gupta & L. Zadeh (eds), *Soft Computing and Intelligent Systems: Theory and Applications*, Academic Press, London, pp. 2–33.

3

Reliable Long-Term Navigation in Indoor Environments

Mattias Wahde, David Sandberg and Krister Wolff
Department of Applied Mechanics, Chalmers University of Technology, Göteborg Sweden

1. Introduction

Long-term robustness is a crucial property of robots intended for real-world tasks such as, for example, transportation in indoor environments (e.g. warehouses, industries, hospitals, airports etc.). In order to be useful, such robots must be able to operate over long distances without much human supervision, something that sets high demands both on the actual robot (hardware) and its artificial brain[1] (software). This paper is focused on the development of artificial robotic brains for reliable long-term navigation (for use in, for example, transportation) in indoor environments.

Reliable decision-making is an essential tool for achieving long-term robustness. The ability to make correct decisions, in real-time and often based on incomplete and noisy information, is important not only for mobile robots but also for animals, including humans (McFarland, 1998; Prescott et al., 2007). One may argue that the entire sub-field of behavior-based robotics emerged, at least in part, as a result of the perceived failure of classical artificial intelligence to address real-time decision-making based on a robot's imperfect knowledge of the world. Starting with the *subsumption method* (Brooks, 1986), many different methods for decision-making, often referred to as methods for *behavior selection* or *action selection*, have been suggested in the literature on behavior-based robotics (see, for example, Bryson (2007); Pirjanian (1999) for reviews of such methods).

In actual applications, a common approach is to combine a *reactive layer* of decision-making, using mainly behavior-based concepts, with a *deliberative layer* using, for example, concepts from classical artificial intelligence, such as high-level reasoning) (Arkin, 1998). Several approaches of this kind have been suggested (see, for example, Arkin (1987) and Gat (1991)) and applied in different robots (see, for example, Sakagami et al. (2002)).

In the *utility function* (UF) method for decision-making (Wahde, 2003; 2009), which will be used here, a somewhat different approach is taken in which, for the purposes of decision-making, no distinction is made between reactive and deliberative aspects of the robotic brain. In this method, an artificial robotic brain is built from a repertoire of *brain processes* as well as a single decision-making system responsible for activating and de-activating brain processes

[1] The term *artificial (robotic) brain* is here used instead of the more common *control system* since the latter term, in the authors' view, signifies the low-level parts (such as motor control to achieve a certain speed) of robot intelligence, whereas more high-level parts, such as decision-making are better described by the term used here.

(several of which may run in parallel). Two kinds of brain processes are defined, namely *(motor) behaviors*, which make use of the robot's motors and *cognitive processes*, which do not. Both kinds of processes may include both reactive and deliberative aspects. Note also that, regardless of content, the brain processes used in the UF method are all written in a unified manner, described in Sect. 4 below.

As its name implies, the UF method, described in Sect. 3 below, is based on the concept of utility, formalized by von Neumann & Morgenstern (1944). The method is mainly intended for complex motor tasks (e.g. transportation of objects in arenas with moving obstacles) requiring both reactivity and deliberation. While the method has already been tested (both in simulation and in real robots) for applications involving navigation over distances of 10-20 m or less (Wahde, 2009), this paper will present a more challenging test of the method, involving operation over long distances. The task considered here will be a delivery task, in which a robot moves to a sequence of target points in a large arena containing both stationary and moving obstacles. The approach considered here involves long-term simulations, preceded by extensive validation of the simulator, using a real robot.

2. Robot

The differentially steered, two-wheeled robot (developed in the authors' group) used in this investigation is shown in Fig. 1. The robot has a circular shape with a radius of 0.20 m. Note that an indentation has been made for the wheels. The height (from the ground to the top of the laser range finder (LRF)) is around 0.84 m. The weight is around 14.5 kg. The robot's two actuators are Faulhaber 3863A024C DC motors, each with a Faulhaber 38/1S-14:1 planetary gearhead. Each motor is controlled by a Parallax HB-25 motor controller. The robot is equipped with two 8 Ah power sources, with voltages of 12 V and 7.2 V, respectively.

In addition to the two drive wheels, the robot has two Castor wheels, one in the back and one in the front. The front Castor wheel is equipped with a suspension system, in order to avoid situations where the drive wheels lose contact with the ground, due to bumps in the surface on which the robot moves. The wheels are perforated by 24 holes which, together with a Boe-Bot Digital Encoder Kit detector on each wheel, are used for measuring the rotation of the wheels for use in odometry. In addition, the robot is equipped with a Hokuyo URG-04LX LRF with a range of 4 m and a 240 degree sweep angle. For proximity detection, the robot has two forward-pointing Sharp GP2D12 IR sensors (oriented towards ± 30 degrees, respectively, from the robot's front direction), and one IR sensor, of the same type, pointing straight backwards. The robot is equipped with two Basic Stamp II microcontrollers (which, although slow, are sufficient for the task considered here), one that reads the wheel encoder signals, and one that (i) sends signals to the motor controllers and (ii) receives signals from the three IR sensors. The two Basic Stamps are, in turn, connected via a USB hub to a laptop that can be placed on top of the robot (under the beam that holds the LRF). The LRF is also connected to the laptop, via the same USB hub.

3. Decision-making structure

The decision-making structure is based on the UF method, the most recent version of which is described by Wahde (2009). The method is mainly intended for use in tasks such as navigation, transportation, or mapping. In this method, the artificial brain is built from (i) a set of brain processes and (ii) a decision-making system based on the concept of utility, allowing the robot

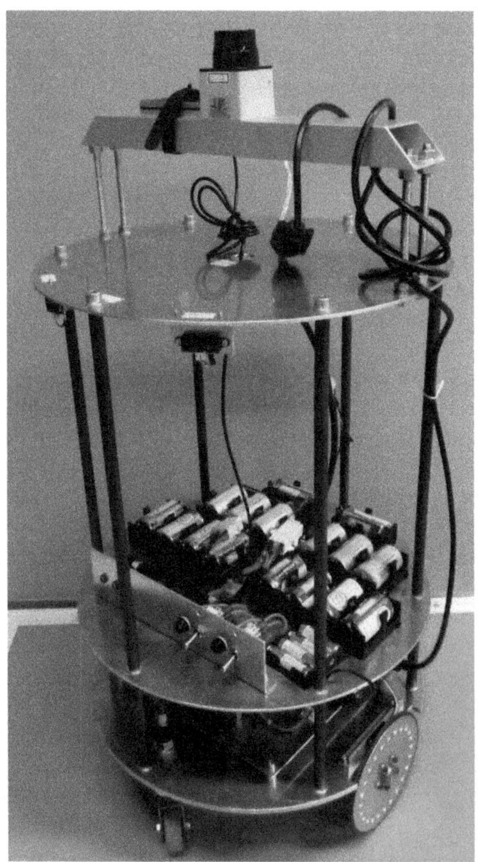

Fig. 1. Left panel: The differentially steered robot considered here. The laser range finder is mounted on top. Right panel: The simulated version of the robot.

to activate and de-activate the various brain processes. Here, only a brief description will be given. For a more detailed description, see Wahde (2009).

Brain processes are divided into several types. The two categories used here are *cognitive processes* (that do not make use of the robot's motors) and *motor behaviors* that *do* make use of the motors. In the UF method, any number of cognitive process can be active simultaneously, whereas exactly one motor behavior is active at any time, the rationale being that a robot of the kind considered here (with two wheels as its only actuators) can only carry out one motor action at any given time.

Each brain process (regardless of its type) is associated with a *utility function* whose task it is to determine the merit of running the brain process, in any given situation. The utility functions, in turn, depend on scalar *state variables*. A simple example of a state variable z may be the reading of an IR sensor mounted at the front of a robot. The value of z is an (admittedly incomplete) measure of the risks involved in moving forward. A more complex example is the *moving obstacle danger level* (see Subsect. 4.4 below), which uses consecutive LRF scans to obtain a scalar value measuring the risk of collision with moving obstacles.

State variable values are measured continuously, but asynchronously (since not all sensors are updated with the same frequency), and the most recent values available are used as inputs to the utility functions from which the utility value u_i for brain process i is obtained using the equation

$$\tau_i \dot{u}_i + u_i = \sigma_i \left(\sum_{k=1}^{m} a_{ik} z_k + b_i + \Gamma_i \right), \; i = 1, \ldots, n, \tag{1}$$

where n is the number of brain processes, τ_i are time constants determining the reaction time of the robot (typically set to around 0.1 s), m is the number of state variables (denoted z_k), and a_{ik} and b_i are tunable parameters. The squashing function $\sigma_i(x)$ is taken as $\tanh(c_i x)$ where c_i is a positive constant (typically set to 1). Thus, the squashing functions σ_i serve to keep the utility values in the range $[-1, 1]$.

Once the values of the utility functions have been computed, decision-making is simple in the UF method, and works as follows: (i) The motor behavior with largest utility (among all motor behaviors) is active and (ii) any cognitive process with positive utility is active. All other brain processes are inactive.

The parameters Γ_i (referred to as *gamma parameters*), which are normally equal to zero, allow direct activation or de-activation of a brain process. Ideally, the state variables z_k should provide the robot will the information needed to make an informed decision regarding which brain processes to use in any situation encountered. However, in practice, the parameters Γ_i are needed in certain situations. Consider, for example, a simple case in which the utility function for an obstacle avoidance behavior depends on a single state variable z (multiplied by a positive constant a) that measures obstacle proximity (using, for example, IR or sonar sensors). Now, if the obstacle avoidance behavior is activated, the robot's first action is commonly to turn away from the obstacle. When this happens, the value z will then drop, so that u also drops, albeit with a slight delay depending on the value of the time constant in Eq. (1). At this point, the obstacle avoidance behavior is likely to be de-activated again, before actually having properly avoided the obstacle. The gamma parameters, which can be set directly by the brain processes, have been introduced to prevent such problems. Thus, for example, when the obstacle avoidance behavior is activated, it can set its *own* gamma parameter to a positive value, thus normally avoiding de-activation when the state variable z drops as described above. Whenever the obstacle avoidance behavior has run its course, it can set the gamma parameter to a large negative value, thus effectively de-activating itself. Note that the decision-making system is active continuously, so that, in the example above, obstacle avoidance *can* be de-activated (even after raising its own gamma parameter) should another brain process reach an even higher utility value. Once a gamma parameter has been set to any value (positive or negative) other than zero, its magnitude falls of exponentially with time, with a time constant (τ_i^{Γ}) specific to the brain process at hand. This time constant typically takes a larger value than the time constant in the utility function.

As mentioned above, the UF method allows several (cognitive) brain processes to run in parallel with the (single) active motor behavior. Thus, for example, while moving (using a motor behavior), a robot using the UF method would simultaneously be able to run two cognitive processes, one for generating odometric pose estimates from encoder readings and one for processing laser scans in order to recalibrate its odometric readings, if needed.

As is evident from Eq. (1), for a given set of state variables, the actual decisions taken by the robot will depend on the parameters τ_i, a_{ij}, b_i, and c_i as well as the values of the gamma parameters (if used). Thus, in order to make a robot carry out a specific task correctly, the

user must set these parameters to appropriate values. Note that, for many cognitive processes (e.g. odometry, which normally should be running continuously), all that is needed is for the utility values to be positive at all times, which can easily be arranged by setting all the a-parameters for that process to zero, and the b-parameter to any positive value.
In many cases, the remaining parameters (especially those pertaining to the motor behaviors) can be set by hand using trial-and-error. In more complex situations, one can use a stochastic optimization method such as, for example, a genetic algorithm or particle swarm optimization to find appropriate parameters values. In those situations, one must first define a suitable objective function, for example the distance travelled in a given amount of time, subject to the condition that collisions should be avoided. Optimization runs of this kind are normally carried out in the simulator (described in Subsect. 5.1 below) rather than a real robot, since the number of evaluations needed during optimization can become quite large.

4. Brain processes

Just as the decision-making system, brain processes used in the UF method also have a unified structure; the main part of each brain process is a finite-state machine (FSM) consisting of *states* (in which various computations are carried out and, in the case of motor behaviors, actions are taken) and *conditional transitions* between states. In any brain process, the FSM executes the *current state* until a transition condition forces it to jump to another state, which is then executed etc. An illustration of a brain process FSM is shown in Fig. 8 below.
For the task considered here, namely reliable long-term navigation, six brain processes have been used, namely *Grid navigation* (denoted B_1), *Odometry* (B_2), *Odometric calibration* (B_3), *Moving obstacle detection* (B_4), *Moving obstacle avoidance* (B_5), and *Long-term memory* (B_6). The six brain processes will be described next.

4.1 Grid navigation

The navigation method used here is contained in a brain process called *Grid navigation*. As the name implies, the navigation method (which is intended for indoor, planar environments) relies on a grid. Dividing a given arena into a set of convex cells is a common problem in robotics, and it is typically approached using *Voronoi diagrams* (or their dual, *Delaunay triangulation*) (Okabe et al., 2000) or *Meadow maps* (Singh & Agarwal, 2010). However, these tessellations tend to generate jagged paths, with unnecessary turns that increase the length of the robot's path (a problem that can be overcome using *path relaxation*; see Thorpe (1984)). Nevertheless, in this paper, a different method for automatic grid generation will be used, in which the map of the arena is first contracted to generate a margin for the robot's movement. Next, the arena is divided into convex cells in a preprocessing step similar to that described by Singh & Wagh (1987). Note, however, that the method described in this paper also can handle non-rectangular arenas (and obstacles). Finally, the cells are joined in order to generate larger convex cells in which the robot can move freely. Once such a grid is available, a path is generated using Dijkstra's algorithm (see Sect. 4.1.2 below), optimizing the placement of waypoints (on the edges of the cells) in order to minimize the path length. The method for grid generation, which is summarized in Table 1, will now be described.

4.1.1 Grid generation method

It is assumed that a map is available, in the form of a set of closed *map curves* consisting, in turn, of a sequence of connected (i.e. sharing a common point) *map curve segments*, each

Step 1 Generate the contracted map.
Step 2 Generate the preliminary map curve intersection grid.
Step 3 Process non-axis parallel map curve segments.
Step 4 Generate the map curve intersection grid by removing cells outside the grid.
Step 5 Generate the convex navigation grid by joining cells.

Table 1. The algorithm for generating the convex navigation grid. See the main text for a detailed description of the various steps.

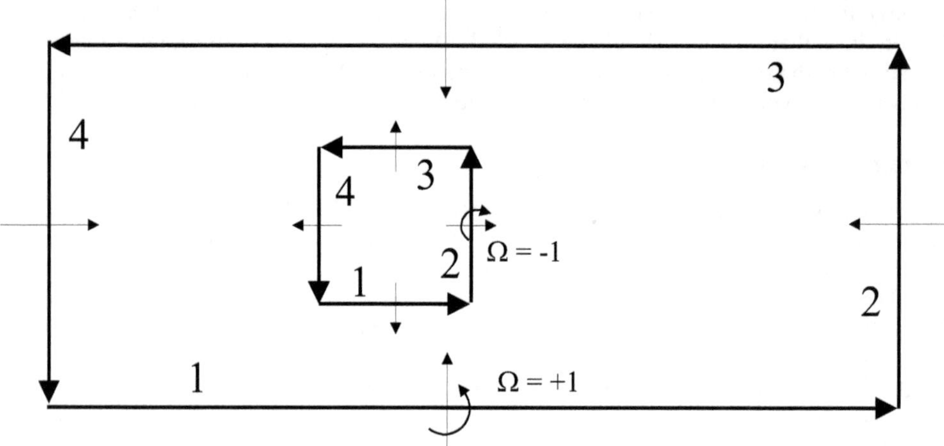

Fig. 2. The definition of in-directions in a rectangular arena with a square obstacle. In this simple arena, there are two map curves, each with four enumerated map curve segments. The thick arrow on each map curve segment points from the starting point to the end point of the segment. The thin arrows point towards the in-direction. The corresponding value of Ω (see Eq. (2)) is equal to $+1$ for the outer map curve (wall) and -1 for the inner map curve (obstacle). Note that the sign of Ω depends on the (arbitrary) choice of enumeration (clockwise or counterclockwise) for the map curve segments.

defined by a starting point $\mathbf{P}_a = (x_a, y_a)$ and an end point $\mathbf{P}_b = (x_b, y_b)$. Thus, each map curve is a simple (i.e. without self-intersections) polygon. In addition, for each map curve segment, an *in-direction* is defined. The in-direction Ω is given by the sign (either -1 or 1) of the z-component of the cross product between the vector $\mathbf{p} = \mathbf{P}_b - \mathbf{P}_a$ and a unit vector $\mathbf{d}$ (orthogonal to $\mathbf{p}$) pointing towards the side of the map curve segment which is accessible to the robot. Thus,

$$\Omega = \text{sgn}\left((\mathbf{p} \times \mathbf{d}) \cdot \hat{z}\right). \tag{2}$$

The in-direction concept is illustrated in Fig. 2. Note that, for any given map curve, all segments will have the same in-direction.

The method divides the accessible area of a robot's environment into a set of convex grid cells. Due to the convexity of the grid cells the robot can move freely within any grid cell, a property that facilitates path optimization, as described below. The grid generation method operates as follows: First (Step 1 in the algorithm described in Table 1), since the robot has a certain size, the actual map is shrunk to generated a *contracted map*. The margin μ (a tunable parameter) used when generating the contracted map should be such that, if the robot is positioned on a

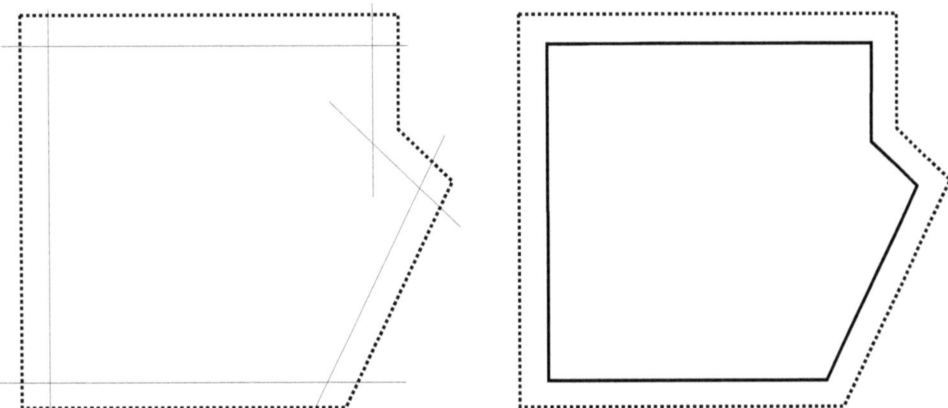

Fig. 3. The map contraction procedure. The left panel shows the original map (dotted line), along with the line segments obtained using the procedure described in the main text. In the right panel, the intersection points between those lines have been determined, resulting in the contracted map (solid line).

line in the contracted map, it should not touch any wall or other fixed obstacle. The procedure is illustrated in Fig. 3. The algorithm for generating the contracted map is straightforward: Consider a given line segment $\mathbf{p} = (p_x, p_y) = \mathbf{P}_b - \mathbf{P}_b = (x_b - x_a, y_b - y_a)$. Let $\mathbf{c}$ be a vector of length μ orthogonal to $\mathbf{p}$ and pointing towards the in-direction. Thus

$$\mathbf{c} = \mu\Omega \frac{(-p_y, p_x)}{\sqrt{p_x^2 + p_y^2}}. \tag{3}$$

It is now possible to obtain two points on the contracted map curve segment as

$$\mathbf{Q}_a = \mathbf{P}_a + \mathbf{c} \tag{4}$$

and

$$\mathbf{Q}_b = \mathbf{P}_b + \mathbf{c}. \tag{5}$$

Note, however, that these two points are normally not the end points of the contracted map curve segment. In order to determine the end points, one must first carry out the process described above for all map curve segments in a given map curve. Once that process has been completed, the end points of the contracted map curve segments can be determined as the intersection points between the lines passing through the point pairs generated during the contraction process just described. If these lines are extended to infinity, any given line may intersect many other lines. In that case, the points chosen as end points of the contracted map curve segments are taken as those intersection points that are nearest to the points $\mathbf{Q}_a$ and $\mathbf{Q}_b$, respectively. The process is illustrated in Fig. 3, where a map consisting of a single map curve, with six map curve segments, is contracted.

Note that, in its simplest form, the map contraction method requires that no contracted map *curve* should intersect any of the other contracted map curves. The cases in which this happens

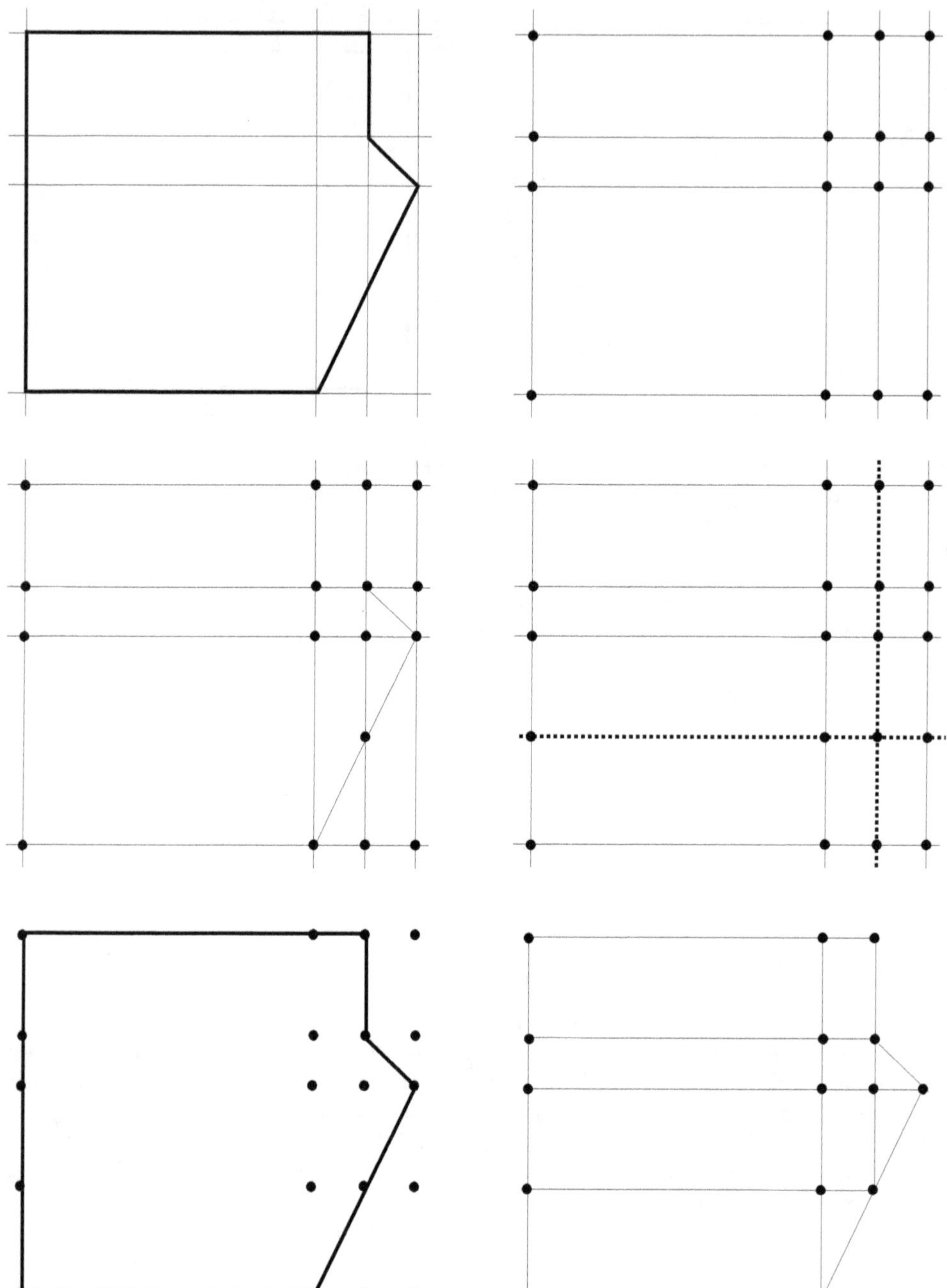

Fig. 4. The grid generation algorithm, applied to the contracted map from Fig. 3. See the main text for a detailed description of the various steps.

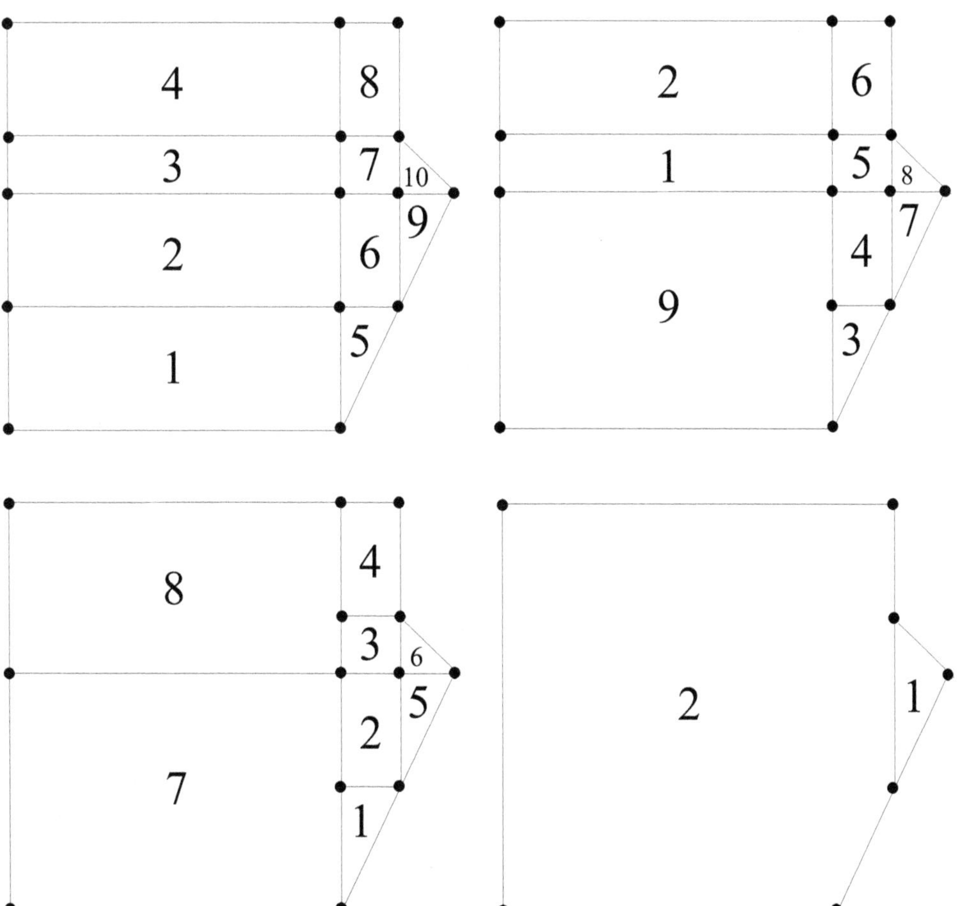

Fig. 5. The grid simplification algorithm, applied to the grid shown in Fig. 4. During grid simplification, polygonal cells are combined to form larger (but still convex) cells. Note that, in each step, the enumeration changes after the removal of the two cells that are joined. The upper left panel shows the initial grid, whereas the upper right panel shows the grid after the first step. In the bottom row, the left panel shows the grid after the second step, and the right panel shows the final result, after several steps.

can be handled, but the procedure for doing so will not be described here. Instead, we shall limit ourselves to considering maps in which the contracted map curves do not intersect.
Once the contracted map has been found, the algorithm (Step 2) proceeds to generate a temporary grid (referred to as the *preliminary map curve intersection grid*), consisting of the intersections between all axis-parallel (horizontal or vertical) map curve segments expanded to infinite length, as illustrated in the first row of Fig. 4. Next (Step 3), all non-axis parallel map curve segments are processed to check for intersections between those lines and the sides of the cells in the preliminary map curve intersection grid. For any intersection found, the

corresponding point is added to the grid, as illustrated in the middle row (left panel) of Fig. 4. Lines are then drawn horizontally and vertically from the newly added points, and the points corresponding to any intersections between those two lines and the cells of the preliminary map curve intersection grid are added as well; see the right panel in the middle row of Fig. 4. Some of the cells will be located outside the (contracted) map, as shown in the left panel in the bottom row of Fig. 4. Those cells are removed (Step 4), resulting in the *map curve intersection grid*, shown in the final panel of Fig. 4.

Note that the cells in the map curve intersection grid are convex by construction. In principle, one could just remove the inaccessible cells and use the remaining cells during navigation. However, in a realistic arena (more complex than the simple arena used for illustrating the procedure) the map curve intersection grid typically contains very many cells, thus slowing down the path generation. Furthermore, with a large number of cells, the paths generated will typically have an unwanted zig-zag nature. Thus, in Step 5 of the algorithm, cells in the map curve intersection grid are joined to form larger cells, with the condition that two cells are only joined if the result is again a convex polygon, something that can easily be checked since, in a convex polygon, all cross products between consecutive sides (i.e. sides that share a common point) have the same sign. In the current version of the algorithm, no attempt has been made to generate a minimal (in number) set of convex polygons. Instead, the cells are simply processed in order, removing joined cells and adding newly formed cells, until there are no cells left such that their combination would be convex. The first steps of this process, as well as the final result, are shown in Fig. 5. The result is referred to as the *convex navigation grid*.

4.1.2 Path planning

The path planning procedure uses the convex navigation grid (which can be generated once and for all for any given (fixed) arena). Given a starting point (i.e. the robot's estimate of its current location) and a desired goal point, a path is generated using Dijkstra's algorithm (Dijkstra, 1959), augmented with a process for selecting navigation waypoints on the edges of the grid cells through which the path passes. As in the standard Dijkstra algorithm, cells are considered in an expanding pattern from the starting cell (i.e. the cell in which the robot is currently located). For any considered cell c, all neighbors a_i of c (i.e. those cells that, partially or fully, share a common side with c) are considered. For each a_i, a randomly generated waypoint q_{c,a_i} is selected (and stored) along the common side shared by the two cells (see Fig. 6). The distance from a_i to the starting point can then be computed and stored.

However, the path planning algorithm should also be able to handle situations in which a moving obstacle (which, of course, would not be included in the convex navigation grid) is found. If, for a line segment l connecting c to a_i, an intersection is found between l and a circle (with an added margin) around a detected moving obstacle (see below), a large penalty is added in the computation of the distance from a_i to the starting point, thus effectively rendering that path useless. Note that the robot, of course, has no global knowledge of the positions of moving obstacles: Only those moving obstacles that it can actually *detect* are included in the analysis. Note also that the path planning method will ignore any detected moving obstacle at a distance of at least Δ (a tunable parameter) from the robot. This is so, since there is no point to account for a faraway moving obstacle (which may not even be headed in the robot's direction) while planning the path.

Fig. 6. Path planning: The upper left panel shows the convex navigation grid for the arena described in Sect. 6 and the remaining panels show the paths considered during path planning. In this case, the robot was located in the lower left corner of the arena, and its navigation target was located near the upper right corner. The first path considered is shown in the upper middle panel. After some time, the optimization procedure found a shorter path (upper right panel), along a different route. In the remaining steps (lower panels) the path planning algorithm further optimized the path (length minimization) by adjusting the placement of the waypoints (on the edges of the convex grid cells). Note that, in this figure, the actual navigation path is shown, i.e. not only the waypoints on the cell edges, but also all the intermediate (interpolated) waypoints.

The process is carried out for all grid cells, since the geometrically shortest path may involve a potential collision with a moving obstacle, as described above, in which case another path must be found, as shown in Fig. 7. A navigation path candidate is obtained once all grid cells have been considered.

However, as noted above, the exact locations of the waypoints q_{c,a_i} (on the edges of the navigation grid cells) are randomly chosen. Thus, an optimization procedure is applied, during which the entire path generation process is iterated n times, using different randomly selected waypoints along the cell sides (see also Fig. 6), and the shortest path thus found is taken as the final path. Once this path has been obtained, the actual *navigation path* is formed by adding interpolated waypoints between the waypoints obtained during path generation, so that the distance between consecutive waypoints is, at most, d_{wp} (a user-specified parameter).

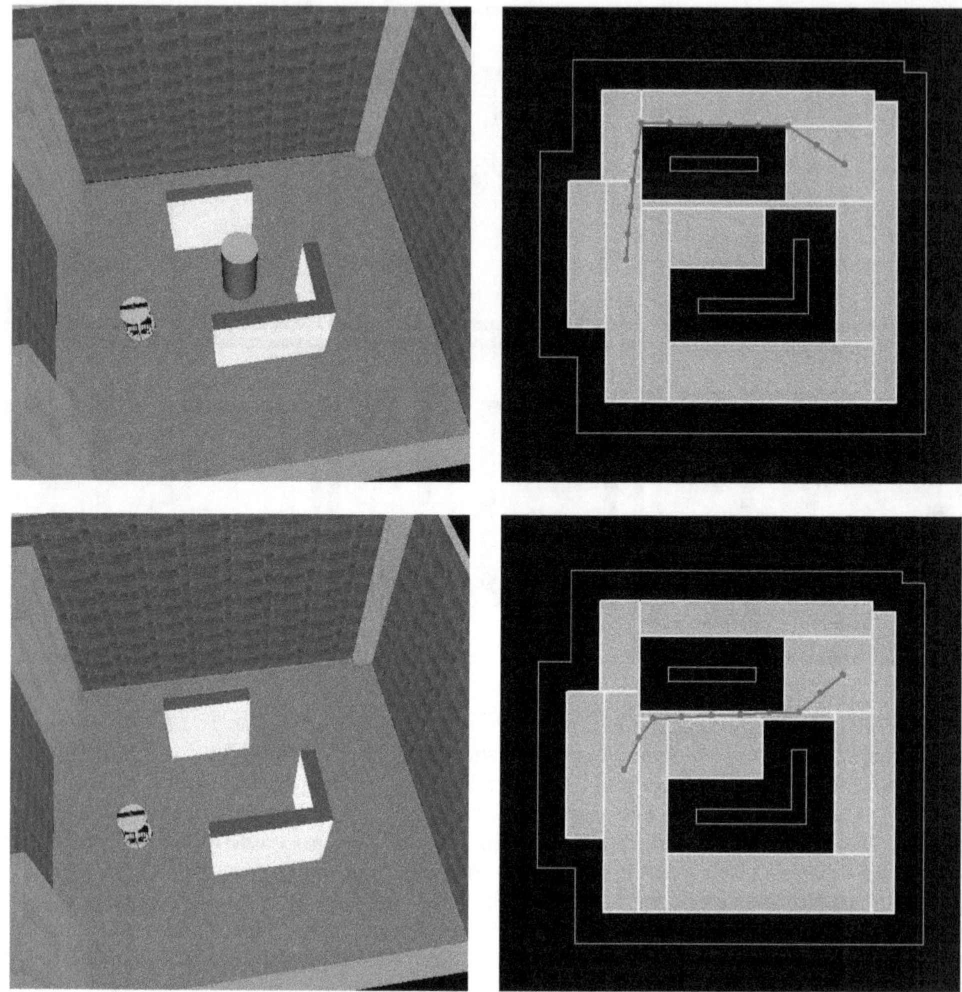

Fig. 7. Path planning with (upper panels) and without (lower panels) moving obstacles present. The moving obstacle is represented as a gray cylinder.

4.1.3 Navigation

The navigation motor behavior, illustrated in Fig. 8, consists of an FSM with 9 states. In State 1, the robot stops (path planning takes place at standstill), and jumps to State 2 in which the robot determines whether or not (based on its odometric position estimate) it is actually located in the convex navigation grid. This check is necessary, since the robot may have left the grid (slightly) as a result of carrying out moving obstacle avoidance; see below. If the robot finds that it is located *outside* the grid, it jumps to State 3 in which it checks, using the LRF[2], whether a certain sector (angular width γ_c, radius r_c) in front of the robot is clear, i.e. does

[2] It is assumed that the height of the moving obstacles is such that they are detectable by the (two-dimensional) LRF.

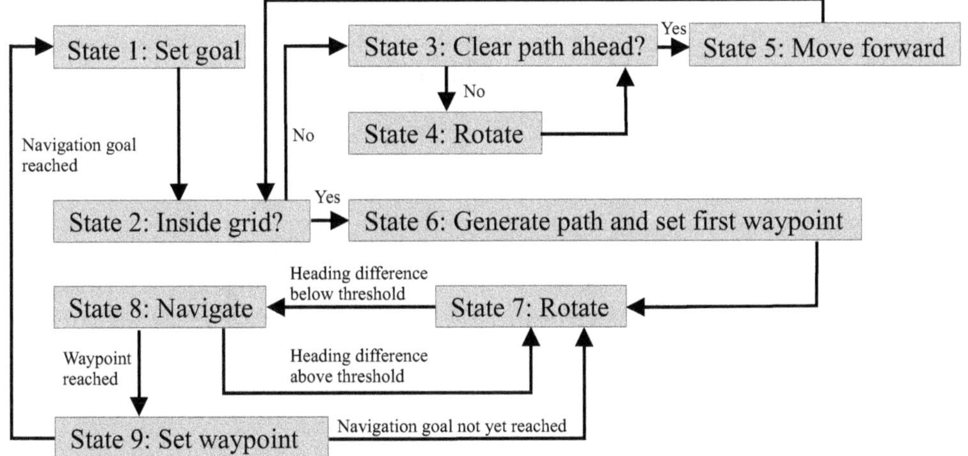

Fig. 8. The FSM implementing grid navigation.

not contain any obstacle. If this is so, the robot jumps to State 5 of the FSM, in which it sets the motor signals to equal, positive values, making the robot move forward, essentially in a straight line[3]. It then jumps back to State 2, again checking if it has reached the grid etc. If instead the robot finds that the path ahead is not clear, it jumps to State 4, in which it sets the motor signals to equal values but with opposite signs, so that it begins rotating without moving its center of mass. It then jumps back to State 3, to check whether the path ahead is clear.

If, in State 2, the robot finds that it is located *inside* the grid, it jumps to State 6, in which it determines the starting point (from the current odometric estimate) and the navigation goal (provided by the user), then generates an optimized path using the procedure described in Subsect. 4.1.2 and, finally, sets the current waypoint as the second waypoint in the path (the first point being its current location). It then jumps to State 7, in which it checks whether (the absolute value of) the difference[4] in estimated heading (using odometry) and desired heading (obtained using the direction to the current waypoint) exceeds a threshold δ. If so, it begins a pure rotation (as in State 4, see above). Once the difference in heading drops below the threshold, the robot jumps to State 8, in which the left and right motor signals are set as

$$v_{\mathrm{L}} = V_{\mathrm{nav}} - \Delta V \tag{6}$$

and

$$v_{\mathrm{R}} = V_{\mathrm{nav}} + \Delta V, \tag{7}$$

respectively, where

$$\Delta V = K_p V_{\mathrm{nav}} \Delta\varphi, \tag{8}$$

where $\Delta\varphi$ is the difference in heading angle, and K_p is a constant. While in State 8, the robot continuously checks the difference between estimated and desired heading. If it should

[3] Even though the motor settings are equal, the actual path may differ slightly from a straight line, due to actuator noise and other similar sources of error.

[4] The difference $\Delta\phi$ between two heading angles φ_1 and φ_2 is defined as $\varphi_1 - \varphi_2 \bmod 2\pi$.

exceed the threshold δ, the robot returns to State 7. If instead the robot reaches within a distance d_p (around 0.1 - 0.2 m) of the current waypoint, the robot jumps to state 9 where it checks whether it has reached the navigation goal (i.e. the final waypoint). If this is the case, the robot picks the next navigation goal from the list, and returns to State 1.
If not, the robot selects the next waypoint in the current path. Now, the simplest way to do so would be to just increase the index of the waypoint by one. Indeed, this is normally what happens. However, in certain situations, it may be favorable to skip one waypoint, namely in cases where the path changes direction abruptly. In such cases, passing all waypoints will lead to an unnecessarily jerky motion. Thus, in State 9, before selecting the next waypoint, the robot considers the angle between the vectors $\mathbf{v}_1 = \mathbf{p}_{i+1} - \mathbf{p}_i$ and $\mathbf{v}_2 = \mathbf{p}_{i+2} - \mathbf{p}_{i+1}$, where $\mathbf{p}_i$ is the waypoint that the robot just has reached, and $\mathbf{p}_{i+1}$ and $\mathbf{p}_{i+2}$ are the two next waypoints (this procedure is of course skipped if $\mathbf{p}_i$ or $\mathbf{p}_{i+1}$ is the final waypoint along the path). If this angle exceeds a certain threshold *and* the path ahead of the robot is clear (measured as in State 3, but possibly with different thresholds defining the sector in front of the robot), the robot skips waypoint $i+1$ and instead targets waypoint $i+2$.

4.2 Odometry

The *Odometry* brain process is straightforward: Given the estimated wheel rotations since the last update (based on the pulse counts from the wheel encoders), the robot uses the simple kinematic model of a differentially steered two-wheeled robot (with two additional supports for balance) to obtain an estimate of its velocity (v_x, v_y). Using standard forward kinematics, the robot also obtains estimates of its position (x, y) and angle of heading φ.

4.3 Odometric calibration

The purpose of the *odometric calibration* brain process is to correct the inevitable errors in the pose obtained from the *odometry* brain process. Odometric calibration relies on scan matching and builds upon an earlier version (called *laser localization*) described by Sandberg et al. (2009). However, whereas the earlier version required the robot to stop before correcting the odometric errors, the new version operates continuously and in concert with other brain processes. Thus, using the brain process taxonomy described in Sect. 3, the new version is a cognitive process rather than a motor behavior.
The FSM of the odometric calibration brain process operates as follows: In State 1, the robot checks whether the (modulus of the) difference in motor signals (between the left and right motor) exceeds a given threshold $T_{\Delta v}$. If it does, the robot remains in State 1, the rationale being that scan matching is less likely to give a reliable result if the robot is currently executing a sharp turn rather than moving forward (or backward). This condition is less of a restriction than one might think: Since the grid navigation method attempts to find the shortest path from the starting point to the goal point, the resulting path typically consists of long straight lines, with occasional turns where necessary.
When the wheel speed difference drops below the threshold, the robot jumps to State 2 of the FSM, where it stores the most recent laser scan and then carries out a laser scan match. The procedure for the scan match has been described in full detail by Sandberg et al. (2009), and will not be given here. Suffice it to say that the scan matching procedure compares the stored laser scan to a virtual scan taken in the map, by directly computing (essentially) the root mean square error between the laser ray readings. Thus, unlike many other methods, this method does *not* rely on the (brittle and error-prone) identification of specific landmarks. Furthermore, unlike the version presented by Sandberg et al. (2009), the current version is able to filter out

moving obstacles. It does so by noting that there is an upper limit on the pose errors, since the odometric calibration is executed at least a few times per second. Thus, laser rays for which the discrepancy between the expected and actual reading is too large to be caused by pose errors are simply ignored in the scan matching procedure; see also Subsect. 5.3 below.
If the scan match error is below a threshold T_ϵ, the robot concludes that its pose is correct and jumps back to State 1. However, if the error exceeds the threshold, a search procedure is initiated in State 3 of the FSM. Here, the robot carries out a search in a region of size $L_x \times L_y \times L_\varphi$ in pose space, around the estimated pose (obtained from odometry) recorded when the stored laser scan was taken.
Since the search procedure takes some time (though less than one second) to carry out, once it has been completed the robot must correct for the movement that has occurred since the stored laser scan was taken. This correction is carried out using odometry, which is sufficiently accurate to provide a good estimate of the *change* in the robot's pose during the search procedure (even though the *absolute* pose may be incorrect, which is the reason for running odometric calibration in the first place!)

4.4 Moving obstacle detection

As mentioned above, in the case considered here, the only long-range sensor is an LRF mounted on top of the robot. Thus, a method for detecting moving obstacles, i.e. any obstacle that is not part of the map, will have to rely on differences between the expected and actual LRF readings at the robot's current pose. Needless to say, this requires, in turn, that the robot's pose estimate, generated by the *Odometry* and *Odometric calibration* brain processes, is rather accurate. Assuming that this is the case, the FSM implementing the *Moving obstacle detection* (MOD) brain process works as follows: The FSM cycles between Steps 1 and 2, where Step 2 is a simple waiting state (the duration of which is typically set to around 0.02 to 0.10 s). In State 1, the robot compares the current LRF readings with the expected readings (given its pose in the map). Essentially, those readings for which the difference between the actual and expected readings exceed a certain threshold T_{mo} are considered to represent moving obstacles. Thus, even in cases where the robot's pose is slightly incorrect (due to odometric drift), it will still be able reliably to detect moving obstacles, provided that T_{mo} is set to a sufficiently large value. On the other hand, the threshold cannot be set *too* large, since the robot would not be able to detect, for example, moving obstacles just in front of a wall. In the investigation considered here, the value of T_{mo} was equal to 0.30 m.
Normally, when a moving obstacle is present within the range of the LRF, a set of consecutive LRF rays will be found to impinge on the obstacle. Given the first and last of those rays, estimates can be obtained for the position and radius of the obstacle. However, in order to minimize the risk of spurious detections, a list of moving obstacle *candidates* is maintained, such that only those candidates that fulfill a number of conditions are eventually placed in the actual list of moving obstacles. To be specific, detected obstacle candidates are removed if (i) they are located beyond a certain distance $d_{\text{mo}}^{\text{max}}$ or (ii) the inferred radius is smaller than a threshold $r_{\text{mo}}^{\text{min}}$. An obstacle candidate that survives these two checks is artificially increased in size by a given margin (typically 50%). On the other hand, for a detected obstacle candidate such that the inferred radius (after adding the margin) exceeds a threshold $r_{\text{mo}}^{\text{max}}$ (typically as a result of pose errors), the actual radius is set equal to $r_{\text{mo}}^{\text{max}}$.
Despite the precautions listed above, the robot will, from time to time, make spurious detections. Hence, a probability measure p_i for each moving obstacle i is introduced as well. A newly detected moving obstacle candidate is given probability $p_i = p_0$. If the

obstacle candidate is found again (in the next time step), its probability is modified as $p_i \leftarrow \max(\alpha p_i, 1)$, where $\alpha > 1$ is a constant. If, on the other hand, the previously detected obstacle candidate is not found, the probability is modified as $p_i \leftarrow \beta p_i$, where $\beta < 1$ is a constant. Whenever the probability p_i drops below a threshold p_{min}, the corresponding obstacle is removed from the list of candidates. Finally, the remaining candidates are added to the list of actual moving obstacles, when they have been detected (i.e. with probability above the minimum threshold) for at least ΔT_{mo} s (a user-specified parameter, typically set to around 0.5 s).

In order to avoid multiple detections of the same moving obstacle (remembering that the MOD method runs many times per second), detected moving obstacle candidates that are within a distance $\delta_{\text{mo}}^{\text{max}}$ of a previously detected obstacle candidate are identified with that obstacle candidate. Obstacle candidates that do not fulfill this condition are added as separate obstacle candidates. In addition to generating a list of moving obstacles, the brain process also determines which of the moving obstacles is closest to the robot.

Furthermore, a scalar variable, referred to as the *moving obstacle danger level* (denoted μ) is computed (also in State 1 of the FSM). This variable is used as a state variable for the decision-making system, and influences the activation (or de-activation) of the moving obstacle avoidance motor behavior, which will be described below. Needless to say, a scalar variable for assessing the threat posed by a moving obstacle can be generated in a variety of ways, without any limit on the potential complexity of the computations involved. One may also, of course, consider using additional long-range sensors (e.g. cameras) in this context. Here, however, a rather simple definition has been used, in which the robot only considers moving obstacles in the frontal[5] half-plane (relative direction from $-\pi/2$ to $\pi/2$). The robot first computes the change in distance to the nearest obstacle between two consecutive executions of State 1. If the distance is increasing, the current danger level m is set to zero. If not, the current danger level is determined (details will not be given here) based on the distance and angle to the moving obstacle. Next, the actual danger level μ is obtained as $\mu \leftarrow (1-\alpha)\mu + \alpha m$, where α is a constant in the range $]0,1]$.

4.5 Moving obstacle avoidance

The *Moving obstacle avoidance* (MOA) motor behavior is responsible for handling emergency collision avoidance[6]. Once the MOA motor behavior has been activated[7], it devotes all its computational capacity to avoiding the nearest moving obstacle; it makes no attempt to assess the danger posed by the obstacle, a task that is instead handled by the decision-making system, using the moving obstacle danger level computed in the MOD brain process. Just as in the case of the MOD brain process, MOA can also be implemented in complex and elaborate ways.

Here, however, a rather simple MOA method has been used, in which the region in front of the robot is divided into four zones, as illustrated in Fig. 9. Should the MOA motor behavior be activated, the actions taken by the robot depend on the zone in which the moving obstacle

[5] The robot thus only handles frontal collisions; it is assumed that a moving obstacle, for example a person, will not actively try to collide with the robot from behind.

[6] Note that the grid navigation motor behavior also handles *some* collision avoidance, in the sense that it will not deliberately plan a path that intersects a nearby moving obstacle.

[7] Note that, whenever the MOA motor behavior is activated, the grid navigation motor behavior is turned off, and vice versa. This is so since, in the UF method, only one motor behavior is active at any given time; see Sect. 3.

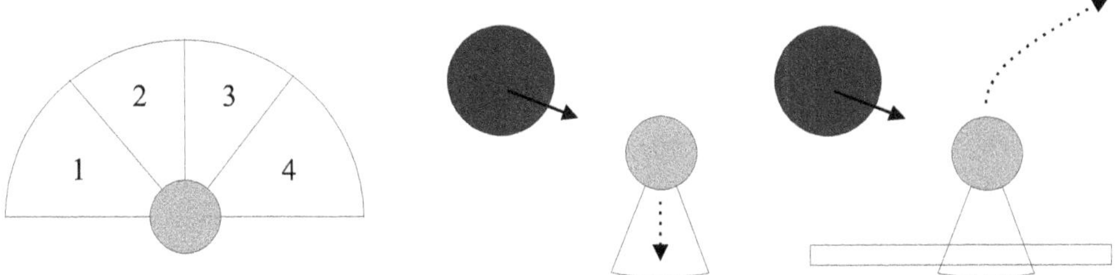

Fig. 9. Left panel: The zones used in the moving obstacle avoidance motor behavior. The robot is shown as a light gray disc. Middle and right panels: The two evasive maneuvers defined for cases in which the moving obstacle (dark gray disc) is located in Zone 1. If the robot is able to move backwards (based on the readings from the rear IR sensor) it does so (middle panel). On the other hand, if an obstacle behind the robot prevents it from moving backwards (right panel), it attempts instead to evade the obstacle by moving forward and to the right.

is located. For example, if the moving obstacle is located in Zone 1, the robot checks (using the rear IR sensor) whether it can move backwards without collisions. If so, it sets the motor signals to equal, negative values, thus (asymptotically) moving straight backwards, until it has moved a certain, user-specified distance, at which point the robot stops as the MOA motor behavior is awaiting de-activation.

If the robot is unable to move backwards (either initially or after some time), it instead attempts to circumnavigate the moving obstacle by turning right, setting the left motor signal to a large positive value, and the right motor signal to a small negative value. Once the robot has turned a certain angle (relative to the initial angle of heading), it instead sets the motor signals to equal positive values, thus (asymptotically) moving forward in a straight line, until it has moved a certain distance. At some point, the obstacle will disappear from the field of view of the LRF (either because it has passed the robot, or as a result of the robot's evasive action), at which point the obstacle danger level (μ) drops, and the MOA motor behavior is de-activated shortly thereafter, so that grid navigation can be resumed. A similar, but somewhat more elaborate, procedure is used if the obstacle is located in Zone 2. Zones 3 and 4 are handled as Zones 2 and 1, respectively, but with the motor signal signs etc. reversed.

The utility function for the MOA motor behavior uses the obstacle danger level (see Subsect. 4.4) as a state variable. Hence, when the perceived obstacle danger level rises above a threshold, which is dependent on the parameter values used in Eq. (1), the MOA motor behavior is activated. At this point, the robot attempts to avoid the moving obstacle, as described above. In cases where it does so by turning, the perceived obstacle danger level typically drops. Thus, in order to avoid immediate de-activation, the MOA motor behavior uses a gamma parameter to raise its own utility once activated (in State 1); see also Sect. 3 above.

4.6 Long-term memory

The *Long-term memory* (LTM) brain process simply stores (and makes available to the other brain processes) the resources necessary for the robot to carry out its navigation task. In the case considered here, the LTM stores the map of the arena (needed during odometric calibration) and the convex navigation grid (needed for navigation).

5. Implementation and validation

5.1 The general-purpose robotics simulator

A simulator referred to as the *General-purpose robotics simulator* (GPRSim) has been written with the main purpose of evaluating the UF decision-making method. The simulator allows a user to set up an arbitrary arena, with or without moving obstacles, and a robot, including both its physical properties (e.g. actuators, sensors, physical parameters such as mass, moment of inertia etc.) and its artificial brain (i.e. a set of brain process, as described above, and the decision-making system implementing the UF method, also described above). In order to make the simulations as realistic as possible, the simulator introduces noise at all relevant levels (e.g. in sensors and actuators). Moreover, simulated sensor readings are taken only with the frequency allowed by the physical counterpart.

5.2 Simulations vs. reality

Simulations are becoming increasingly important in many branches of science, and robotics is no exception. However, in the early days of behavior-based robotics many researchers expressed doubts as to the possibility of transferring results obtained in simulations onto real robots. In particular, Brooks took an especially dim view of the possibility of such transfer, as discussed in Brooks (1992). Similar critique of simulations has been voiced by, among others, Jakobi et al. (1995) and Miglino et al. (1996). However, their critique mainly targeted naive simulation approaches that use idealized sensors and actuators in block-world-like environments. In fact, both Jakobi et al. (1995) and Brooks (1992) also note that simulations are indeed useful, provided that great care is taken to simulate, for example, actuators and sensors in a realistic manner and to validate the simulation results using real robots. Furthermore, even though the discrepancy between simulations and reality is likely to increase as ever more complex robotic tasks are considered (as argued by Lipson (2001)), it is also likely that carrying out proper tests of such robots will, in fact, *require* simulations to be used, particularly if the simulations can be executed faster than real time. In addition, careful simulations allow researchers to test various hardware configurations before embarking on the often costly and time-consuming task of constructing the real counterpart.

In our view, simulations are an essential tool in mobile robot research, but should only be applied in cases where one can argue convincingly that (i) the various hardware components (e.g. sensors and actuators) can be simulated in a reliable way, and (ii) the actions taken do not require extreme accuracy regarding time or space. As an example regarding simulation of hardware components, consider the sensors used on our robot; in the application considered here almost all relevant sensing is carried out using the LRF, a sensor that is, in fact, very simple to simulate, in view of its high accuracy: Our LRF has a typical error of $\pm$ 1 mm out to distances of around 4 m. On the other hand, if a *camera* had been used on the real robot, it is unlikely that a reliable simulation could have been carried out. Even though GPRSim allows an almost photo-realistic representation of the environment, the readings from a simulated camera would be unlikely to capture all aspects (such as shadows, variations in light levels etc.) of vision. Regarding time and space, the implementation of the navigation method (described above) introduces, by construction, a margin between the robot and the arena objects. Thus, while the actions of the simulated robot may differ somewhat from those of a real robot (and, indeed, also differ between different runs in the simulator, depending on noise in actuators and sensors), there is little risk of *qualitative* differences in the results obtained.

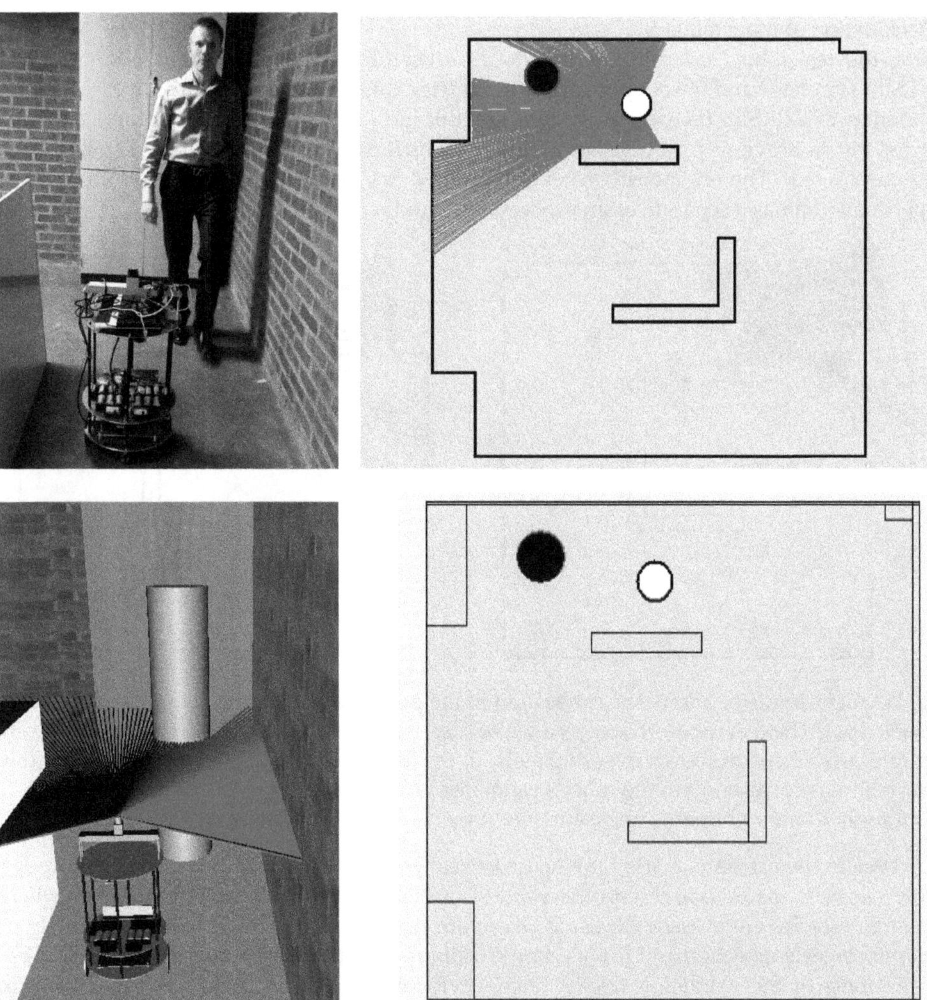

Fig. 10. Upper panels: Detection of a moving obstacle by the real robot. Left: The person approaching the robot. Right: a screenshot showing the LRF rays (for the real robot). The robot is shown as a white disc at its estimated pose, and the detected moving obstacle as a black disc. Lower panels: The same situation, shown for the simulator. Here, the laser rays (which obviously are not visible in the real robot) have been plotted in the screenshot rather than in the map view shown in the right panel.

Nevertheless, before using any simulator for realistic robotics experiments, careful validation and system identification must be carried out, a process that will now be described briefly, in the case of our simulator.

5.3 Validation of the simulator

Using the real robot, several experiments have been carried out in order to validate the GPRSim simulator and to set appropriate parameter values for the simulator. In fact, careful validation of GPRSim (as well as system identification of sensors and actuators) has been carried out in previous (unpublished) work, in which navigation over distances of 10-20 m was considered. The odometric calibration method has been carefully tested both in those experiments and as a separate brain process; see Sandberg et al. (2009).

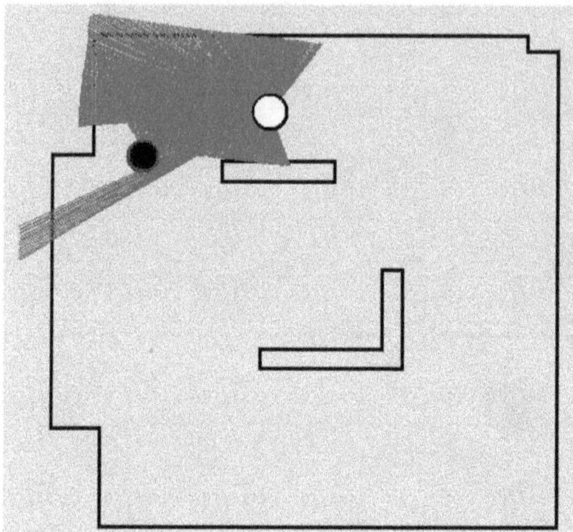

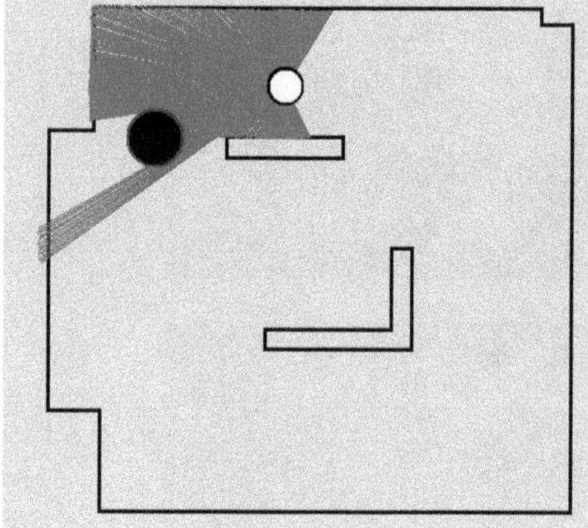

Fig. 11. An example of odometric calibration in the presence of a moving obstacle (shown as a black disc). The real robot, shown as a white disc, was used. The same laser readings (from a single laser scan) are shown in both panels. In the left panel, the origin and direction of the laser readings correspond to the robot's estimated (and incorrect) pose before odometric calibration. In the right panel, the estimated pose has been corrected.

However, in the previous work, moving obstacles were not included. Thus, here, a thorough effort has been made to ascertain the validity of the simulation results in situations where moving obstacles are present. Basically, there are two things that must be tested namely (i) moving obstacle detection and (ii) odometric calibration in the presence of moving obstacles. An example of the validation results obtained for moving obstacle detection is shown in Fig. 10. The upper panels show the results obtained in the real robot, and the lower panel show the results from GPRSim, for a given robot pose and obstacle configuration. As can be seen, the robot (shown as a white disc) detected the moving obstacle (shown as a black disc) in both cases.

Fig. 11 shows the results (obtained in the real robot) of a validation test regarding odometric calibration. In this particular test, the actual position of the robot was $(x, y) = (2.45, 4.23)$ m. The heading angle was equal to π radians (so that the robot was directed towards the negative x-direction). For the purposes of testing the odometric calibration process, an error was introduced in the *estimated* pose; the estimated position was set to $(x, y) = (2.25, 4.20)$ m, and the estimated heading to 3.0 radians. The left panel shows the robot and the moving obstacle (black disc), as well as the actual LRF readings, plotted with the origin and direction provided by the robot's *estimated* pose. As can be seen, the map is not very well aligned with the readings. The right panel shows the same laser readings, again with origin and direction given by the robot's estimated pose, but this time using the estimate obtained *after* odometric calibration. At this point, the alignment is almost perfect, despite the presence of the moving obstacle. Repeating the experiment 10 times, the estimated position (after odometric calibration) was found to be $(x, y) = (2.40 \pm 0.02, 4.22 \pm 0.01)$ m and the estimated heading was 3.11 ± 0.02 radians, where the errors correspond to one standard deviation. The corresponding values found in the simulator were $(x, y) = (2.38 \pm 0.03, 4.23 \pm 0.01)$ m, whereas the estimated heading was found to be 3.13 ± 0.02 radians.Several experiments of this kind were carried out. In all cases, both the simulated robot and the real robot were able to correct their estimated pose using odometric calibration, with essentially identical level of precision.

6. Results

The (simulated) arena used for studying long-term robustness is shown from above in the upper left panel of Fig. 12. The size of the arena, which represents a warehouse environment, is 10×10 m. The large object near the lower right corner represents an elevator shaft (inaccessible to the robot).

Regarding the decision-making system, a single state variable was used, namely the obstacle danger level described in Subsect. 4.4 above. The four cognitive brain processes (*Odometry*, *Odometric calibration*, *Moving obstacle detection*, and *Long-term memory*) were allowed to run continuously. Hence, in their utility functions, the parameters a_{ij} were equal to zero, and b_i was positive. The main task of the decision-making system was thus to select between the two motor behaviors, namely *Grid navigation* (B_1) and *Moving obstacle avoidance* (B_5), using the obstacle danger level as the state variable (z_1). Since only the relative utility values matter (see Sect. 3), the parameters a_{1j} were set to zero, b_{1j} was set to a positive value (0.25), and the time constant τ_1 was set to 0.10 s, thus making u_1 rapidly approach a constant value of around 0.245. After extensive testing, suitable parameters for B_5 were found to be: $a_{51} = 1.00$, $b_5 = -0.45$, and $\tau_1 = 0.10$ s. The gamma parameter Γ_5 (used as described in Subsect. 4.5 above) was set to 3.0 in State 1 of B_5. The corresponding time constant (for exponential decay of the gamma parameter) was set to 1.0 s. The sigmoid parameters c_i were set to 1 for all brain processes.

Several runs were carried out with GPRSim, using the robot configuration described above. Here, the results of two runs will be presented.

6.1 Run 1: Stationary arena

As a first step, a long run (*Run 1*) was carried out, without any moving obstacle present, with the intention of testing the interplay between navigation, odometry, and odometric calibration. The robot was released in the long corridor near the left edge of the arena, and was then required to reach a set of target points in a given order. A total of 11 different target points were used. When the final point was reached, the robot's next target was set as the first target point, thus repeating the cycle. The right panel of Fig. 12 shows the locations of the 11 target points. The robot's trajectory during the first 100 meters of movement (256 s) is shown in the lower left panel of the figure. Both the actual trajectory (thick green line) and the odometric trajectory (thin red line) are shown. As can be seen, the robot successfully planned and followed the paths between consecutive target points. Over the first 100 m of the robot's

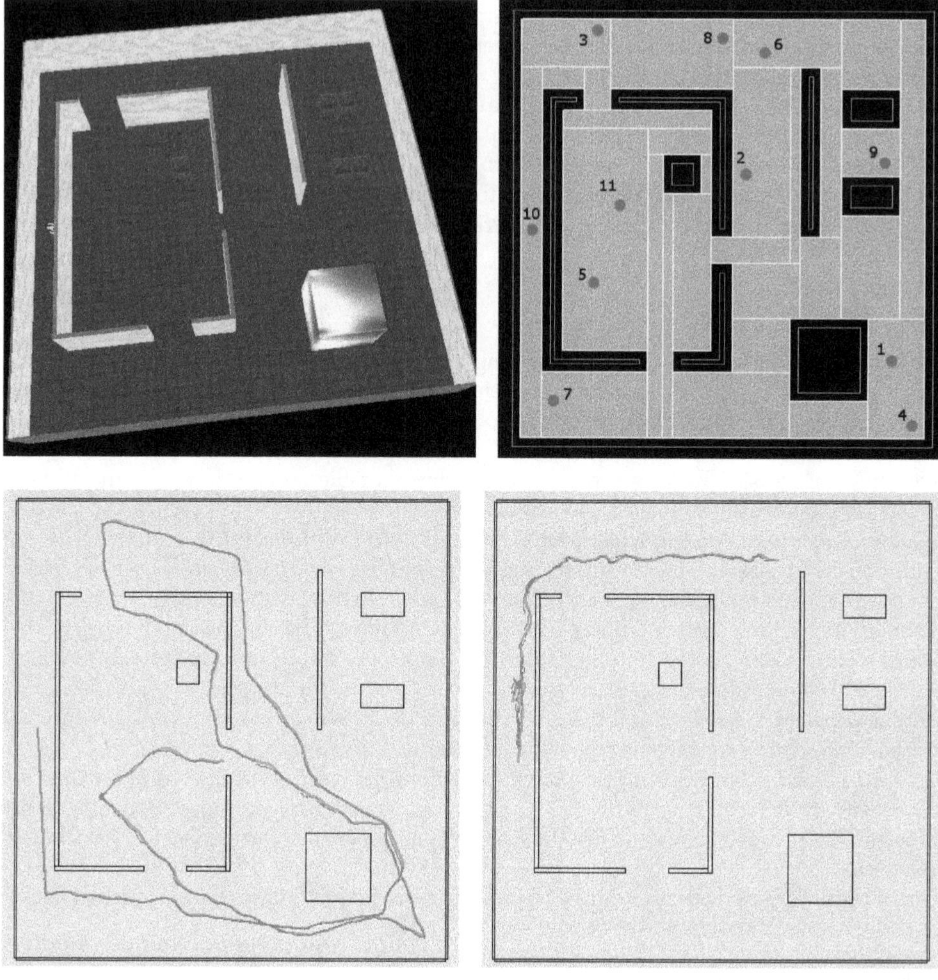

Fig. 12. Upper left panel: The arena used for the simulations. Upper right panel: The 11 target points, with enumeration. Note that, after reaching the 11th target point, the robot's next target becomes the first target point etc. The robot is initially located near target point 10, as can be seen in the upper left panel. Lower left panel: The trajectory for the first 100 m of navigation. The actual trajectory is shown as a thick green line, and the odometric trajectory as a thin red line. Lower right panel: A later part of the trajectory showing the robot approaching 200 m of distance covered. As can be seen, in the long corridor, the robot suffered a rather large pose error at one point. The error was swiftly corrected, however, so that the robot could continue its task.

movement, the average position error[8] was around 0.063 m, and the average heading error

[8] The average position error is computed by averaging the *modulus* of the position error over time. Similarly the average heading error is obtained by averaging the *modulus* of the heading error (modulo 2π) over time.

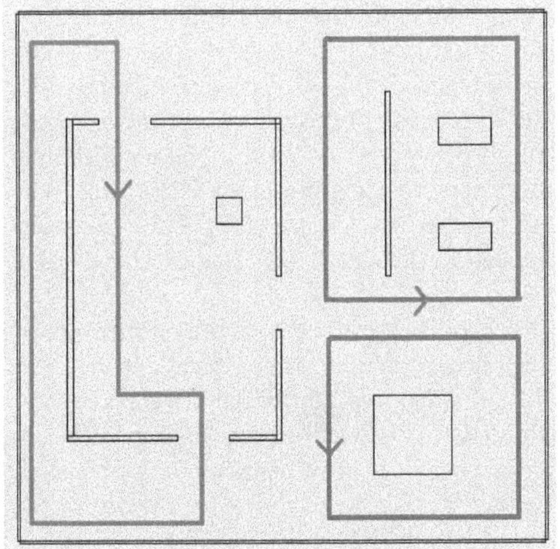

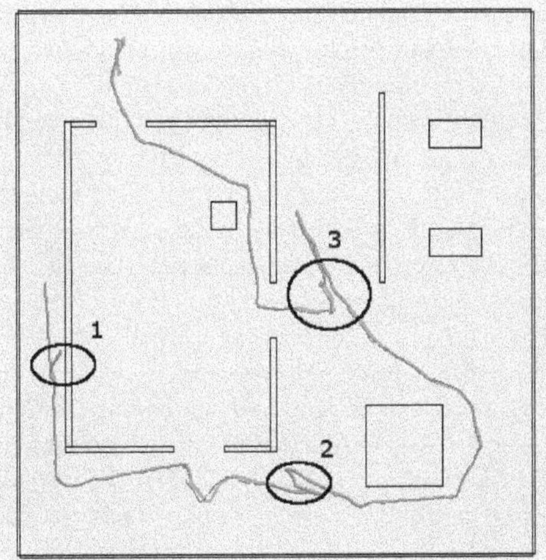

Fig. 13. Left panel: The trajectories of the moving obstacles in Run 2. Right panel: The robot's trajectory during the first 50 m of movement. The ellipses highlight the robot's three evasive maneuvers.

0.031 radians (1.8 degrees). Note, however, that most of the contribution to the error comes from a few, rather rare occasions, where the robot's pose accuracy was temporarily lost. In all cases, the robot quickly recovered (through continuous odometric calibration) to find a better pose estimate. An example of such an error correction is shown in the lower right panel of Fig. 12.

6.2 Run 2: Arena with moving obstacles

In the second run (*Run* 2) presented here, the robot moved in the same arena as in Run 1, and with the same sequence of target points. However, in Run 2, there were three moving obstacles present in the arena. The approximate trajectories of the moving obstacles are illustrated in the left panel of Fig. 13. The right panel shows the first 50 m of the robot's movement. Here, the average position error was around 0.068 m and the average heading error around 0.025 radians (1.4 degrees). The ellipses highlight the three cases where the robot was forced to make an evasive maneuver (during the first 50 m of movement) in order to avoid a collision with a moving obstacle. Note that the odometric calibration process managed to maintain a low pose error even during the evasive maneuver. The run was extended beyond 100 m, with only a small increase in pose error; after 100 m, the average position error was 0.096 m and the average heading error 0.038 radians (2.2 degrees). In this run, the first 100 m took 352 s to traverse (compared to the 256 s in Run 1), due to the evasive maneuvers. Note that, after the third evasive maneuver, the robot modified the original path towards target point 3 (cf. the lower left panel of Fig. 12), having realized that a shorter path was available after evading the moving obstacle.

7. Discussion and conclusion

Even though the results presented in Sect. 6 show that the method used here is capable of achieving robust navigation over long distances in the presence of moving obstacles, a possible criticism of our work is that there is no guarantee that the robot will not suddenly fail, for example by generating an incorrect heading estimate, such that its position error will grow quickly. Indeed, in very long runs, exactly such problems do occur, albeit rarely. Our approach to overcoming those problems would follow the philosophy of error correction rather than error avoidance. An example of this philosophy can be found in the topic of gait generation for humanoid robots: Rather than trying to generate a humanoid robot that will never fall, it is better to make sure that the robot can get up if it *does* fall. This approach has been applied in, for example, the HRP-2 robot, see Kaneko et al. (2002), and many other humanoid robots subsequently developed.

Similarly, rather than trying to further tune the brain process parameters or the parameters of the decision-making system, we would, for full long-term reliability, advocate an approach in which the robotic brain would be supplemented by (at least) two additional brain processes. These are: (i) An *Emergency stop* motor behavior which would cause the robot to stop if it finds itself in a situation where a collision with a stationary object is imminent (moving obstacles being handled by the *Moving obstacle avoidance* behavior). Such situations would occur, for example, if the odometric calibration suffers a catastrophic failure so that the error in the robot's estimated pose grows without bound. Once the robot has stopped it should turn towards the most favorable direction for calibration (which would hardly be the direction towards the stationary object and neither any direction in which *no* part of the map would be visible). Once a suitable direction has been found, the robot should activate the second added brain process, namely *Wake-up*. This cognitive process would attempt a global search (again matching the current laser scan to virtual scans taken in the map) for the robot's current pose. With these two brain processes, which are to be added in future work, the robot would be able to correct even catastrophic errors.

As for the currently available brain processes, the *Moving obstacle avoidance* brain process could be improved to handle also situations in which *several* moving obstacles are approaching the robot simultaneously. However, to some extent, the brain process already does this since, before letting the robot move, it generally checks whether or not there is a clear path in the intended direction of movement; see also the right panel of Fig. 9. In cases where multiple moving obstacles are approaching, typically the best course of action (at least for a rather slow robot) is to do nothing.

Another obvious part of our future work is to carry out long-term field tests with the real robot. Such field tests could, for example, involve a delivery task in an office environment, a hospital, or a factory.

To conclude, it has been demonstrated that the approach of using modularized brain processes in a given, unified format coupled with decision-making based on utility functions, is capable of robustly solving robot motor tasks such as navigation and delivery, allowing the robot to operate autonomously over great distances. Even so, the robot may occasionally suffer catastrophic errors which, on the other hand, we suggest can be dealt with by the inclusion of only a few additional brain processes.

8. References

Arkin, R. C. (1987). *Towards Cosmopolitan Robots: Intelligent Navigation in Extended Man-Made Environments*, PhD thesis, University of Massachusetts.

Arkin, R. C. (1998). *Behavior-Based Robotics*, MIT Press.

Brooks, R. A. (1986). A robust layered control system for a mobile robot, *IEEE J. of Robotics and Automation* RA-2(1): 14–23.

Brooks, R. A. (1992). Artificial life and real robots, *Proceedings of the First European Conference on Artificial Life*, pp. 3–10.

Bryson, J. J. (2007). Mechanisms of action selection: Introduction to the special issue, *Adaptive Behavior* 15: 5–8.

Dijkstra, E. W. (1959). A note on two problems in connexion with graphs, *Numerische Mathematik* 1: 269–271.

Gat, E. (1991). *Reliable Goal-directed Reactive Control for Real-world Autonomous Mobile Robots*, PhD thesis, Virginia Polytechnic Institute and State University.

Jakobi, N., Husbands, P. & Harvey, I. (1995). Noise and the reality gap: The use of simulation in evolutionary robotics, *in* F. Morán, A. Moreno, J. J. Merelo & P. Chacón (eds), *Proceedings of the Third European Conference On Artifical Life, LNCS 929*, Springer, pp. 704–720.

Kaneko, K., Kanehiro, F., Kajita, S. et al. (2002). Design of prototype humanoid robotics platform for HRP, *Proceedings of the IEEE/RSJ International Conference on Intelligent Robots and Systems (IROS2002)*, pp. 2431–2436.

Lipson, H. (2001). Uncontrolled engineering: A review of nolfi and floreano's evolutionary robotics (book review), *Artificial Life* 7: 419–424.

McFarland, D. (1998). *Animal Behaviour: Psychobiology, Ethology and Evolution, 3rd Ed.*, Prentice Hall.

Miglino, O., Lund, H. H. & Nolfi, S. (1996). Evolving mobile robots in simulated and real environments, *Artificial Life* 2: 417–434.

Okabe, A., Boots, B., Sugihara, K. & Chiu, S. N. (2000). *Spatial Tessellations: Concepts and Applications of Voronoi Diagrams*, 2nd edn, John Wiley and Sons, Ltd.

Pirjanian, P. (1999). Behavior coordination mechanisms - state-of-the-art, *Technical report*, Institute for Robotics and Intelligent Systems, University of Southern California.

Prescott, T. J., Bryson, J. J. & Seth, A. K. (2007). Introduction. modelling natual action selection, *Philos. Trans. R. Soc. Lond. B* 362: 1521–1529.

Sakagami, Y., Watanabe, R., Aoyama, C., Matsunaga, S., Higaki, N. & Fujimura, K. (2002). The intelligent ASIMO: System overview and integration, *Proceedings of the 2002 IEEE/RSJ International Conference on Intelligent Robots and Systems*, pp. 2478–2483.

Sandberg, D., Wolff, K. & Wahde, M. (2009). A robot localization method based on laser scan matching, *in* J.-H. Kim, S. S. Ge, P. Vadakkepat, N. Jesse et al. (eds), *Proocedings of FIRA2009, LNCS 5744*, Springer, pp. 179–186.

Singh, J. S. & Wagh, M. D. (1987). Robot path planning using intersecting convex shapes: Analysis and simulation, *IEEE Journal of Robotics and Automation* RA-3: 101–108.

Singh, S. & Agarwal, G. (2010). Complete graph technique based optimization in meadow method of robotic path planning, *International Journal of Engineering Science and Technology* 2: 4951–4958.

Thorpe, C. E. (1984). Path relaxation: Path planning for a mobile robot, *Proceedings of AAAI-84*, pp. 318–321.

von Neumann, J. & Morgenstern, O. (1944). *Theory of Games and Economic Behavior*, Princeton University Press.

Wahde, M. (2003). A method for behavioural organization for autonomous robots based on evolutionary optimization of utility functions, *Journal of Systems and Control Engineering (IMechI)* 217: 249–258.

Wahde, M. (2009). A general-purpose method for decision-making in autonomous robots, *in* B.-C. Chien, T.-P. Hong, S.-M. Chen & M. Ali (eds), *Next-Generation Applied Intelligence, LNAI 5579*, Springer, pp. 1–10.

4

Navigation of Quantum-Controlled Mobile Robots

Eugene Kagan and Irad Ben-Gal
Tel-Aviv University
Israel

1. Introduction

The actions of autonomous mobile robots in stochastic medium imply certain intellectual behavior, which allows fulfilling the mission in spite of the environmental uncertainty and the robot's influence on the characteristics of the medium. To provide such a behavior, the controllers of the robots are considered as probabilistic automata with decision-making and, in some cases, learning abilities. General studies in this direction began in the 1960s (Fu & Li, 1969; Tsetlin, 1973) and resulted in practical methods of on-line decision-making and navigation of mobile robots (Unsal, 1998; Kagan & Ben-Gal, 2008).

Along with the indicated studies, in recent years, the methods of mobile robot's navigation and control are considered in the framework of quantum computation (Nielsen & Chuang, 2000), which gave rise to the concept of quantum mobile robot (Benioff, 1998; Dong, et al., 2006). Such approach allowed including both an environmental influence on the robot's actions and the changes of the environment by the robot by the use of the same model, and the ability to apply the methods of quantum communication and decision-making (Levitin, 1969; Helstrom, 1976; Davies, 1978; Holevo, 2001) to the mobile robot's control.

Following Benioff, *quantum robots* are "mobile systems that have a quantum computer and any other needed ancillary systems on board… Quantum robots move in and interact (locally) with environments of quantum systems" (Benioff, 1998). If, in contrast, the quantum robots interact with a non-quantum environment, then they are considered as *quantum-controlled mobile robots*. According to Perkowski, these robots are such that "their controls are quantum but sensors and effectors are classical" (Raghuvanshi, et al., 2007). In the other words, in the quantum-controlled mobile robot, the input data obtained by classical (non-quantum) sensors are processed by the use of quantum-mechanical methods, and the results are output to classical (non-quantum) effectors.

In this chapter, we present a brief practical introduction into quantum computation and information theory and consider the methods of path planning and navigation of quantum-controlled mobile robots based on quantum decision-making.

2. Quantum information theory and models of quantum computations

We begin with a brief introduction into quantum information theory and stress the facts, which are required for the tasks of mobile robots' navigation. An application of quantum-mechanical methods for the mobile robot's control is based on the statistical interpretation of quantum mechanics (Ballentine, 2006); for a complete review of quantum computation and information theory, see, e.g., (Nielsen & Chuang, 2000).

2.1 Basic notation and properties of quantum-mechanical systems

In general, in the considerations of the finite quantum-mechanical systems, it is postulated (Ballentine, 2006) that the state of the quantum-mechanical system is represented by a Hermitian complex matrix $\sigma = \left\|\sigma_{jk}\right\|$, $\sigma_{jk} = \sigma_{kj}^*$, where $(\cdot)^*$ stands for complex conjugate, with the unit sum of the diagonal elements, $\mathrm{tr}(\sigma) = 1$; and that the observed value of the quantum-mechanical system is specified by the eigenvalues of the matrix σ. Since matrix σ is Hermitial, its eigenvalues are real numbers. If matrix σ is diagonal, then such a representation of the state is equivalent to the representation of the state of stochastic classical system, in which diagonal elements σ_{jj} form a probability vector (Holevo, 2001).

Let $\vec{s} = (s_1, s_2, \ldots)$ be a vector of complex variables $s_j = a_j + ib_j$, $j = 1, 2, \ldots$, and let $s_j^* = a_j - ib_j$ be its complex conjugate. According to the Dirac's "bra[c]ket" notation, row-vector $\langle s| = (s_1, s_2, \ldots)$ is called *bra-vector* and a column-vector $|s\rangle = (s_1^*, s_2^*, \ldots)^T$, where $(\cdot)^T$ stands for the transposition of the vector, is called *ket-vector*. Vector $\vec{s}$ or, equivalently, a Hermitian matrix $\rho = |s\rangle\langle s|$, are interpreted as a state of quantum object; vector $\vec{s}$ is called *state vector* and matrix ρ is called *state matrix* or *density matrix*.

Let $\vec{s}_1 = (s_{11}, s_{12}, \ldots)$ and $\vec{s}_2 = (s_{21}, s_{22}, \ldots)$ be two state vectors. The inner product $\langle s_2|s_1\rangle = (s_{21}, s_{22}, \ldots) \cdot (s_{11}^*, s_{12}^*, \ldots)^T = s_{21} \cdot s_{11}^* + s_{22} \cdot s_{12}^* + \ldots$ is called *amplitude* of the event that the object moves from the state $\vec{s}_1$ to the state $\vec{s}_2$. The *probability* of this event is defined by a square of the absolute value of the amplitude $\langle s_2|s_1\rangle$, i.e. $P(s_2 \mid s_1) = \left|\langle s_2|s_1\rangle\right|^2$, where $\left|\langle s_2|s_1\rangle\right| = |\alpha + i\beta| = \sqrt{\alpha^2 + \beta^2}$. By definition, regarding the probability of any state $\vec{s}$, it is assumed that $0 \le P(s \mid s) \le 1$. For example, let $\langle s_1| = (1, 0)$ and $\langle s_2| = \left(1/\sqrt{2}, 1/\sqrt{2}\right)$, where both $P(s_1 \mid s_1) = 1$ and $P(s_2 \mid s_2) = 1$. However, the probability of transition from the state $\vec{s}_1$ to the state $\vec{s}_2$ is $P(s_2 \mid s_1) = \left|\langle s_2|s_1\rangle\right|^2 = 1/2$.

The *measurement* of the state of the object is described by an *observation operator* O, which in the finite case is determined by a squire matrix $\mathrm{O} = \left\|o_{jk}\right\|$. The result of the observation of the state $\vec{s}$ is defined by a multiplication $\langle s|\mathrm{O}|s\rangle = (s_1, s_2, \ldots) \cdot \left\|o_{jk}\right\| \cdot (s_1^*, s_2^*, \ldots)^T$, and the state, which is obtained after an observation is defined as $|s'\rangle = \mathrm{O}|s\rangle$. For example, assume that the measurement is conducted by the use of the operator $\mathrm{O} = \begin{pmatrix} 1 & 0 \\ 0 & -1 \end{pmatrix}$. Then, the observation of the states $\langle s_1| = (1, 0)$ and $\langle s_2| = \left(1/\sqrt{2}, 1/\sqrt{2}\right)$ results in $\langle s_1|\mathrm{O}|s_1\rangle = 1$ and $\langle s_2|\mathrm{O}|s_2\rangle = 0$, i.e. operator O unambiguously detects the states $\vec{s}_1$ and $\vec{s}_2$. Moreover, an observation of the state $\vec{s}_1$ does not change this state, i.e. $\mathrm{O}|s_1\rangle = (1, 0)^T$, while the observation of the state $\vec{s}_2$ results in the new state $\mathrm{O}|s_2\rangle = \left(1/\sqrt{2}, -1/\sqrt{2}\right)^T$. From such a property of observation, it follows that in contrast to the classical systems, the actual state of the quantum-mechanical system obtains a value, which was measured by the observer, and further evolution of the system starts from this value. In the other words, the evolution the quantum-mechanical system depends on the fact of its observation.

An actual evolution of the quantum-mechanical system is governed by the evolution operators, which are applied to the state matrix σ or state vector $\vec{s}$. Below, we consider the states and operators, which are used in quantum information theory.

2.2 Concepts of the quantum information theory

The elementary state, which is considered in quantum information theory (Nielsen & Chuang, 2000), is called *qubit* (quantum bit) and is represented by a two-element complex vector $|s\rangle = (s_1, s_2)^T$, $0 \le |s_1|^2 + |s_2|^2 \le 1$. Among such vectors, two vectors are specified $|0\rangle = (1,0)^T$ and $|1\rangle = (0,1)^T$, which correspond to the bit values "0" and "1" known in classical information theory (Cover & Thomas, 1991). In general, vectors $|0\rangle$ and $|1\rangle$ determine the states "spin up" and "spin down" of electron, i.e. $|0\rangle \equiv "\uparrow"$ and $|1\rangle \equiv "\downarrow"$.

Given vectors $|0\rangle = (1,0)^T$ and $|1\rangle = (0,1)^T$, any qubit $|s\rangle = (s_1, s_2)^T$ is represented as $|s\rangle = (s_1, s_2)^T = s_1|0\rangle + s_2|1\rangle$. In particular, if there are defined two states $|"\rightarrow"\rangle = \left(1/\sqrt{2}, 1/\sqrt{2}\right)^T$ and $|"\leftarrow"\rangle = \left(1/\sqrt{2}, -1/\sqrt{2}\right)^T$, which represent the electron states "spin right" and "spin left", then $|"\rightarrow"\rangle = \left(1/\sqrt{2}\right)\cdot|1\rangle + \left(1/\sqrt{2}\right)\cdot|0\rangle$ and $|"\leftarrow"\rangle = \left(1/\sqrt{2}\right)\cdot|1\rangle - \left(1/\sqrt{2}\right)\cdot|0\rangle$. Moreover, since it holds true that $|0\rangle = \left(1/\sqrt{2}\right)\cdot|"\rightarrow"\rangle + \left(1/\sqrt{2}\right)\cdot|"\leftarrow"\rangle$ and $|1\rangle = \left(1/\sqrt{2}\right)\cdot|"\rightarrow"\rangle - \left(1/\sqrt{2}\right)\cdot|"\leftarrow"\rangle$, the pairs of the vectors $|0\rangle$ and $|1\rangle$, and $|"\rightarrow"\rangle$ and $|"\leftarrow"\rangle$ can be used interchangeably.

In general, the evolution of the qubits is governed by the use of the following operators:

Pauli operators: $\mathrm{I} = \begin{pmatrix} 1 & 0 \\ 0 & 1 \end{pmatrix}$, $\mathrm{X} = \begin{pmatrix} 0 & 1 \\ 1 & 0 \end{pmatrix}$, $\mathrm{Y} = \begin{pmatrix} 0 & -i \\ i & 0 \end{pmatrix}$, and $\mathrm{Z} = \begin{pmatrix} 1 & 0 \\ 0 & -1 \end{pmatrix}$;

- Hadamard operator: $\mathrm{H} = \frac{1}{\sqrt{2}}\begin{pmatrix} 1 & 1 \\ 1 & -1 \end{pmatrix}$, Phase shift operator: $\mathrm{S} = \begin{pmatrix} 1 & 0 \\ 0 & e^{i\pi/8} \end{pmatrix}$;
- Controlled NOT (CNOT) operator: $\mathrm{CNOT} = \begin{pmatrix} 1 & 0 & 0 & 0 \\ 0 & 1 & 0 & 0 \\ 0 & 0 & 0 & 1 \\ 0 & 0 & 1 & 0 \end{pmatrix} = \begin{pmatrix} \mathrm{I} & 0 \\ 0 & \mathrm{X} \end{pmatrix}$.

The Pauli operators are the most known qubits operators that are in use in general quantum mechanics, while the other three operators are more specific for quantum information theory. According to the Kitaev-Solovey theorem (Nielsen & Chuang, 2000), an algebra $\mathcal{U} = \{\{|0\rangle, |1\rangle\}, \mathrm{CNOT}, \mathrm{H}, \mathrm{S}\}$, which consists of the qubits $|0\rangle$ and $|1\rangle$, and CNOT, Hadamard and phase shift operators, forms a universal algebra that models any operation of the Boolean algebra $\mathfrak{B} = \{\{0,1\}, \neg, \&, \vee\}$. Notice that the qubit operators are reversible. In fact, direct calculations show that $\mathrm{NOT}(\mathrm{NOT}(\vec{s})) = \mathrm{X}\cdot\mathrm{X}\cdot|s\rangle = |s\rangle$, $\mathrm{H}\cdot\mathrm{H}\cdot|s\rangle = |s\rangle$ and so on.

To illustrate the actions of the simplest qubit operators and their relation with classical logic operations, let us present the corresponding quantum gates and their pseuodocode.

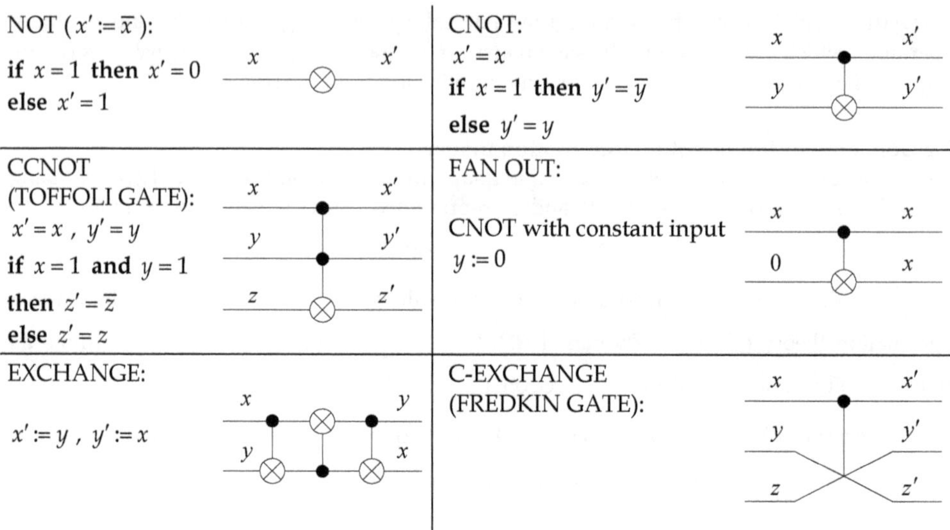

The other types of the qubit gates, e.g. phase shift operator S and its derivatives, cannot be represented by the classical operators and require quantum computing devices. In such computing, it is assumed that each matrix operation is conducted in one computation step providing a power of quantum computing. The indicated dependence of quantum states on the observation process allows an implementation of such operations by the use of adaptive computation schemes. Below, we demonstrate a relation between quantum operations and evolving algebras and consider appropriate probabilistic decision-making.

3. Probabilistic automata and mobile robots control

In this section we describe a simulation example, which illustrates an implementation of qubit operators for the mobile robot's navigation. Then, based on this example, we consider the model of probabilistic control and algorithmic learning methods based on evolving algebras.

3.1 Control of the mobile robot by qubit operators

Let us consider a mobile robot, which moves on the plain, and assume that the inner states of the robots correspond to its direction on a plane. Let $S = \{"\uparrow", "\downarrow", "\rightarrow", "\leftarrow"\}$ be a set of pure or errorless states such that each state corresponds to a certain direction of the robot. The set $V = \{"\textit{step forward}", "\textit{step backward}", "\textit{turn left}", "\textit{turn right}", "\textit{stay still}"\}$ includes the actions that are available to the robot, where the steps "*forward*" and "*backward*" are restricted by a certain fixed distance. Being at time moment t in the state $s^t \in S$ and choosing the action $v^t \in V$, the robot receives to its input a finite scalar value $\varepsilon^t = \varepsilon\left(s^t, v^t\right)$, which depends on the robot's position in the environment. Notice that the elements of the set V form a group with multiplication acting on the set of states S. The steps "*forward*", "*backward*" and "*stay still*" action do not change the state of the robot but change its position relatively to the environment that changes the obtained value ε.

Assume that the states of the robot are described by four qubits: $"\uparrow" = |s_1\rangle = (1,0)^T$, $"\downarrow" = |s_2\rangle = (0,1)^T$, $"\rightarrow" = |s_3\rangle = \left(1/\sqrt{2}, 1/\sqrt{2}\right)^T$ and $"\leftarrow" = |s_4\rangle = \left(1/\sqrt{2}, -1/\sqrt{2}\right)^T$, and the actions "*turn right*" and "*turn left*" are, respectively, conducted by the use of the Hadamard $\mathrm{H} = \frac{1}{\sqrt{2}}\begin{pmatrix} 1 & 1 \\ 1 & -1 \end{pmatrix}$ and reverse Hadamard $\mathrm{H}^R = \frac{1}{\sqrt{2}}\begin{pmatrix} -1 & 1 \\ 1 & 1 \end{pmatrix}$ operators. Then for the Hadamard operator it holds true that $\mathrm{H}|s_1\rangle = |s_3\rangle$, $\mathrm{H}|s_3\rangle = |s_1\rangle$, $\mathrm{H}|s_2\rangle = |s_4\rangle$, and $\mathrm{H}|s_4\rangle = |s_2\rangle$; and similarly, for the reverse Hadamard it holds true that $\mathrm{H}^R|s_1\rangle = |s_4\rangle$, $\mathrm{H}^R|s_4\rangle = |s_1\rangle$, $\mathrm{H}^R|s_2\rangle = |s_3\rangle$ and $\mathrm{H}^R|s_3\rangle = |s_2\rangle$, where the states $|s\rangle = a|0\rangle + b|1\rangle$ and $-|s\rangle = (-1)(a,b)^T$ conventionally are not distinguished.

Let us consider the states and the actions of the robot in particular. If there is no influence on the environment, then the relations between the states can be represented as follows. Assume that the robot is in the state $|s_1\rangle$. Then according to the above-indicated equality $|s_1\rangle = \left(1/\sqrt{2}\right)\cdot|s_3\rangle + \left(1/\sqrt{2}\right)\cdot|s_4\rangle$, and the probability that at the next time moment the robot will be in the state $|s_3\rangle$ is $\rho(s_3 \mid s_1) = \Pr\left\{s^{t+1} = |s_3\rangle \mid s^t = |s_1\rangle\right\} = \left|1/\sqrt{2}\right|^2 = 1/2$ and the probability that it will be at the state $|s_4\rangle$ is also $\rho(s_4 \mid s_1) = \Pr\left\{s^{t+1} = |s_4\rangle \mid s^t = |s_1\rangle\right\} = \left|1/\sqrt{2}\right|^2 = 1/2$. The similar equalities hold true for the other remaining states $|s_2\rangle$, $|s_3\rangle$ and $|s_4\rangle$. In general, for complex amplitudes s_{jk}, it follows that $|s_1\rangle = s_{13}|s_3\rangle + s_{14}|s_4\rangle$, $|s_2\rangle = s_{23}|s_3\rangle - s_{24}|s_4\rangle$, $|s_3\rangle = s_{31}|s_1\rangle + s_{32}|s_2\rangle$, and $|s_4\rangle = s_{41}|s_1\rangle - s_{42}|s_2\rangle$; thus $\rho(s_k \mid s_j) = \Pr\left\{s^{t+1} = |s_k\rangle \mid s^t = |s_j\rangle\right\} = \left|s_{jk}\right|^2$.

Now let us take into account an influence of the environment. Recall that the quantum state of the qubit is equivalently defined both by the state-vector $|s_i\rangle$ and by the density matrix. Let $|s_1\rangle = (1,0)^T$ and $|s_2\rangle = (0,1)^T$, and let $|s_3\rangle = (s_{31}, s_{32})^T$, $|s_4\rangle = (s_{41}, s_{42})^T$. Then, by definition, the states are defined by the density matrices $\sigma_1 = |s_1\rangle\langle s_1| = \begin{pmatrix} 1 & 0 \\ 0 & 0 \end{pmatrix}$, $\sigma_2 = |s_2\rangle\langle s_2| = \begin{pmatrix} 0 & 0 \\ 0 & 1 \end{pmatrix}$, $\sigma_3 = |s_3\rangle\langle s_3| = \begin{pmatrix} |s_{31}|^2 & s_{31}s_{32}^* \\ s_{31}^*s_{32} & |s_{32}|^2 \end{pmatrix}$, and $\sigma_4 = |s_4\rangle\langle s_4| = \begin{pmatrix} |s_{41}|^2 & s_{41}s_{42}^* \\ s_{41}^*s_{42} & |s_{42}|^2 \end{pmatrix}$, while $\mathrm{tr}(\sigma_i) = 1$. Since the non-diagonal elements of the matrices can obtain arbitrary values, let us use these elements for specifying the relation with the environmental variable ε. In particular, assume that $\sigma_3(\varepsilon) = \sigma_3 + \begin{pmatrix} 0 & i\cdot\varepsilon \\ -i\cdot\varepsilon & 0 \end{pmatrix}$. Then, the application of the Hadamard operators H and H^R to the state $\sigma_3(\varepsilon)$ after normalization results in the states $|s'\rangle = \left(1/\sqrt{2} - i\left(1/\sqrt{2}\right)\varepsilon, i\left(1/\sqrt{2}\right)\varepsilon\right)^T$ and $|s''\rangle = \left(-i\left(1/\sqrt{2}\right)\varepsilon, 1/\sqrt{2} + i\left(1/\sqrt{2}\right)\varepsilon\right)^T$. If, e.g., $\varepsilon = 1/3$, then $|s'\rangle = (0.72 - 0.24i, 0.24i)^T$ and

$|s''\rangle = (-0.24i, 0.72+0.24i)^T$ instead of the states $|s_1\rangle = (1,0)^T$ and $|s_2\rangle = (0,1)^T$, which are obtained in the errorless case. The transition probabilities are $\rho(s_1 \mid s') = \rho(s_2 \mid s'') = 0.76$ and $\rho(s_2 \mid s') = \rho(s_1 \mid s'') = 0.24$. The presented method of control is illustrated by the example, shown in Fig. 1. In the simulation (Rybalov et al., 2010), the robot was programmed to follow the trajectory by the use of a compass sensor, and its turns were controlled by Hadamard and reverse Hadamard operators.

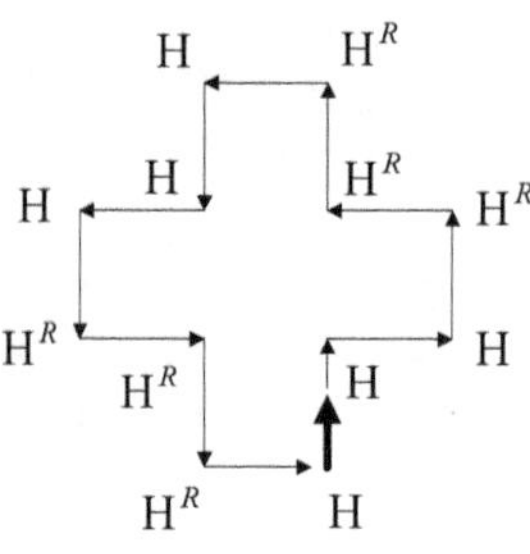

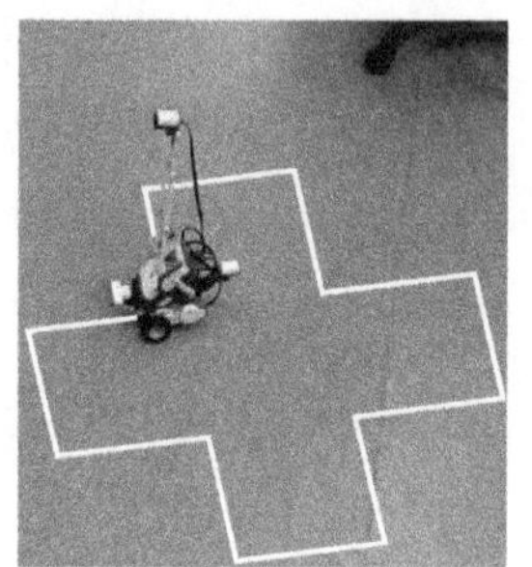

Fig. 1. From left to right: the planned trajectory with corresponding Hadamard and reverse Hadamard operator; the resulting trajectory with the directions of turns; and the mobile robot following the trajectory by the use of the compass sensor.

The presented approach to control the robot is a particular case of the dynamics of open quantum dynamical system (Holevo, 2001). In general, the dynamics of such systems is determined by transitions $|s\rangle \to \mathrm{U}|s\rangle$, where U is an appropriate Hermitian operator.

3.2 Probabilistic model and algorithmic learning

Let us take into account the observation process and present a brief description of the algorithmic model of the robot's control system with variable evolution structure. As it is usual for automata models, let X be a set of input values, Y be a set of output values, and S be a set of inner states. The automaton is defined by two characteristic functions $f: X \times S \to Y$ and $g: X \times S \to S$, such that discrete time dynamics is defined as $y^t = f(x^t, s^t)$ and $s^{t+1} = g(x^t, s^t)$, where $x^t \in X$, $y^t \in Y$, $s^t, s^{t+1} \in S$, $t = 0, 1, 2, \ldots$ In the case of *learning automata*, the probability distributions are defined over the sets X, Y and S, and functions f and g act on such distributions (Fu & Li, 1969; Tsetlin, 1973), while the metrical or topological structures of the sets are constant. By the other approaches, the *program structure learning* is specified by a convergence to the appropriate transition function g, or by the choice of a metric or topology over the set S, called *data structure learning*.
Algorithmically, the variable data structure is defined as follows (Gurevich, 1991). Let G be a global namespace with three distinguished elements $1 = "true"$, $0 = "false"$ and $\Diamond = "undef"$. A map $\varphi: G^r \to G$ is called basic function of arity r, while a basic function of arity $r = 0$ is considered as a distinguished element, and basic functions $\psi: G^r \to \{1, 0, \Diamond\}$ are considered as terms. The evolving algebra $\mathfrak{A} = \{G, \varphi_0, \varphi_1, \ldots\}$ is defined by the following updates:

- local update: $\varphi(a_0, a_1, \ldots, a_r) \leftarrow a$, which specifies the value to the basic function,
- guarded update: if $\psi = 1$ then $\varphi(a_0, a_1, \ldots, a_r) \leftarrow a$,

where $a_0, a_1, \ldots, a_r, a \in G$ and ψ is a term. Thus, evolving algebra $\mathfrak{A}$ permits its functions to change their domains according to the data flow, as it is required for the learning property. Notice that the implementation of the updates of the algebra $\mathfrak{A}$ by the use of quantum gates results in the universal algebra $\mathfrak{U}$, which defines quantum computations. Nevertheless, similar to the algebra $\mathfrak{U}$, the direct implementation of the evolving algebra $\mathfrak{A}$ is possible only in particular cases of computable operators.

Let us consider an implementation of the evolving algebra for probabilistic control of the mobile robots control (Kagan & Ben-Gal, 2008). Since during its mission the robot acts in a stochastic environment, the input variable $x \in X$ is random; thus, given characteristic f and g, both inner states $s \in S$ and outputs $y \in Y$ are also random variables. Assume that the namespace G, in addition to elements 1, 0 and $\Diamond$, includes all possible realizations of the inputs, inner states and outputs, i.e. $G = \{1, 0, \Diamond\} \cup X \cup Y \cup S$. For the realizations of inputs, we define the terms $\psi_1(x) \in \{1, 0, \Diamond\}$, and for the inner states we define the terms $\psi_2(s) \in \{1, 0, \Diamond\}$ and the local updates $\varphi(s') \leftarrow s''$, $s', s'' \in G$.

Now we can formulate transition function g in the terms of evolving algebra. We say that the pairs $(x', s'), (x'', s'') \in X \times S$ are equal (in the sense of the function g) if $g : (x', s') \mapsto s$ and $g : (x'', s'') \mapsto \overline{s}$. Then, since there are at most m distinguished realizations of the inner states and there may exist the pairs from $X \times S$, such that the map g is not defined, the space $X \times S$ is partitioned into at most $m + 1$ equivalence classes. Denote such equivalence classes by A^g. Then, the transition function g is defined as a guarded update if $(x, s) \in A_j^g$ then $\phi(s) \leftarrow s_j$, which is checked for each pair $(x, s) \in X \times S$, and if the appropriate class A^g is not found, then $\phi(s) \leftarrow \Diamond$ is specified.

The presented algorithmic model allows simulations of quantum-control of the mobile robot and its navigation on the basis of the qubits model of states. Below, we consider an example of such simulations with probabilistic decision-making.

4. Navigation of quantum-controlled mobile robot along predefined path

Let us start with a simulation example. The simulation follows an idea of the experiment of checking a spin of elementary particle by three Stern-Gerlach apparatus, which are defined by a sequence of certain quantum operators (Albert, 1994).

As above, let the inner states of the robot be specified by four qubits $"\uparrow" = |s_1\rangle = (1, 0)^T$, $"\downarrow" = |s_2\rangle = (0, 1)^T$, $"\rightarrow" = |s_3\rangle = (1/\sqrt{2}, 1/\sqrt{2})^T$ and $"\leftarrow" = |s_4\rangle = (1/\sqrt{2}, -1/\sqrt{2})^T$. In addition, assume that there are two types of detectors defined by the above-defined Pauli operators X and Z, so that $\langle s_3|\mathrm{X}|s_3\rangle = +1$, $\langle s_4|\mathrm{X}|s_4\rangle = -1$, $\langle s_1|\mathrm{Z}|s_1\rangle = +1$ and $\langle s_2|\mathrm{Z}|s_2\rangle = -1$. The robot starts with a random initial state $|s^0\rangle$ and arrives to the first detector Z. Detector Z checks the state $|s^0\rangle$ and the robot obtains a new state $|s^1\rangle$. According to the maximum of probabilities $P(s_3 \mid s^1)$ and $P(s_4 \mid s^1)$, the robot chooses the left or right trajectory and arrives to the second detector X. Then, after the similar actions and obtaining the state $|s^2\rangle$, the robot continues to the third detector Z, which checks the robot's state and results in the state $|s^3\rangle$. The fragment of the experiment with the robot following its path is shown in Fig. 2.

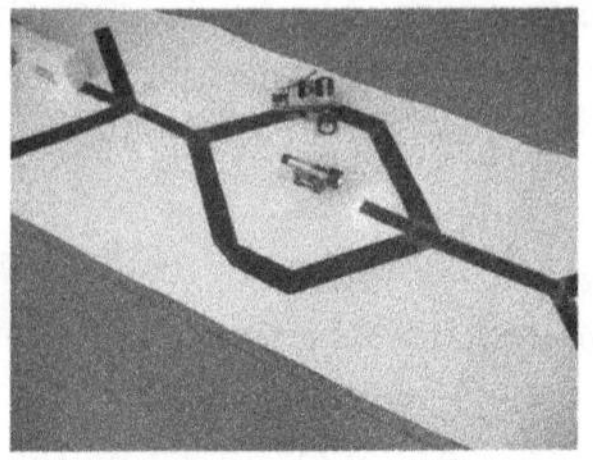
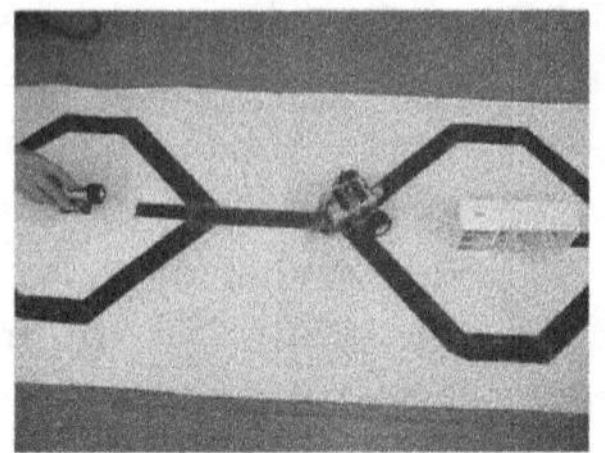

Fig. 2. The mobile robot follows a path (from right to left) with three simulated detectors using touch sensor ("Z detector") and light sensors ("X detector").

Let us consider the actions of the robot in particular. As indicated above, the robot starts with an initial state $|s^0\rangle = (s_1^0, s_2^0)^T$, where $s_1^0, s_2^0 \in [0,1]$ are random values such that $0 < (s_1^0)^2 + (s_2^0)^2 \leq 1$. Then the first detector Z results in the state $|s^1\rangle = \mathrm{Z}|s^0\rangle$ and the decision-making is conducted regarding the further left or right trajectory, which is based on the probabilities $P(s_3 \mid s^1) = |\langle s^1 | s_3 \rangle|^2$ and $P(s_4 \mid s^1) = |\langle s^1 | s_4 \rangle|^2$. If $P(s_3 \mid s^1) > P(s_4 \mid s^1)$, then the robot turns left, and if $P(s_3 \mid s^1) < P(s_4 \mid s^1)$, then the robot turns right (the ties are broken randomly). Following the chosen trajectory, the robot arrives to the second detector X. The check with this detector results in the state $|s^2\rangle = \mathrm{X}|s^1\rangle$, and the decision regarding the further trajectory is obtained on the basis of the probabilities $P(s_1 \mid s^2) = |\langle s^2 | s_1 \rangle|^2$ and $P(s_2 \mid s^2) = |\langle s^2 | s_2 \rangle|^2$. Similar to the above, if $P(s_1 \mid s^2) > P(s_2 \mid s^2)$, then the robot turns left, and if $P(s_1 \mid s^2) < P(s_2 \mid s^2)$, then the robot turns right (the ties are broken randomly). The third check is again conducted by the Z detector, which results in the state $|s^3\rangle = \mathrm{Z}|s^2\rangle$, and the decision-making is conducted by the same manner as for the state $|s^1\rangle$.

Now let us present a general description of the process (Kagan & Ben-Gal, 2008), that implements the above-indicated equivalence classes A^f and A^g. As indicated above, the evolution of the quantum-mechanical system with observations does not depend on the previous states and starts from the value of the state, which is obtained by the measurement. Thus, the outputs $y \in Y$ of the system are specified by a Markov process, which is controlled by input states $x \in X$ and inner states $s \in S$. Then the probability $p(y) = p(A^f(y))$ of the equivalence class $A^f(y) = f^{-1}(y)$ is defined as a sum

$p(y)=\sum_{(x,s)\in A^f(y)} p(x,s)$ of the probabilities of the pairs $(x,s)\in X\times S$. Similarly, the probability $p(s)$ of the inner state $s\in S$ of the system is defined by the use of equivalence class $A^g(s)=g^{-1}(s)$ as a sum $p(s)=\sum_{(x,s)\in A^g(s)} p(x,s)$.

Recall that, by definition, the equivalence classes $A^f(y)$, $y\in Y$, and A^g, $A^g(s)$, form partitions α^f and α^g of the space $X\times S$. Then, the relation between the dynamics of inner states is determined by the relation between the partitions α^f and α^g. For example, let $d_{Orn}\left(\alpha^g,\alpha^f\right)=\sum_{i=0}^{k-1}\left|p\left(A_i^g\right)-p\left(A_i^f\right)\right|$, where $k=\max\left\{\left|\alpha^g\right|,\left|\alpha^f\right|\right\}$, and if $\left|\alpha^g\right|>\left|\alpha^f\right|$, then α^f is completed by empty sets, while if $\left|\alpha^g\right|<\left|\alpha^f\right|$, then empty sets are added to α^g, being Ornstein distance between the partitions (Ornstein, 1974). Then, for the finite time case, $t=0,1,2,\ldots,T-1$, the Ornstein distance between partitions α_t^f and α_t^g, is defined as $d_{Orn}\left(\left\{\alpha_t^g\right\}_0^{T-1},\left\{\alpha_t^f\right\}_0^{T-1}\right)=\left(1/(T-1)\right)\sum_{t=0}^{T-1} d_{Orn}\left(\alpha_t^g,\alpha_t^f\right)$. Since the structure of the partitions α_t^f and α_t^g is constant, the distance represents the relation between probability measures defined by these partitions. Thus the goal of the robots navigation is to find a process for governing the inner states $s\in S$ such that the distance $d_{Orn}\left(\left\{\alpha_t^g\right\}_0^{T-1},\left\{\alpha_t^f\right\}_0^{T-1}\right)$ reaches its minimum over the considered time interval. Below we will consider general informational algorithms of local search, which can be implemented for such a task.

5. Information theoretic decision-making and path planning

The section presents information theoretic methods for quantum inspired decision-making and general path-planning algorithms. We start with a motivating example of informational decision-making, then we consider the logic of quantum mechanics and informational distance between the partitions of the events space. Finally, we present the navigation algorithms, which are based on the representation of the states' evolution by the use of partitions.

5.1 Decision-making by the use of quantum-mechanical information measure

An application of informational criteria for decision-making and path planning of the quantum-controlled mobile robots is motivated by the criteria of classical information theory (Cover & Thomas, 1991). Recall that in the classical case, an informational distance between the probability vectors $\vec{p}=(p_1,\ldots,p_n)$ and $\vec{q}=(q_1,\ldots,q_n)$, $\sum_{j=1}^{n} p_j=1$, $\sum_{j=1}^{n} q_j=1$, $p_j\geq 0$, $q_j>0$, $j=1,2,\ldots,n$, is specified by the *relative Shannon entropy* or *Kullback-Leibler distance* $KL(\vec{p}\,||\,\vec{q})=\sum_{j=1}^{n} p_j\log\left(p_j/q_j\right)=\sum_{j=1}^{n} p_j\log p_j-\sum_{j=1}^{n} p_j\log q_j$, where by convention it is assumed that $0\log 0=0$. The distance $KL(\vec{p}\,||\,\vec{q})$ satisfies both $KL(\vec{p}\,||\,\vec{q})\geq 0$ and $KL(\vec{p}\,||\,\vec{q})=0$ if and only if $p_j=q_j$ for all $j=1,2,\ldots,n$, and, in general

$KL(\vec{p}\,||\,\vec{q}) \neq KL(\vec{q}\,||\,\vec{p})$. Vectors $\vec{p}$ and $\vec{q}$ represent the states of the stochastic system, and distance $KL(\vec{p}\,||\,\vec{q})$ characterizes the information-theoretic difference between these states. In contrast, in quantum-mechanical systems, the state is represented by the above-presented Hermitian density matrix σ with $\mathrm{tr}(\sigma)=1$. The informational measures for such states are defined on the basis of the von Neumann entropy (Nielsen & Chuang, 2000) $VN(\sigma) = -\mathrm{tr}(\sigma \log \sigma) = -\sum_{j=1}^{n} \lambda_j \log \lambda_j$, where λ_j are the eigenvalues of the matrix σ. Then the *relative von Neumann entropy* of the state σ' relative to the state σ'' is defined as $VN(\sigma'\,||\,\sigma'') = \mathrm{tr}(\sigma' \log \sigma') - \mathrm{tr}(\sigma' \log \sigma'') = -\sum_{j=1}^{n} \lambda'_j \log \lambda'_j - \sum_{j=1}^{n} \lambda'_j \log \lambda''_j$, where λ' and λ'' are eigenvalues of the matrices σ' and σ'', correspondingly.

Let σ^t be a state of the system at time moment t, and consider its representation $\sigma^t = \sum_{j=1}^{n} \lambda_j^t E_j$, where $\lambda_1^t < \lambda_2^t < \ldots < \lambda_n^t$ are eigenvalues of the matrix σ^t and $E = \{E_1, \ldots, E_n\}$ is a set of matrices such that $E_j E_k = \delta_{jk} E_k$ and $\sum_{j=1}^{n} E_j = I$, where $\delta_{jk} = 1$ if $j = k$, and $\delta_{jk} = 0$ otherwise. According to the dynamics of the quantum system, if the system is in the state σ^t, then its next state σ^{t+1} is specified by the use of the selected operator E_j according to the projection postulate $\sigma^{t+1} = E_j \sigma^t E_j / \mathrm{tr}(\sigma^t E_j)$. This postulate represents the above indicated influence of the measurement, i.e. of the application of the operator E_j, to the state of the system, and the generalized the Bayesian rule to the evolution of quantum-mechanical systems. The decision-making problem required a definition of such projection that given a state σ^t, the next state σ^{t+1} is optimal in a certain sense. Since for a state σ^t, there exist several sets E of matrices with the above indicated properties, the decision-making includes two stages (Davies, 1978; Holevo, 2001) and requires finding the set $E = \{E_1, \ldots, E_n\}$, and then selecting an operator E_j from the set E according to optimality criteria.

One of the methods, which is widely used in classical information theory, implies a choice of such probability vector $\vec{p}^{t+1}$, that given a vector $\vec{p}^t$, the Kullback-Leibler distance $KL(\vec{p}^t\,||\,\vec{p}^{t+1})$ reaches its maximum. Likewise, in the simulations, we implemented the choice of the set E and operator $E_j \in E$ such that it maximizes the relative von Neumann entropy $VN(\sigma^t\,||\,\sigma^{t+1})$. In the simulations (Kagan et al., 2008), the mission of the robot was to navigate in the environment and to find the objects, which randomly change their location. The amplitudes, which defined the states, were derived from the distances between the objects and as they were measured by the ultra-sonic sensor. The scheme of the simulation and the fragment of the robot's movement are shown in Fig. 3.

The robot scans the environment, chooses such an object that maximizes the von Neumann relative entropy, and moves to this object. After the collision, the object was moved to a new random location. The comparison between quantum and classical decision-making demonstrated the difference in nearly 50% of the decisions, and in the search for variable number of objects, quantum decision-making demonstrated nearly 10% fewer decision errors than the classical one. Such results motivated an application of information theoretic methods for navigation of quantum-controlled mobile robots. In the next section we consider the algorithms which follow this direction.

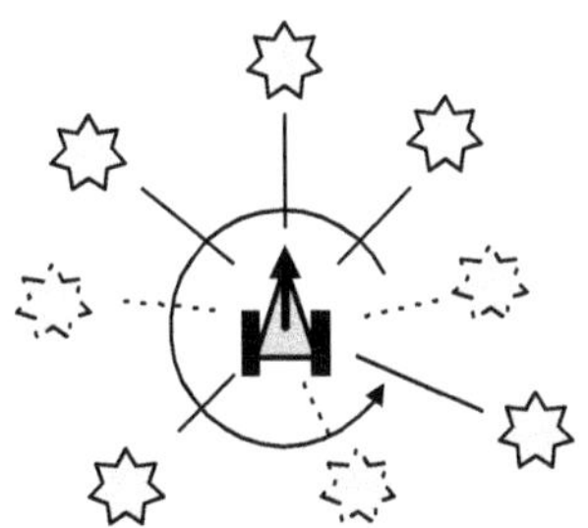

Fig. 3. The mobile robot with ultra-sonic sensor acting in the stochastic environment: after collision, the obstacle randomly changes its location in the environment.

5.2 Logic of quantum mechanics and informational distance

Now let us consider the logic of quantum mechanics and the structure of quantum events over which the informational algorithms act. Such a formal scheme of quantum events, which is called *quantum logic* (Cohen, 1989), was introduced by Birkhoff and von Neumann (Birkhoff & Neumann, 1936) as an attempt of to find an axiomatic description of quantum mechanics.

A quantum logic is a lattice Λ of events A, which contains the smallest element $\varnothing$, the greatest element I, relation $\subset$, unary operation $'$, and binary operations $\cup$ and $\cap$. It is assumed that for the events $A \in \Lambda$ the following usual set properties hold true: a) For any element $A \in \Lambda$ there exist an event $A' \in \Lambda$ such that $(A')' = A$; b) $A \cap A' = \varnothing$ and $A \cup A' = I$; c) For any pair of events $A, B \in \Lambda$, $A \subset B$ implies $B' \subset A'$, and $A \subset B$ implies $B = A \cup (B \cap A')$; d) For any countable sequence $A_1, A_2, \ldots$ of events from Λ their union is in Λ, i.e., $\left(\cup_j A_j\right) \in \Lambda$. Notice that is not required that the events $A \in \Lambda$ are subsets of the same set. That leads to the fact that under the requirements of quantum logic the distributive rule may not hold and, in general, $B \cap (A \cup A') \geq (B \cap A) \cup (B \cap A')$. In contrast, in the probabilistic scheme, it is postulated that the events are the subsets of the set of elementary outcomes, so the distribution rule is satisfied.

In quantum logic Λ, events $A \in \Lambda$ and $B \in \Lambda$ are called *orthogonal*, and denoted by $A \perp B$, if $A \subset B'$, and a finite set $\alpha = \{A_1, \ldots, A_n\}$, of events $A_j \in \Lambda$, is called $\vee$-*orthogonal system* if $\left(\cup_{j=1}^{k} A_j\right) \perp A_{k+1}$, $k = 1, 2, \ldots, n-1$. The state μ over the quantum logic Λ is defined as a map $\mu : \Lambda \to [0,1]$ such that $\mu(I) = 1$ and for any $\vee$-orthogonal system $\alpha = \{A_1, \ldots, A_n\}$ it holds true that $\mu\left(\cup_{j=1}^{n} A_j\right) = \sum_{j=1}^{n} \mu\left(A_j\right)$. Given a state $\mu : \Lambda \to [0,1]$, an $\vee$-orthogonal system α is a *partition* of the logic Λ with respect to the state μ if $\mu\left(\cup_{j=1}^{n} A_j\right) = 1$, $A_j \in \alpha$ (Yuan, 2005).

Following classical ergodic theory (Rokhlin, 1967), *entropy* of the partition α is defined as $H_\mu(\alpha) = -\sum_{A \in \alpha} \mu(A) \log \mu(A)$, and *conditional entropy* of partition α relatively to partition β is $H_\mu(\alpha \mid \beta) = -\sum_{B \in \beta} \sum_{A \in \alpha} \mu(A, B) \log \mu(A \mid B)$, where α and β are partitions of quantum logic Λ (Yuan, 2005; Zhao & Ma, 2007). In addition, similarly to the ergodic theory (Rokhlin,

1967), a Rokhlin distance is defined as $d_\mu(\alpha,\beta)=H_\mu(\alpha|\beta)+H_\mu(\beta|\alpha)$ between the partitions α and β of the quantum logic Λ, which preserves the metric properties (Khare & Roy, 2008), as it holds for the probabilistic scheme.

5.3 Actions of the robot over the quantum events

The indicated properties of the partitions and states of quantum logic allow an application of informational A*-type search algorithms acting over partitions space (Kagan & Ben-Gal, 2006) for navigation of quantum-controlled mobile robots. Let χ be a set of all possible partitions of the logic Λ given a state μ. By $d_\mu(\alpha,\beta)$, $\alpha,\beta\in\chi$, we denote a Rokhlin distance between the partitions α and β, and by $\tilde{d}_\mu(\alpha,\beta)$ an estimated distance such that $\tilde{d}_\mu(\alpha,\beta)\le d_\mu(\alpha,\beta)$; as a distance $\tilde{d}_\mu$ the above-defined Ornstein distance $d_{Orn}(\alpha,\beta)$ can be applied. Let $r\in R^+$ be a constant value. The non-empty neighborhood $N(\alpha,r)\subset\chi$ of partition $\alpha\in\chi$ is a set of partitions such that $\alpha\notin N(\alpha,r)$ and for every partition $\beta\in N(\alpha,r)$ it holds true that $\tilde{d}_\mu(\alpha,\beta)\le r\le d_\mu(\alpha,\beta)$. In the algorithm, α stands for a partition specified by the robot, and τ stands for a partition specified by the environment. The robot is located in the environment, which is specified by a partition $\tau_{cur}=\vartheta$ and starts with the initial partition $\alpha_{cur}=\theta$. Then, the actions of the robot are defined by the following sequence:

The robots actions given the environment τ_{cur}:

- Choose the next partition $\alpha_{next}\leftarrow\arg\min_{\alpha\in N(\alpha_{cur},r)}\left\{\tilde{d}_\mu(\alpha,\tau_{cur})\right\}$;
- Update estimation $\tilde{d}_\mu(\alpha_{cur},\tau_{cur})\leftarrow\max\left\{\tilde{d}_\mu(\alpha_{cur},\tau_{cur}),r+\min_{\alpha\in N(\alpha_{cur},r)}\tilde{d}_\mu(\alpha,\tau_{cur})\right\}$;
- Set current partition $\alpha_{cur}\leftarrow\alpha_{next}$.

The robots actions while the environment changes ($\tau_{cur}\leftarrow\tau$):

- Update estimation $\tilde{d}_\mu(\alpha_{cur},\tau)\leftarrow\max\left\{\tilde{d}_\mu(\alpha_{cur},\tau),\tilde{d}_\mu(\alpha_{cur},\tau_{cur})-r\right\}$.

Let us clarify the application of the algorithms to the quantum logic Λ by half steps, $t=0,1/2,1,3/2,\ldots$, which correspond to the robot's and the environment's actions. Let the initial map by $\mu^0:\Lambda\to[0,1]$. Regarding the robot's states, it means a choice of the basis for the states; in the above-given examples, such a choice corresponds to the definition of the qubits for the states $"\uparrow"=|s_1\rangle$ and $"\downarrow"=|s_2\rangle$, or that is equivalent to the states $"\rightarrow"=|s_3\rangle$ and $"\leftarrow"=|s_4\rangle$. By the use of the map μ^0 the robot chooses the state $|s^0\rangle$ and a certain partition which specifies the next map $\mu^{1/2}$. According to the obtained map $\mu^{1/2}$, the environment changes and the map μ^1 is specified. Now the robot chooses the $|s^1\rangle$ and according to the chosen partition specifies the map $\mu^{3/2}$. The environment, in its turn, changes according to the map $\mu^{3/2}$, and so on.

The presented actions over the partitions of quantum logic provide a representation, which differs from the one use in the quantum learning algorithms (Chen & Dong, 2008; Dong, et al., 2010) and implements information theoretic decision-making using the Rokhlin distance. However, the meaning of the actions is similar, and the goal of the robot is to determine the actions such that the inner states of the robot correspond to the environmental states, which, in their turn, are changed by the robot.

6. Notes on fuzzy logic representation of quantum control

Let us return to the representation of the robot's states by qubits $|s\rangle = a|0\rangle + b|1\rangle = a(1,0)^T + b(0,1)^T$, where both amplitudes a and b are real numbers. In such a case, the state-vector $|s\rangle$ can be represented by two membership functions $\mu_a : X \to [0,1]$ and $\mu_b : X \to [0,1]$ for some universal set X, which are defined as (Hannachi, et al., 2007) $\mu_a = (2/\pi)\arcsin\sqrt{\left(sign(a)a^2 - sign(b)b^2 + 1\right)/2}$ and $\mu_b = (2/\pi)\arcsin\sqrt{\left(sign(a)a^2 + sign(b)b^2 + 1\right)/2}$, with backward transformations $a = sign(\mu_a + \mu_b - 1)\sqrt{\left|\sin^2 \mu_a\pi/2 + \sin^2 \mu_b\pi/2 - 1\right|}$ and $b = sign(\mu_b - \mu_a)\sqrt{\left|\sin^2 \mu_b\pi/2 - \sin^2 \mu_a\pi/2\right|}$. Over the pairs of membership functions μ_a and μ_b, fuzzy analogs of quantum mechanical operators are defined (Hannachi, et al., 2007). Let us consider the Hadamard operators, which represent the turns of the robot.

Fuzzy analog $\tilde{\mathrm{H}}$ of the Hadamard operator H is the following (Hannachi, et al., 2007): $\tilde{\mathrm{H}}(\mu_a, \mu_b) = \left(\max(0, \min(1, 1/2 - \mu_a + \mu_b)), \min(1, \max(0, \mu_a + \mu_b - 1/2))\right)$. Similarly, reverse fuzzy Hadamard operator is defined by the following transformation (Rybalov et al., 2010): $\tilde{\mathrm{H}}^R(\mu_a, \mu_b) = \left(\min(1, \max(0, 3/2 - \mu_a - \mu_b)), \max(0, \min(1, 1/2 - \mu_a + \mu_b))\right)$.

Straightforward calculations show that the direct and reverse fuzzy Hadamard operators $\tilde{\mathrm{H}}$ and $\tilde{\mathrm{H}}^R$ are reversible and preserve the properties of corresponding quantum Hadamard operators. Trajectories of the robot that acts according to the fuzzy Hadamard operators are illusrated by Fig. 4; the simulattions have been conducted likewise in the example shown in Fig. 1.

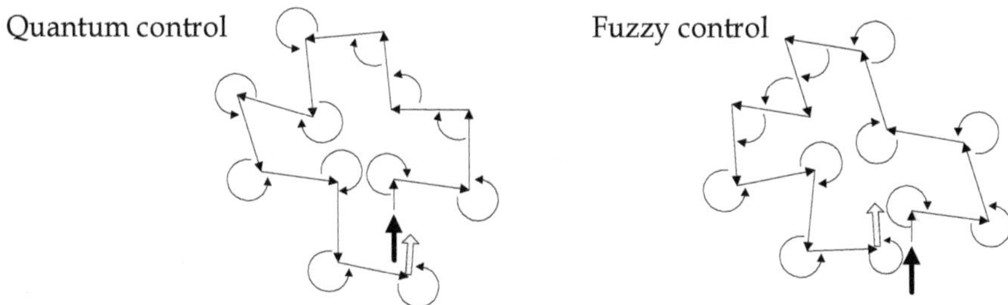

Fig. 4. Trajectories and turns of the mobile robot according to quantum and fuzzy control.

According to simulations, the turns of quantum and fuzzy controlled robot are different; however, the states of the robot and the results of its actions are statistically equivalent. Such preliminary results show that in case of real amplitudes, fuzzy logic models of quantum control may be applied.

7. Conclusion

In the chapter we presented a brief introduction into the methods of navigation of quantum-controlled mobile robots and considered the ideas of its implementation by the use of probabilistic and information theoretic techniques. The described methods represent such a property of the quantum-controlled mobile robots that the state of quantum-mechanical system includes the state of the environment as a part of its inner state.

In particular, the state of the mobile robot in the environment was defined by the use of the density matrix, which, in addition to the inner state of the robot, included the state of the environment. Such a specification of the state allowed calculations of both the robot's influence on the environment and the environmental influence on the robot by the use of the unified techniques.

The decision-making methods, which define the robot's behavior, implemented the indicated representation of the state and were based on the probabilistic and informational schemes. These schemes generalize the known maximum probability and maximum information criteria while taking into account additional information regarding the robot's influence on the environment, and correspond to the statistical considerations of quantum-mechanical methods (Malley & Hornstein, 1993; Barndorff-Nielsen & Gill, 2003).

In general, the actions of quantum-controlled mobile robot were specified by the choices of certain partitions of quantum logic. The choices were based on informational distances following the line of informational search algorithms (Kagan & Ben-Gal, 2006). As indicated above, such a method gives an alternative view to quantum learning and path-planning algorithms (Chen & Dong, 2008; Dong, et al., 2010).

The presented methods were simulated by the use of small mobile robots, while the complete realization of quantum control requires quantum-mechanical on-board computers. However, as it follows from preliminary considerations (Rybalov, et al., 2010), fuzzy control of the mobile robot demonstrates similar results as probabilistic and informational schemes of quantum control; thus in some cases fuzzy logic models of quantum control may be applied.

8. References

Albert, D. Z. (1994). *Quantum Mechanics and Experience*. Harvard University Press, Cambridge, Massachusetts and London, England.

Ballentine, L. E (2006). *Quantum Mechanics. A Modern Development*. Word Scientific, Singapore.

Barndorff-Nielsen, O. E. & Gill, R. D. (2003). On Quantum Statistical Inference. *J. Royal Statistical Society B*, Vol. 65, No. 4, pp. 775-816.

Benioff, P. (1998). Quantum Robots and Environments. *Phys. Rev. A*, Vol. 58, pp. 893-904.

Birkhoff, G. & Neumann, J. von. (1936). The Logic of Quantum Mechanics. *Annals Math.*, Vol. 37, No. 4, pp. 823-843.

Chen, C.-L. & Dong, D.-Y. (2008). Superposition-Inspired Reinforcement Learning and Quantum Reinforcement Learning. In *Reinforcement Learning: Theory and Applications*, C. Weber, M. Elshaw, N. M. Mayer, (Eds.), InTech Education and Publishing, Vienna, Austria, pp. 59-84.

Cohen, D. W. (1989). *An Introduction to Hilbert Space and Quantum Logic*. Springer-Verlag, New York.

Cover, T. M. & Thomas, J. A. (1991). *Elements of Information Theory*. John Wiley & Sons, New York.

Davies, E. B. (1978). Information and Quantum Measurement. *IEEE Trans. Inform. Theory*, Vol. 24, No. 5, pp. 596-599.

Dong, D.-Y.; Chen, C.-L.; Zhang, C.-B. & Chen, Z.-H. (2006). Quantum Robot: Structure, Algorithms and Applications. *Robotica*, Vol. 24, No. 4, pp. 513-521.

Dong, D.; Chen, C.; Chu, J. & Tarn, T.-J. (2010). Robust Quantum-Inspired Reinforcement Learning for Robot Navigation. *IEEE/ASME Trans. Mechatronics*, To appear.

Fu, K. S. & Li, T. J. (1969). Formulation of Learning Automata and Automata Games. *Information Science*, Vol. 1, No. 3, pp. 237-256.

Gurevich, Y. (1991). Evolving Algebras: An Attempt to Discover Semantics. *Bull. European Assoc. Theor. Comp. Science*, Vol. 43, pp. 264-284.

Hannachi, M. S.; Hatakeyama, Y. & Hirota, K. (2007). Emulating Qubits with Fuzzy Logic. *J. Advanced Computational Intelligence and Intelligent Informatics*, Vol. 11, No. 2, pp. 242-249.

Helstrom, C. W. (1976). *Quantum Detection and Estimation Theory*. Academic Press, New York.

Holevo, A. S. (2001). *Statistical Structure of Quantum Theory*. Springer-Verlag, Berlin.

Kagan, E. & Ben-Gal, I. (2008). Application of Probabilistic Self-Stabilization Algorithms to Robot's Control. *Proc. 15-th Israeli Conf. IE&M'08*, Tel-Aviv, Israel.

Kagan, E. ; Salmona, E. & Ben-Gal, I. (2008). Probabilistic Mobile Robot with Quantum Decision-Making. *Proc. IEEE 25-th Conv. EEEI*. Eilat, Israel.

Kagan, E. & Ben-Gal, I. (2006). An Informational Search for a Moving Target. *Proc. IEEE 24-th Convention of EEEI*. Eilat, Israel.

Khare, M. & Roy, S. (2008). Conditional Entropy and the Rokhlin Metric on an Orthomodular Lattice with Bayesian State. *Int. J. Theor. Phys.*, Vol. 47, pp. 1386-1396.

Levitin, L. B. (1969). On a Quantum Measure of the Quantity of Information. *Proc. Fourth All-Union Conf. Problems of Information Transmission and Coding*. IPPI AN USSR, Moscow, pp. 111-115 (In Russian). *English translation*: A. Blaquieve, et al., eds. (1987). *Information Complexity and Control in Quantum Physics*. Springer-Verlag, New York, pp. 15-47.

Malley, J. D. & Hornstein, J. (1993). Quantum Statistical Inference. *Statistical Sciences*, Vol. 8, No. 4, pp. 433-457.

Nielsen, M. A. & Chuang, I. L. (2000). *Quantum Computation and Quantum Information*. Cambridge University Press, Cambridge, England.

Ornstein, D. S. (1974). *Ergodic Theory, Randomness, and Dynamical Systems*. Yale University Press, New Haven, 1974.

Raghuvanshi, A.; Fan, Y.; Woyke, M. & Perkowski, M. (2007). Quantum Robots for Teenagers, *Proc. IEEE Conf. ISMV'07*, Helsinki, Finland.

Rokhlin, V. A. (1967). Lectures on the Entropy Theory of Measure-Preserving Transformations. *Rus. Math. Surveys*, Vol. 22, pp. 1-52.

Rybalov, A.; Kagan, E.; Manor, Y. & Ben-Gal, I. (2010). Fuzzy Model of Control for Quantum-Controlled Mobile Robots. *Proc. IEEE 26-th Conv. EEEI*. Eilat, Israel.

Tsetlin, M. L. (1973). *Automaton Theory and Modeling of Biological Systems*. Academic Press, New York.

Unsal, C. (1998). *Intelligent Navigation of Autonomous Vehicles in an Automated Highway System: Learning Methods and Interacting Vehicles Approach*. PhD Thesis, Virginia Polytechnic Institute and State University, VI, USA.

Yuan, H.-J. (2005). Entropy of Partitions on Quantum Logic. *Commun. Theor. Phys.*, Vol. 43, No. 3, pp. 437-439.

Zhao, Y.-X. & Ma, Z.-H. (2007). Conditional Entropy of Partitions on Quantum Logic. *Commun. Theor. Phys.*, Vol. 48, No. 1, pp. 11-13.

5

Tracking Control for Reliable Outdoor Navigation Using Curb Detection

Seung-Hun Kim
Korea Electronics Technology Institute, Korea

1. Introduction

In the past several years there has been increasing interest in applications of a mobile robot. Personal service robots perform the missions of guiding tourists in museum, cleaning room and nursing the elderly [1]. Mobile robots have been used for the purpose of patrol, reconnaissance, surveillance and exploring planets, etc [2].
The indoor environment has a variety of features such as walls, doors and furniture that can be used for mapping and navigation of a mobile robot. In contrast to the indoor cases, it is hard to find any specific features in outdoor environment without some artificial landmarks [3]. Fortunately, the existence of curbs on roadways is very useful to build a map and localization in outdoor environment. The detected curb information could be used for not only map building and localization but also navigating safely [4]. And also the mobile robot decides to go or stop using the width of the road calculated from the detected curb.
The present paper deals with the development of a robot which can patrol areas such as industrial complexes and research centers. We have already reported about outdoor navigation of a mobile robot [5]. The extended Kalman filter is applied for the fusion between odometry and Differential Global Positioning System(DGPS) measurement data. It is insufficient for reliable navigation since the error of DGPS measurement data increased near high buildings and trees [6]. Hence, it is necessary to correct the pose of a mobile robot when the position data from the odometry and DGPS are inaccurate. This chapter proposes the curb detection algorithm and calculate the pose error from the curb edge data and then use it to correct the pose of the mobile robot.

2. Outdoor mobile robot system

The outdoor mobile robot system on an ATRV-mini from RWI Inc. is constructed as shown in Fig. 1. It is a four-wheeled robot designed for outdoor use and equipped with odometry. This robot is chosen because it can climb up and down over speed bumps and slopes. It is added a laser range finder to detect obstacles and build a map of the environment. For global localization, a DGPS receiver is added. A camera that can zoom-in and zoom-out, was installed to send the image data to the control station.
Fig. 2 shows the software architecture of the mobile robot system. The system consists of three parts which are the mobile server, the navigation server and the remote server. The mobile sever controls the motor in the ATRV according to the velocity and angular velocity

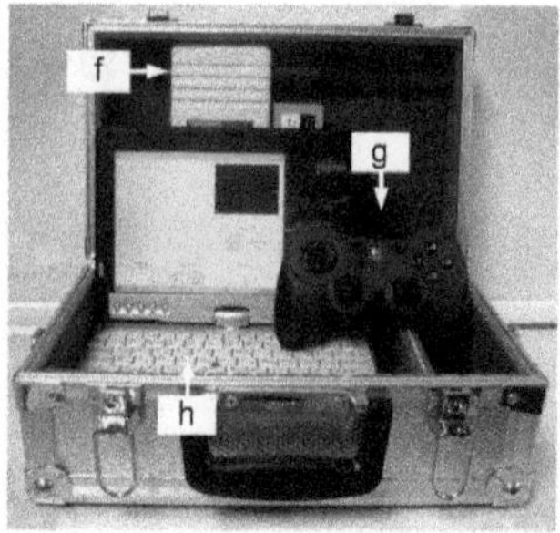

Fig. 1. Mobile robot system configuration.
(a) DGPS antenna (b) PC (c) Image camera (d) LMS200 laser range finder
(e) Attitude/heading sensor (f) Wireless LAN bridges (g) Joypad (h) Control station

from the navigation server and transfers the odometry to the navigation server. The navigation server makes the mobile robot localize using latitude and longitude data from the DGPS and attitude from 3-axis angle sensor. It detects curbs using distance from the tilted laser range finder and sends the distance, image and navigation data to the remote server through the wireless LAN. The remote server operates in two modes, autonomous mode or teleoperated mode, display the image and the navigation information.

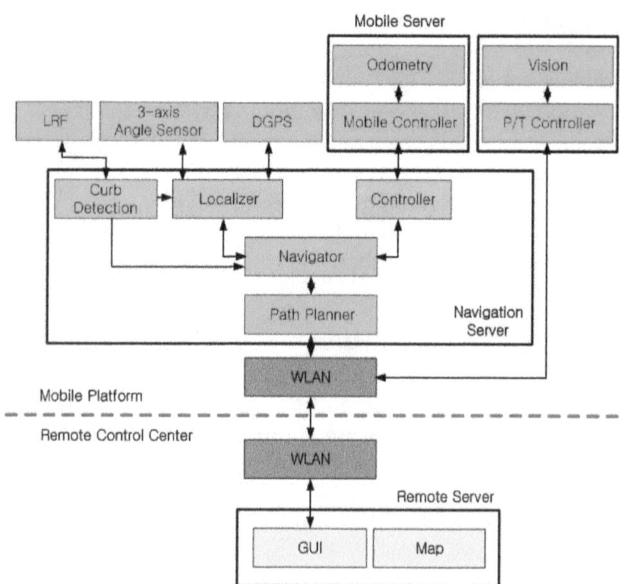

Fig. 2. Software diagram of the mobile robot system.

3. Curb detection

The outdoor mobile robot requires ability to navigate in a dynamic and potentially dangerous environment due to curbs and obstacles. Using the laser range finder, it can measure the distance from the obstacles and calculate the distance between obstacles. These data obtained from the laser range finder can be used in environment mapping and obstacle avoidance.

An important feature of road environment is the existence of curbs as shown in Fig. 3. The curbs could be used in mobile robot navigation.

Fig. 3. Road and curb on the road.

The proposed algorithm is to detect curbs through the laser range finder, which was positioned on the mobile robot and looking down the road with a small tilt angle as shown in Fig. 4. The tilted laser range finder projects three-dimensional shape of curbs on two-dimension space. And then it extracts curb edge points.

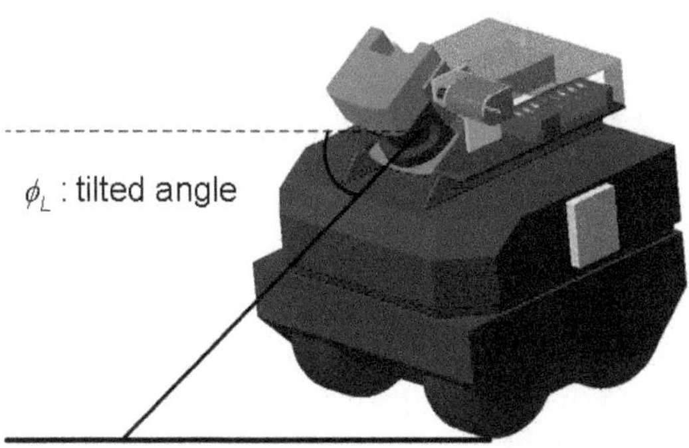

Fig. 4. Tilted laser range finder on the mobile robot.

The mobile robot builds the map of the curbs of roads and the map is used for tracking and localization. (n, x_L, y_L) is obtained by the laser range finder, where n represents n-th data and x_L, y_L are the distance data in the laser range finder's coordinate as shown in Fig. 5. The curb edges on the both sides of the mobile robot is used for mapping and tracking.

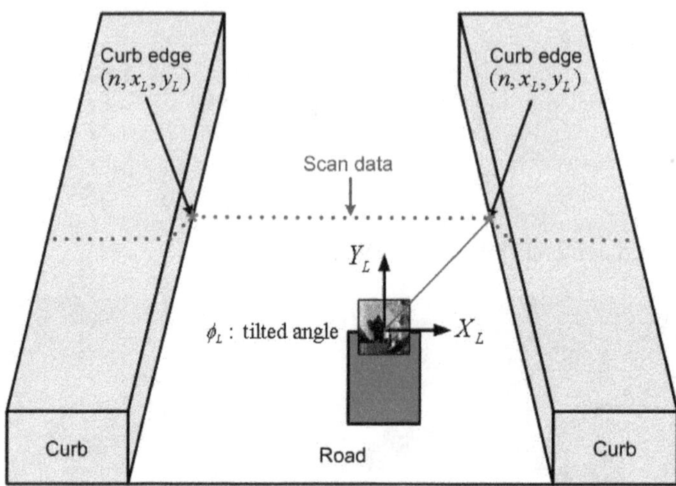

Fig. 5. Curb detection on the road.

Even though the angle between the robot and the curb changes, the feature points of the curb do not change as shown in Fig. 6. The points around 1600mm of the axis Y represent the road surface and the curved points between 0mm and 2000mm of the axis X are the curb of the road.

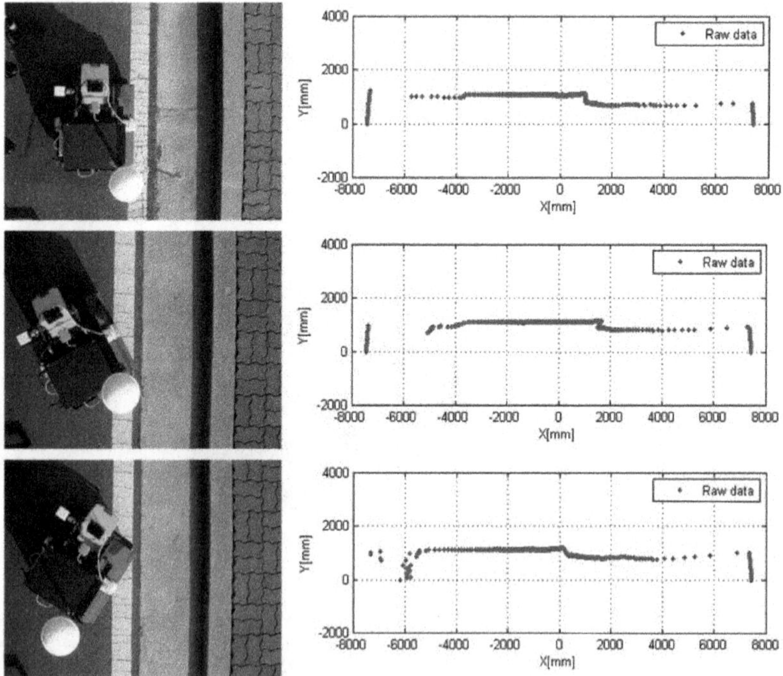

Fig. 6. Angles (0°, 30°, -30°) between the mobile robot and the curb and raw data.

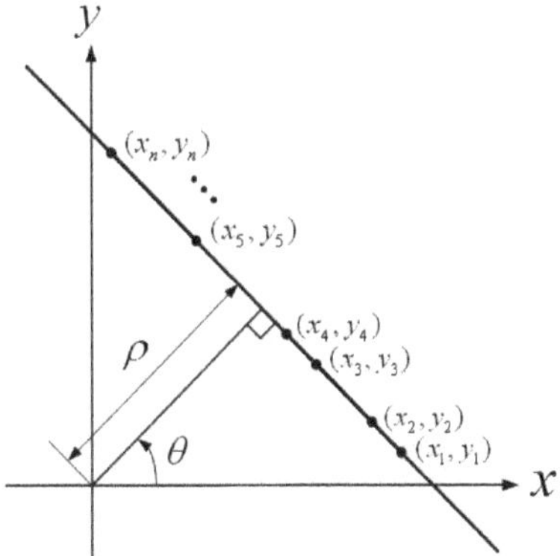

Fig. 7. Projection of collinear points onto a line.

3.1 Extracting the longest straight line

The laser range finder senses the surface of roads due to the small angle between the laser range finder and the mobile robot. Therefore, the longest straight line of the raw data from laser range finder is the surface of roads and the curbs locates the edges of the road. The longest straight line is extracted using Hough Transform [7], [8], [9]. In Hough transform, each line is represented by two parameters, commonly called ρ and θ, which represent the length and angle from the origin of a normal to the line. Using this parameterization, an equation of the line can be written as:

$$\rho = x_n \cos\theta + y_n \sin\theta \tag{1}$$

(x_n, y_n) is the n-th measurement data, where n is 1 to 361 and $\theta \in [0, 2\pi)$. The dominant ρ and θ are obtained by Hough Transform, called ρ_d and θ_d as shown in Fig. 8.

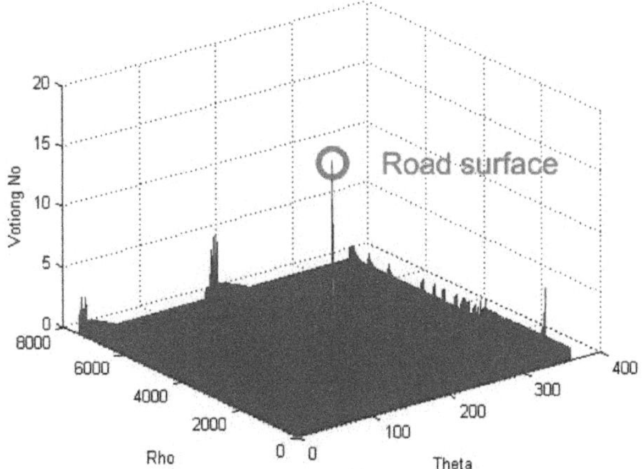

Fig. 8. Dominant ρ_d and θ_d found by Hough Transform.

The straight line in Fig. 9. shows the longest straight line represented by ρ_d and θ_d.

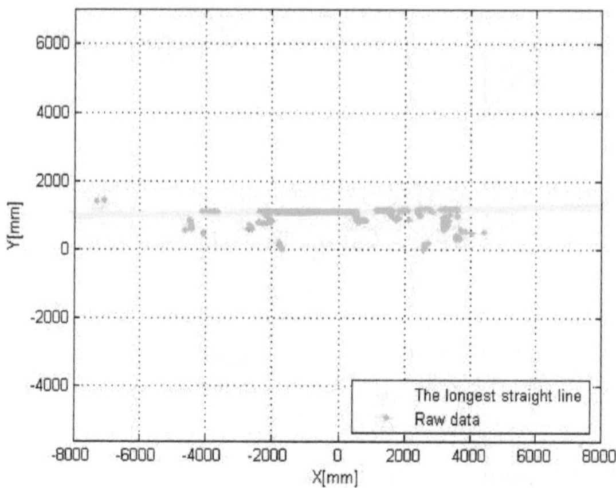

Fig. 9. Raw data and the longest straight line.

3.2 Fitting the points around the longest straight line

It is needed to fit the raw data points corresponding to the straight line for identifying the road surface and the curbs. The road surface and curbs exist on the boundary of the straight line from Fig. 10. To obtain the boundary points of the line, ρ is first calculated by Eq. (1) for each raw scan data with θ_d . And then, choose the points satisfying the following Eq. (2). The points of Fig. 6 show the fitting points with $\Delta\rho$ of 100mm.

$$\rho_d - \Delta\rho \leq \rho \leq \rho_d + \Delta\rho \tag{2}$$

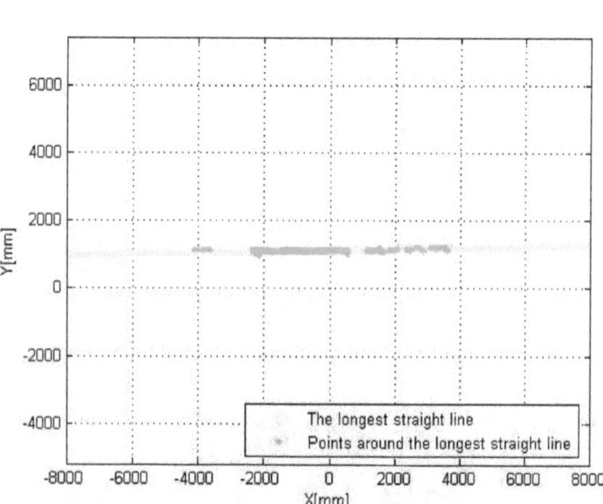

Fig. 10. Points around the longest straight line.

3.3 Grouping the points and finding the road surface

The points around the longest straight line include the road surface and obstacles. To extract the road surface, we classify into six groups (ellipses in Fig. 11). Points of each group are continuous. The biggest group among the six groups indicates the sensed road surface.

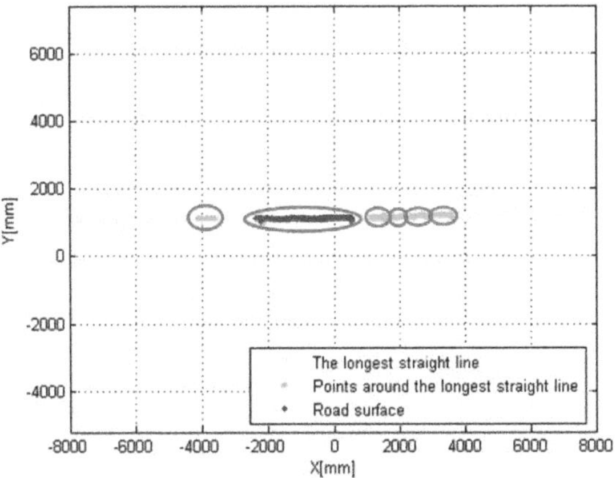

Fig. 11. Grouped points and found road surface.

3.4 Extracting the curb edge points

The scanned data of a flat road surface has two specific characteristics. One is that the scanned data of road surface is on the straight line and the interval between adjacent two points is small. The other is that the slope of the straight line is parallel to the direction X_L of a laser range finder.

When the distance between adjacent two points is continuously within d and the slope between them is less than σ, the two points are considered as being on the straight line (Eq. (3)). The end points (stars in Fig. 12) of the straight line are the road edges and also the start points of the curbs. Therefore the curb edge points could be obtained by finding the end points of the straight line.

$$\sqrt{(x_i - x_{i+1})^2 + (y_i - y_{i+1})^2} \leq d \text{ and } \left| \tan^{-1}\left(\frac{y_i - y_{i+1}}{x_i - x_{i+1}}\right) \right| \geq \sigma \quad (3)$$

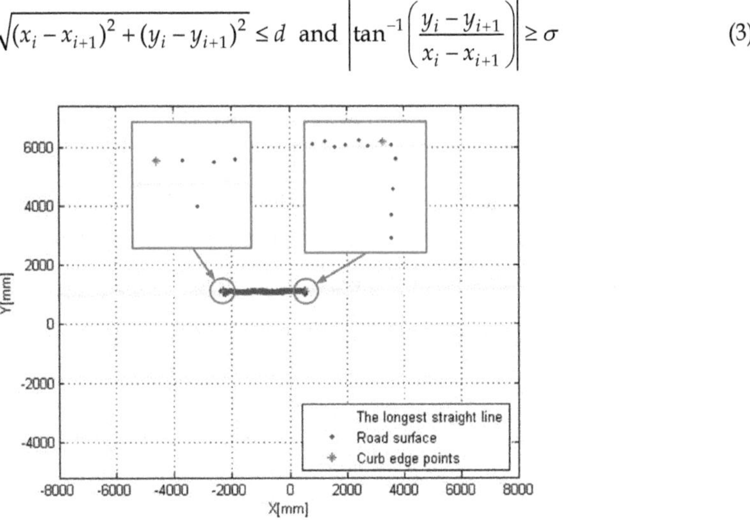

Fig. 12. Extracted curb edge points.

3.5 Parameter estimation of curb

The position of the mobile robot and the extracted curb point can be simultaneously represented in global coordinates. If we can calculate the relative pose of the mobile robot to the curb position, we can correct the mobile robot's pose against its colliding with the curb. Fig. 13 represents the both position in global coordinates.

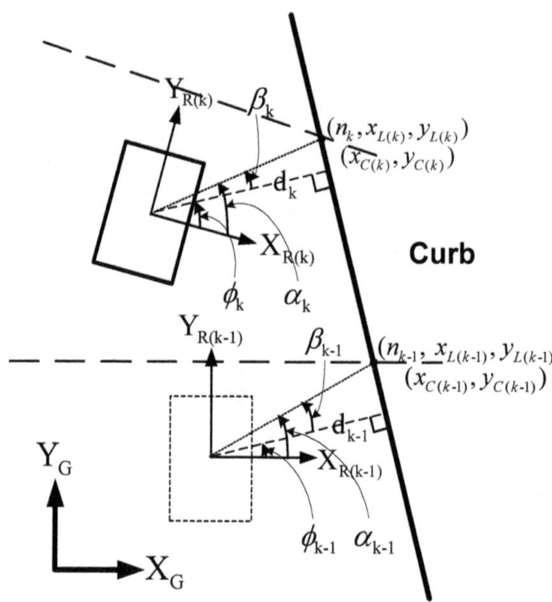

x_k : x position of the robot in global coordinate

y_k : y position of the robot in global coordinate

θ_k : orientation of the robot in global coordinate

$x_{L(k)}$: x position of the curb in robot coordinate

$y_{L(k)}$: y position of the curb in robot coordinate

$x_{C(k)}$: x position of the curb in global coordinate

$y_{C(k)}$: y position of the curb in global coordinate

Fig. 13. Positions of the mobile robot and curbs in global coordinate.

We have obtained $(n_k, x_{L(k)}, y_{L(k)})$ from curb detection method as explained in Section III. Curb position $(x_{C(k)}, y_{C(k)})$ in global coordinate can be obtained from the curb position $(x_{L(k)}, y_{L(k)})$ as shown in Eq. (4).

$$\begin{pmatrix} x_{C(k)} \\ y_{C(k)} \\ 1 \end{pmatrix} = \begin{pmatrix} -\sin\theta_k & -\cos\theta_k & x_k \\ \cos\theta_k & -\sin\theta_k & y_k \\ 0 & 0 & 1 \end{pmatrix} \begin{pmatrix} x_{L(k)} \\ y_{L(k)} \\ 1 \end{pmatrix} \tag{4}$$

3.6 Curb line equation

If we assume that the curb is straight, we can express the curb edges to a line equation such as Eq. (5).

$$ax + by + c = 0 \tag{5}$$

From the subsequent two curb points $(x_{C(k)}, y_{C(k)})$ and $(x_{C(k-1)}, y_{C(k-1)})$, we can find the line equation parameters (a,b,c) as shown in Eq. (6).

$$\begin{aligned}
& y - y_{C(k)} = \frac{y_{C(k)} - y_{C(k-1)}}{x_{C(k)} - x_{C(k-1)}}(x - x_{C(k)}) \\
& (y_{C(k)} - y_{C(k-1)})x + (x_{C(k-1)} - x_{C(k)})y \\
& \quad + (x_{C(k)}y_{C(k-1)} - x_{C(k-1)}y_{C(k)}) = 0 \\
& a = y_{C(k)} - y_{C(k-1)} \\
& b = x_{C(k-1)} - x_{C(k)} \\
& c = x_{C(k)}y_{C(k-1)} - x_{C(k-1)}y_{C(k)}
\end{aligned} \tag{6}$$

3.7 Distance and angle between robot and curb

We can also obtain the distance d_k between the robot and the curb from the line equation and angle ϕ_k between robot and curb by geometric relation as shown in Fig. 11.
Eq. (7) shows the distance d_k.

$$d_k = \frac{|ax_k + by_k + c|}{\sqrt{a^2 + b^2}} \tag{7}$$

The angle ϕ_k can be obtained such as Eq. (8).

$$\begin{aligned}
& \phi_k = \alpha_k - \beta_k \\
& \alpha_k = \text{atan2}(y_{L(k)}, x_{L(k)}) \\
& \beta_k = \text{acos}\left(\frac{d_k}{\sqrt{x^2{}_{L(k)} + y^2{}_{L(k)}}}\right) \\
& \phi_k = \text{atan2}(y_{L(k)}, x_{L(k)}) - \text{acos}\left(\frac{d_k}{\sqrt{x^2{}_{L(k)} + y^2{}_{L(k)}}}\right)
\end{aligned} \tag{8}$$

The proposed curb detection algorithm is experimented to identify the performance of it. The mobile robot was driven along a road with the distance (1m) between the mobile robot and curbs. The tilted angle (ϕ_L in Fig. 4) between the mobile robot and the laser range finder is 25°. The values d and σ as 50mm and $\pi / 4$ are selected in Eq. (4.3). The navigation server sends the data of curb edge points (position data of stars in Fig. 14) to the remote server. The remote server receives the image and curbs edges data from the navigation server and displays the map of curbs and navigation information as shown in Fig. 16. The rectangular and the dots in the map represent the mobile robot and curb edge points, respectively.

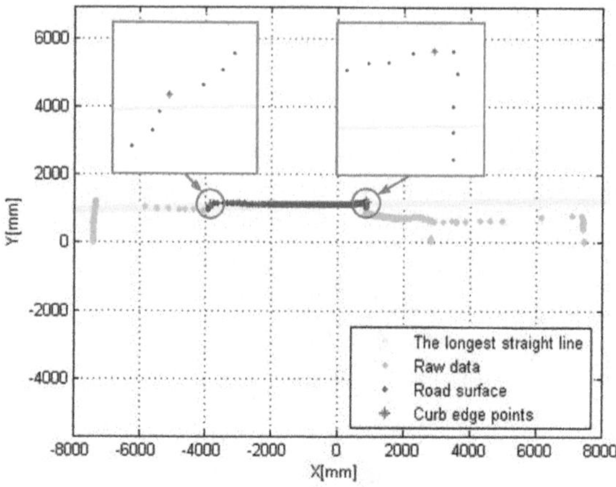

Fig. 14. Extracted curb edge points on the road.

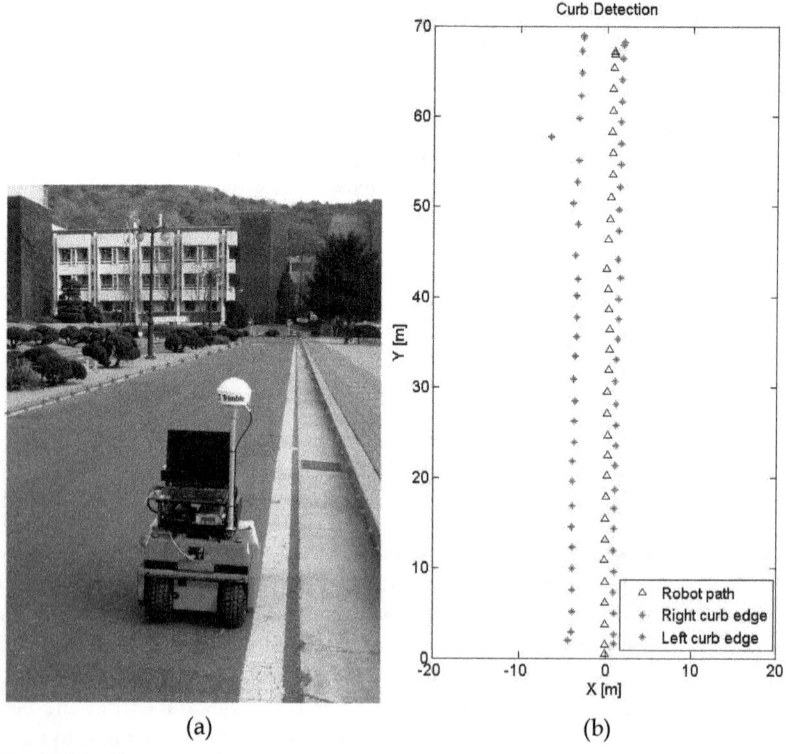

(a) (b)

Fig. 15. Curb detection experiment.
(a) The mobile robot tracking the curbs (b) Robot path and detected curbs.

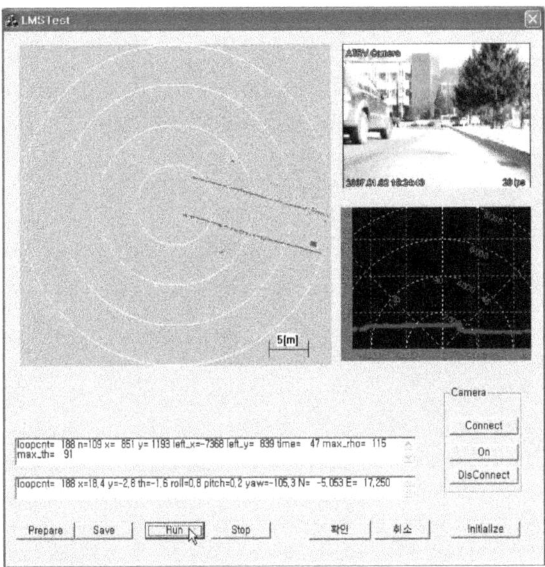

Fig. 16. Screen of the remote server.

The left and right-side stars and the black triangles in Fig. 15(b), 17 show the detected curbs and the robot path, respectively. The left-side stars in Figures are irregular because the right-side curb has regular shape (height: 13cm), but the left-side of the road is not the curb. It is just the structure with low height (2cm). Even though the mobile robot moves zigzag, the detected right-side curb path is smooth. The average error of the curb points is 2cm.

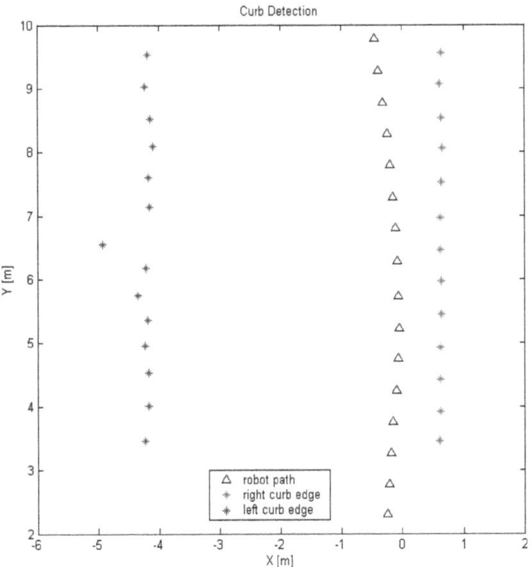

Fig. 17. Robot path and extracted curb positions.

Fig. 18 shows the distance between robot and curb, the curb edge angle α and angle ϕ between robot and curb, respectively. We can know that the proposed curb detection algorithm produces proper results and the distance and angle information calculated from the curb position could be used for reliable and safe navigation.

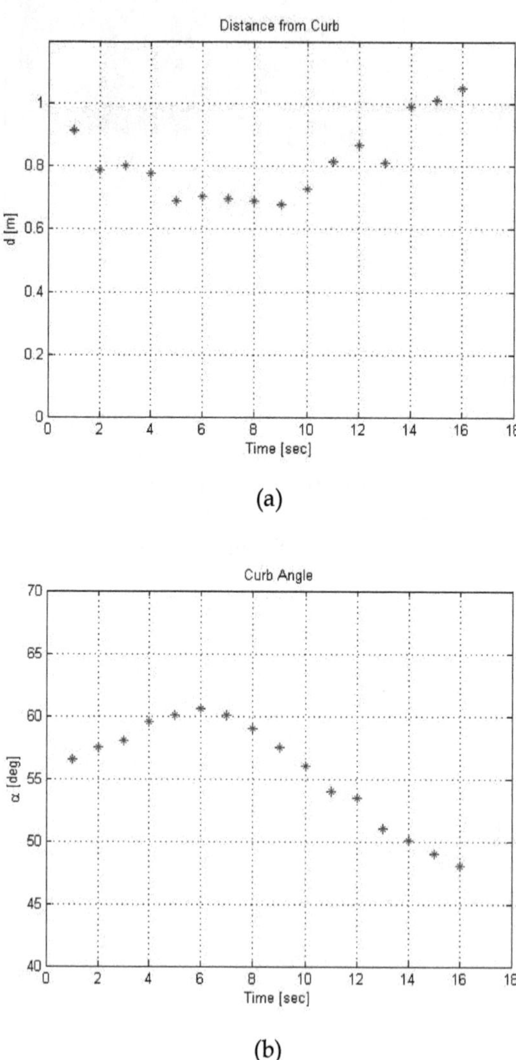

Fig 18. Results for curb detection. (a) Distance d between robot and curb (b) Curb edge angle α

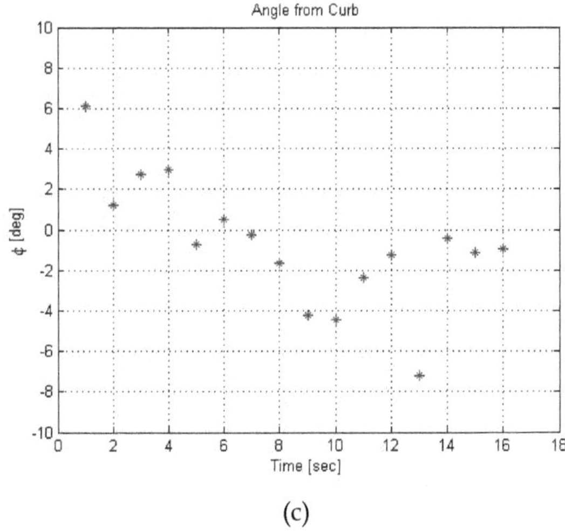

(c)

Fig. 18. Results for curb detection. (c) Angle ϕ between robot and curb.

4. Tracking control using curb position

An ATRV-mini is a nonholonomic mobile robot, equivalent to a four-wheeled differential drive. Kanayama proposed a kinematic control law in Cartesian coordinates for solving the trajectory tracking problems [10]. We add a term related to the distance between the robot and the right-side curb of roads to Kanayama controller.

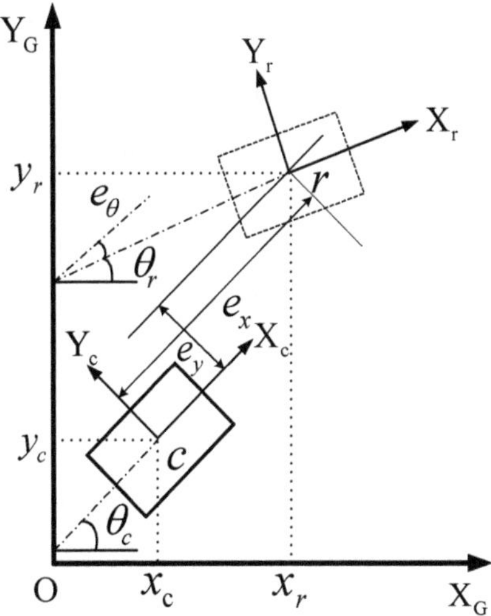

Fig. 19. Mobile robot and reference robot.

Considering the Fig. 19, in which the mobile robot c is represented by the coordinates $\mathbf{p}=[x_c\ y_c\ \theta]^T$ of the center of mass, moving over plane with linear velocity υ and angular velocity ω. The matrix equation that describes the kinematic model of a mobile robot is given by

$$\begin{bmatrix} \dot{x}_c \\ \dot{y}_c \\ \dot{\theta} \end{bmatrix} = \begin{bmatrix} \cos\theta & -\sin\theta \\ \sin\theta & \cos\theta \\ 0 & 1 \end{bmatrix} \begin{bmatrix} \upsilon \\ \omega \end{bmatrix} \tag{9}$$

The posture tracking error vector e_p is calculated in the reference robot frame X_r, Y_r.

$$e_p = R(\theta)(p_r - p)$$
$$e_p = \begin{bmatrix} e_x \\ e_y \\ e_\theta \end{bmatrix} = \begin{bmatrix} \cos\theta & \sin\theta & 0 \\ -\sin\theta & \cos\theta & 0 \\ 0 & 0 & 1 \end{bmatrix} \begin{bmatrix} x_r - x_c \\ y_r - y_c \\ \theta_r - \theta_c \end{bmatrix}, \tag{10}$$

where $R(\theta)$ is the rotation matrix.

Differentiating the Eq. (10), yields

$$\begin{bmatrix} \dot{e}_x \\ \dot{e}_y \\ \dot{e}_\theta \end{bmatrix} = \begin{bmatrix} \omega e_y - \upsilon + \upsilon_r \cos(e_\theta) \\ -\omega e_x + \upsilon_r \sin(e_\theta) \\ \omega_r - \omega \end{bmatrix} \tag{11}$$

The asymptotic convergence of posture error can be obtained

$$V_c = \begin{bmatrix} \upsilon_c \\ \omega_c \end{bmatrix} = \begin{bmatrix} k_x e_x + \upsilon_r \cos(e_\theta) \\ \omega_r + k_y \upsilon_r e_y + k_\theta \upsilon_r \sin(e_\theta) \end{bmatrix}, \tag{12}$$

where k_x, k_y and k_θ are positive constants, defined as controller gains.

For reliable and safe navigation, we design a new controller V_d by adding the extracted curb distance and angle information to the above Kanayama controller.

$$e_d = d_{ref} - d$$
$$V_d = \begin{bmatrix} \upsilon_d \\ \omega_d \end{bmatrix} = \begin{bmatrix} k_x e_x + \upsilon_r \cos(e_\theta) \\ \omega_r + \upsilon_r (k_y e_y + k_\theta \sin(e_\theta) + k_d e_d + k_\phi \sin\phi) \end{bmatrix}, \tag{13}$$

where d_{ref} is the reference distance from the curb for safety and k_d, k_ϕ are positive constants as controller gains.

If the robot is too close to the curb, the added terms of the proposed controller try to prohibit the robot colliding with curbs, and vice versa.

We have performed two kinds of experiments to evaluate the performance of the proposed path tracking controller in outdoor environment. One is that the mobile robot has to follow a straight path. The other is that the mobile robot has to follow a curved path.

4.1 Straight path tracking

We have performed a straight path tracking experiment, which target point is (78m, -22m) and linear velocity is 0.5m/sec. The gains of controller were determined by some simulations and experiments. Fig. 20 shows the reference path and the error of path, position and angle of the robot. Fig. 20(a) shows the reference path and tracking path and the position error of 20cm at the final point, respectively. Even though DGPS position error of 1m has occurred at the point of 12m from X-axis, the mobile robot returned to its reference path by the stable tracking controller.

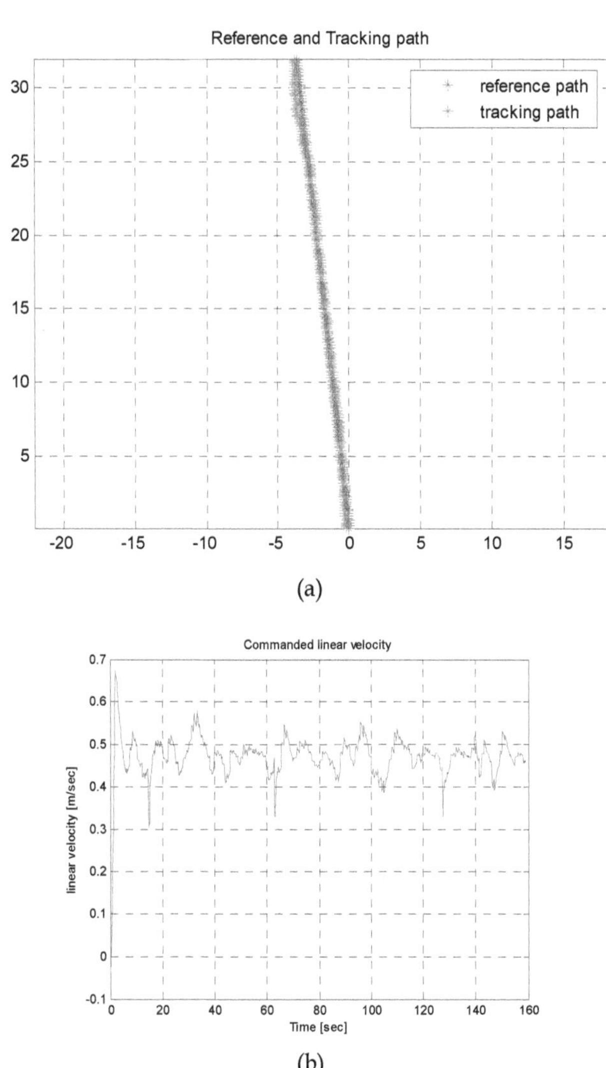

(a)

(b)

Fig. 20. (Continued)

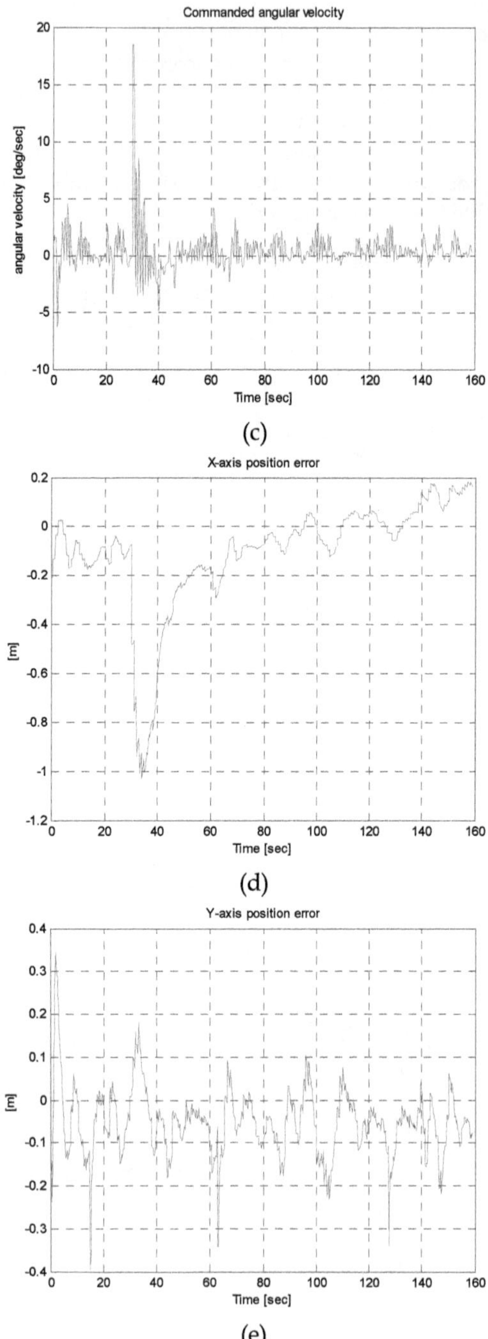

Fig. 20. (Continued)

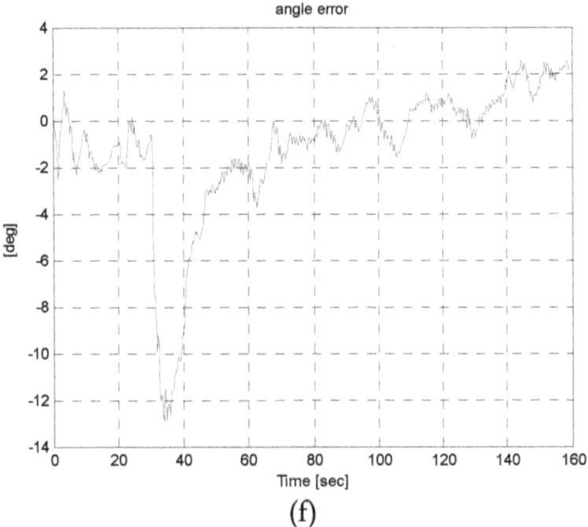

(f)

Fig. 20. Results for straight path tracking. (a) Reference path and tracking path (b) Controller linear velocity (c) Controller angular velocity (d) X-axis position error for robot coordinate (e) Y-axis position error for robot coordinate (f) Angle error for robot coordinate.

4.2 Curved path tracking

We have performed a curved path tracking experiment, which linear velocity is 0.3m/sec and angular velocity is 0.06rad/s. Fig. 21 shows the reference path and the error of path, position and angle of the robot. Fig. 21(a) shows the reference path and tracking path. The X-axis position error and the angular error are 40cm and 7° at the final point, respectively. The reason that the errors are bigger than the straight path tracking experiment is related to the characteristic of skid steering system of our mobile robot platform. It is necessary to optimize the gains of controller to reduce the errors.

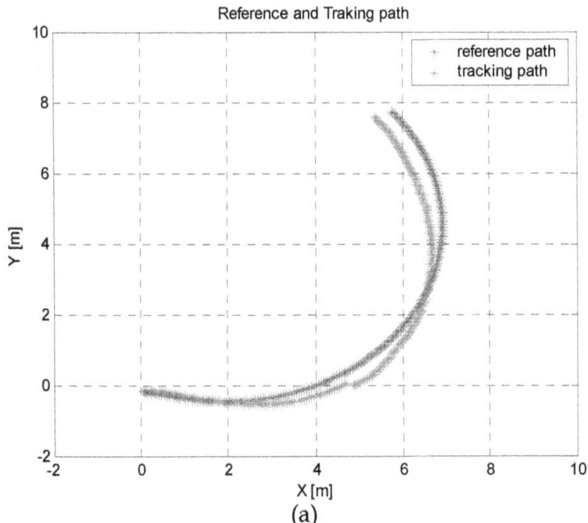

(a)

Fig. 21. (Continued)

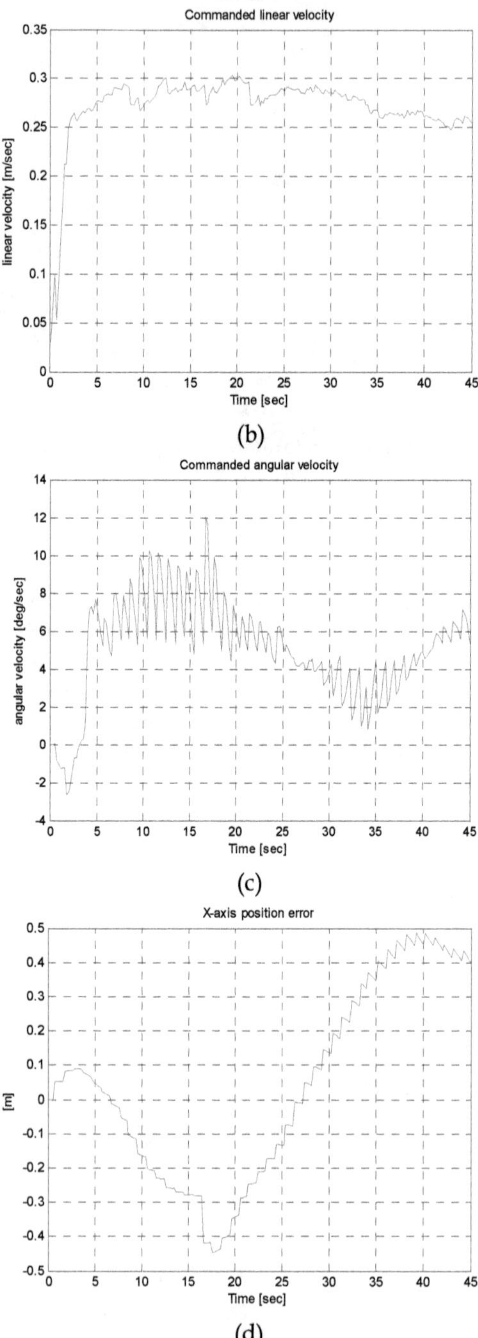

Fig. 21. (Continued)

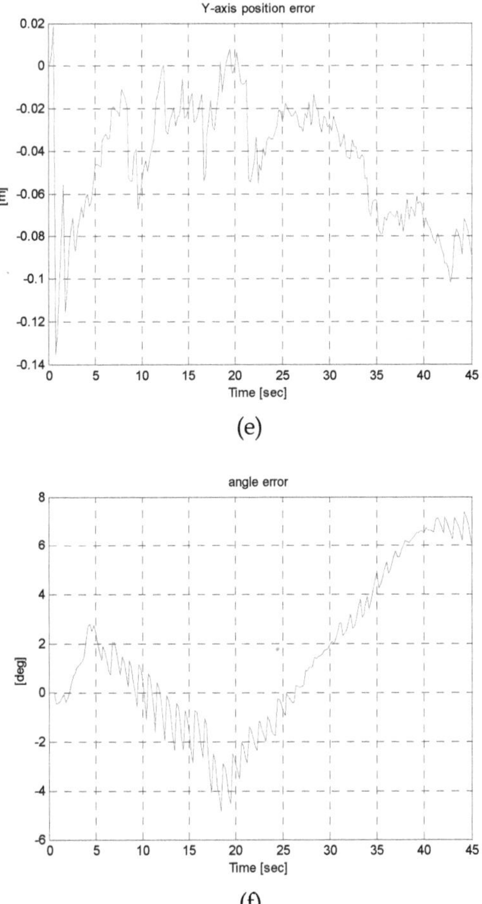

Fig. 21. Results for curved path tracking.
(a) Reference path and tracking path (b) Controller linear velocity (c) Controller angular velocity (d) X-axis position error for robot coordinate (e) Y-axis position error for robot coordinate (f) Angle error for robot coordinate

5. Conclusion

In this chapter, an algorithm is described to detect road curb position through a laser range finder. The curbs map was built and the width of the road was calculated by the detected curbs. The proposed algorithm is to calculate the distance and the angle between the mobile robot and the curb. A new controller is designed by adding the extracted curb distance and angle information to the tracking controller. It helps the mobile robot to navigate safely and reliably. To identify the availability of the proposed algorithm, navigation experiments was performed in outdoor environment and proper results was obtained.

6. References

[1] S. Thrun at el. (2000). Probabilistic Algorithms and the Interactive Museum Tour-Guide Robot Minerva, *International Journal of Robotics Research*, Vol.19, No.11, pp. 972-999

[2] Richard Thrapp at el. (2001). Robust localization algorithms for an autonomous campus tour guide, *Proc. of IEEE Int'l Conf. on Robotics and Automation*, pp. 2065-2071

[3] C.C. Wang and C. Thorpe, "Simultaneous Localization and Mapping with Detection and Tracking of Moving Objects", Proc. of IEEE Int'l Conf. on Robotics and Automation, 2002, pp. 2918-2924

[4] W. S. Wijesoma, K. R. S. Kodagoda, and Arjuna P. Balasuriya, "Road-Boundary Detection and Tracking Using Ladar Sensing", IEEE Trans. On Robotics and Automation, Vol.20, No.3, 2004

[5] Seung-Hun Kim at el. (2006). A hybrid autonomous/teleoperated strategy for reliable mobile robot outdoor navigation, *SICE-ICASE Int. Joint Conf.*, pp. 3120-3125

[6] K. Ohno at el. (2003). Outdoor Navigation of a Mobile Robot between Buildings based on DGPS and Odometry Data Fusion, *Proc. of IEEE Int'l Conf. on Robotics and Automation*, pp. 1978-1984

[7] J. Forsberg at el. (March 1995). Mobile robot navigation using the rangeweighted Hough transform, *IEEE Robotics & Automation Magazine*, vol. 2, pp. 18-26

[8] R. O. Duda. & P. E. Hart. (1972). Use of the Hough transformation to detect lines and curves in pictures, *Communications of the ACM*, vol. 15, pp. 11-15

[9] P. V. C. Hough. (December 1962). Methods and means for recognizing complex patterns, in U.S. Patent 3 069 654

[10] Y. Kanayama at el. (1990). A stable tracking control method for an autonomous mobile robot, *IEEE Int. Conf. on Robotics and Automation*, vol. 1, pp. 384-389

Part 2

Methods for Control

An Embedded Type-2 Fuzzy Controller for a Mobile Robot Application

Leonardo Leottau[1] and Miguel Melgarejo[2]
[1]*Advanced Mining Technology Center (AMTC),*
Department of Electrical Engineering, University of Chile, Santiago
[2]*Laboratory for Automation, Microelectronics and Computational Intelligence (LAMIC),*
Faculty of Engineering, Universidad Distrital Francisco José de Caldas, Bogotá D.C.
[1]*Chile*
[2]*Colombia*

1. Introduction

Fuzzy logic systems (FLS) have been used in different applications with satisfactory performance (Wang, 1997). The human perception cannot be modelled by traditional mathematical techniques, thus, the introduction of fuzzy set (FS) theory in this modelling has been suitable (John & Coupland, 2007). When real-world applications are treated, many sources of uncertainty often appear. Several natures of uncertainties would influence the performance of a system. It is independent from what kind of methodology is used to handle it (Mendel, 2001).

Type-1 Fuzzy logic systems (T1-FLS) have limited capabilities to directly handle data uncertainties (Mendel, 2007). Once a type-1 membership function (MF) has been defined, uncertainty disappears because a T1-MF is totally precise (Hagras, 2007). Type-2 fuzzy logic systems (T2-FLS) make possible to model and handle uncertainties. These are rule based systems in which linguistic variables are described by means of Type-2 fuzzy sets (T2-FSs) that include a footprint of uncertainty (FOU) (Mendel, 2001). It provides a measure of dispersion to capture more about uncertainties (Mendel, 2007). While T2-FSs have non-crisp MFs, T1-FSs have crisp membership grades (MGs) (John & Coupland, 2007).

A representation of the inference model for T2-FLS is depicted in Figure 1 (Mendel, 2007). It begins with fuzzification, which maps crisp points into T2-FSs. Next, inference engine computes the rule base by making logical combinations of antecedent T2-FS, whose results are implicated with consequent T2-FS to form an aggregated output type-2 fuzzy set. Afterwards, Type-Reduction (TR) takes all output sets and performs a centroid calculation of this combined type-2 fuzzy set, which leads to a type-1 fuzzy set called *type-reduced set*. That reduced set is finally defuzzyfied in order to obtain a crisp output (Mendel, 2001; Karnik & Mendel 2001). The computational complexity of this model is reduced if interval type-2 fuzzy sets are used (Mendel, 2001), it is convenient in the context of hardware implementation in order to make softer the computational effort and sped up the inference time.

Type-2 fuzzy hardware is a topic of special interest, since the application of T2-FLS to particular fields that demand mobile electronic solutions would be necessary. Some recent

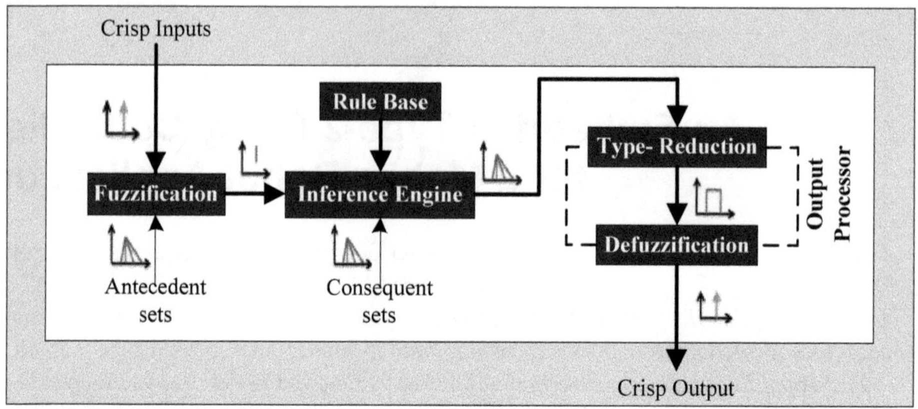

Fig. 1. Type-2 fuzzy system.

applications of T2-FLS have been developed in fields like robotics, communication and control systems among others (Castro et al., 2007; Hagras, 2007; Melgarejo & Peña, 2004, 2007; Torres & Saez, 2008). It is worth to think about the possibility of embedding T2-FLS handling these applications in order to achieve better communication speeds in smaller areas (Leottau & Melgarejo, 2010a).

Control and robotics are one of the most widely used application fields of fuzzy logic. The advantages of type-2 fuzzy logic controllers (T2-FLCs) over type-1 fuzzy logic controllers (T1-FLC) have also been demonstrated and documented in (Castro et al., 2007; Figueroa et al., 2005; Hagras 2004, 2007; Martinez et al., 2008; Torres & Saez, 2008). Although this kind of works presents improvements in the T2-FLCs performance, it is necessary to propose methodologies where these advances can be reflected in design processes as (Hagras 2008, Leottau & Melgarejo, 2010b; Melin & Castillo, 2003; Wu & Tan, 2004) and hardware implementation approaches over embedded devices as (Hagras, 2008; Leottau & Melgarejo, 2010a; Melgarejo & Peña, 2004, 2007). In this way T2-FLS would become widely used in different applicative contexts.

This work presents the implementation of an Interval Type-2 Fuzzy Logic Controller (IT2-FLC) for tracking the trajectory of a mobile robot application. The FLC is design according to the approach proposed in (Leottau & Melgarejo, 2010b), which involves some of the T1-FLC and T2-FLC properties. A hardware implementation of the designed T2-FLC is carried out over a digital signal controller (DSC) embedded platform. Different tests are performed for evaluating the performance of the IT2-FLC with respect to a T1-FLC. Simulation and emulation (i.e. with the embedded controller) results evidence that the IT2-FLC is robust to type reducer changes and exhibits better performance than its T1-FLC counterpart when noise is added to inputs and outputs. The IT2-FLC outperforms the T1-FLC in all tested cases, taking the interval time square error (ITSE) as performance index.

The chapter is organized as follows: Section 2 presents an overview of hardware implementation of IT2-FLCs. Section 3 describes the design approach followed for developing the IT2-FLC and section 4 describes its hardware implementation. Section 5 presents test, obtained results and discussion. Finally, conclusions and future work are presented in Section 6.

2. Hardware Implementation of an IT2-FLC for robot mobile applications

A FLC is a control strategy whose decisions are made by using a fuzzy inference system. Particularly, an IT2-FLC uses IT2-FSs to represent the inputs and/or outputs of the T2-FLS and these sets are completely characterized by their Footprints of Uncertainty (Mendel, 2001). So, this kind of FLC can model the uncertainty by means of the FOUs. In this way, uncertainties in sensors, actuators, operational changes, environmental changes, noise can be considered (Hagras, 2007; Mendel 2001). A FOU is described by its upper and lower MFs. Each MF is composed by as many membership grades (MGs) as discretization levels are considered in antecedents and consequents universes of discourse (Baturone et al., 2000).

This work considers two alternatives for computing the MGs of IT2-FSs: memory based approach and function computing method, which are involved in fuzzyfier and inference engine stages. These two alternatives are the most commonly used in fuzzy hardware implementations (Baturone et al., 2000). Moreover, centroids and center of sets methods are considered for TR stage. As TR algorithm, the Enhanced Karnik-Mendel algorithm is used (Wu & Mendel, 2009). For the sake of clarity, figure 1 depicts the stage structure of an IT2-FLC. These mentioned methods and alternatives are briefly introduced below:

Function Computing Approach (FCA):

This method carries out a direct computation of MF by using numeric algorithms that avoid constructing and storing look up tables (Leottau & Melgarejo, 2010a). It reduces memory usage and facilitates the implementation of MF. However, its execution could require several machine cycles depending on the complexity of MF (Baturone et al., 2000).

Memory Based Approach (MBA):

This approach stores the MGs of every input value into a memory. This strategy is executed considerably fast, because it uses the input value as the pointer to the memory and to directly retrieve the MG (Leottau & Melgarejo, 2010a).

Centroid Type-Reduction (Cent):

The centroid TR combines all the rule-output T2-FSs using union and then finds the centroid of this T2-FS (Karnik & Mendel, 2001). Therefore, this method presents high accuracy but low computational performance (Mendel, 2001). Cent TR is chosen in order to obtain the best accurate output while the hardware platform is forced to the highest computational effort.

Center-of-sets type-reduction (CoS):

In center-of-sets TR, each type-2 consequent set is replaced by its centroid. Then, the weighted average of these centroids is found, being the weight of each centroid the degree of firing of it corresponding rule (Mendel, 2001). CoS TR is chosen because its convenient trade-off between accuracy and computational cost. It is more accurate than Heights method (Mendel, 2001) and computationally less expensive than Cent TR.

Enhanced Karnik Mendel Algorithm (EKM):

The EKM algorithm is an iterative procedure to obtain the generalized centroid of an T2-FS. It uses statistically defined values as initialization points to reduce the amount of iterations that are necessary, in this way it converges monotonically and super-exponentially fast (Wu & Mendel, 2009). In addition, as it has been mentioned in (Leottau & Melgarejo, 2010a), EKM is the fastest algorithm reported up to that date for finding the generalized centroid of a inferred IT2-FS over DSC technology.

3. Proposal to design an IT2-FLC for a truck backer-upper

A proposal for designing an IT2-FLC for tracking the trajectory of a mobile robot application involving some of the T1-FLC and T2-FLC properties is introduced in (Leottau & Melgarejo, 2010b) and briefly explained in this section. T2-FSs are used for the initial modelling taking advantage of its capability to handle linguistic uncertainty because a large number of T1-FSs are embedded. T1-FSs are used in order to make easier a fine tuning, taking into account their small amount of parameters. Finally, returning to T2-FLSs is proposed for handling the uncertainty that appears in the final system performance. The truck backer-upper (Castro et al., 2007; Nguyen & Widrow 1989, Wang, 1997) is a non-linear control problem, whose model is depicted in Figure 2. Backing a trailer truck is a difficult task which requires a great deal of practice. Usually a truck driver backing, going forward, backing again, etc. if forward movements are not permitted, a more difficult problem emerges. Thus, the problem treated in this chapter is to control the steering of a trailer truck for tracking a trajectory, where only are allowed backing movements.

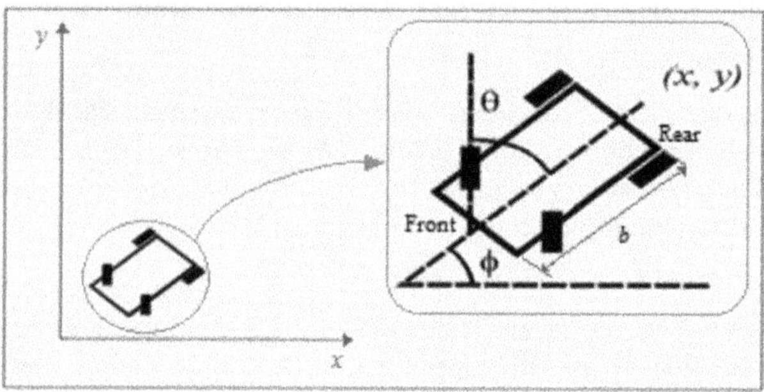

Fig. 2. The simulated truck.

3.1 Modeling the system dynamics: Truck mathematical model

In this step, the control process is studied in order to find and determine the model of the system dynamics. There are different strategies to do it (Kuo & Golnaraghi, 1996; Mendel, 2007; Wang, 1997). One of the most commonly used is by obtaining the mathematical model using physical schemes such as this case (Wang, 1997).

This problem is proposed in (Wang, 1997). The mathematical model that describes the system to control is shown below, where b is the length of the truck.

$$x(t+1) = x(t) + cos[\varphi(t) + \theta(t)] + sin[\theta(t)]sin[\varphi(t)] \tag{1}$$

$$y(t+1) = y(t) + sin(\varphi(t) + \theta(t)) - sin[\theta(t)]cos[\varphi(t)] \tag{2}$$

$$\varphi(t+1) = \varphi(t) - sin^{-1}[2sin(\theta(t))/b] \tag{3}$$

Since this work is focused for academic proposes, b=4 inch is assumed, however as it is handled in (Wang, 1997), units of that length can be considered depending on the operational environment. The truck position is determined by three state variables φ, x and y, where φ is the angle of the truck regarding the horizontal line as shows in Figure 2. The steering angle θ is the control variable to the truck which only moves backward. For the sake of simplicity, y does not have to be considered as state variable, thus position in the vertical axis is not controlled.

3.2 Initial design of an IT2-FLC for the truck

In this step, parameters as feedback variables, inputs (antecedents), outputs (consequents), universe of discourse, control action, rule base and the shape of MFs are defined. Using IT2-FS in this stage is proposed in order to obtain a preliminary design including linguistic uncertainty associated to ambiguity in the knowledge of the control process (Leottau & Melgarejo, 2010b).

Inputs of the controller are $(Ex, E\varphi \; and \; \Delta\varphi)$, where $Ex = Xref - x(t)$ is the error of x position regarding the reference trajectory $Xref$. $E\varphi = \varphi ref - \varphi(t)$ is the error of φ angle regarding the reference φref which is fixed to $\varphi ref{=}\pi/2$. $\Delta\varphi = \varphi(t) - \varphi(t-1)$ is the changing of φ. The output is θ, which is limited to $[-2\pi/9, 2\pi/9]$. The universe of discourse is assumed as $Ex \in [-20, 20]$, $E\varphi$ and $\Delta\varphi \in [-\pi, \pi]$ and $\theta \in [-\pi/3, \pi/3]$.

Based on intuitive reasoning, a preliminary design is obtained. Two antecedent sets for each input are defined, one to determine positive input values and their respective complement set for negative values. Eight rules are obtained by combining the antecedent sets and using four consequent sets are proposed: Positive high (+ +), positive small (+), negative small (-) and negative high (- -). The obtained rule base is shown below:

$$\begin{aligned}
&Rule\ 0\text{: } if\ Ex^{-}\ and\ E\varphi^{-}\ and\ \Delta\varphi^{-}\ then\ \theta^{+}\\
&Rule\ 1\text{: } if\ Ex^{-}\ and\ E\varphi^{-}\ and\ \Delta\varphi^{+}\ then\ \theta^{-}\\
&Rule\ 2\text{: } if\ Ex^{-}\ and\ E\varphi^{+}\ and\ \Delta\varphi^{-}\ then\ \theta^{-}\\
&Rule\ 3\text{: } if\ Ex^{-}\ and\ E\varphi^{+}\ and\ \Delta\varphi^{+}\ then\ \theta^{--}\\
&Rule\ 4\text{: } if\ Ex^{+}\ and\ E\varphi^{-}\ and\ \Delta\varphi^{-}\ then\ \theta^{++}\\
&Rule\ 5\text{: } if\ Ex^{+}\ and\ E\varphi^{-}\ and\ \Delta\varphi^{+}\ then\ \theta^{+}\\
&Rule\ 6\text{: } if\ Ex^{+}\ and\ E\varphi^{+}\ and\ \Delta\varphi^{-}\ then\ \theta^{+}\\
&Rule\ 7\text{: } if\ Ex^{+}\ and\ E\varphi^{+}\ and\ \Delta\varphi^{+}\ then\ \theta^{-}
\end{aligned} \tag{4}$$

Since negative values in antecedent sets are handled as the complement of their respective positive values, in practical terms $Ex^{-}, E\varphi^{-}$ and $\Delta\varphi^{-}$ respectively are equivalent to $\overline{Ex^{+}}$, $\overline{E\varphi^{+}}$ and $\overline{\Delta\varphi^{+}}$.

3.3 Initialization the IT2-FLC for the truck

In order to evaluate the IT2-FLC performance, it is convenient to define a figure of merit. In this way, it is possible to carry out a preliminary tuning, searching for an acceptable response. In this case, an exhaustive tuning is not carried out because to achieve stability is enough (Leottau & Melgarejo, 2010b). It could be carried out using simulations.

Determine an initial wide for FOUs in order to include the uncertainty in the system modelling is the first task of this initialization step. The main sources of uncertainty that would be present in this implementation are:

- Uncertainties in inputs to the FLC. The sensors measurements would be affected by noise levels or by variant conditions of observation (Hagras, 2007), taking into account that these kind of mobile robots applications usually are employed in outdoor environments (Hagras, 2004).

- Uncertainties in control outputs. Actuators characteristics could change with time. The mobile robots mechanics often are susceptible to wear or tear (Hagras, 2007).
- Uncertainties associated with changing operational conditions (Mendel, 2007).

The sources of uncertainty mentioned above are included in the final design. But, as it is mentioned in step 3.2 of this procedure, FOUs in this part of design, makes reference to linguistics uncertainties. This is the main advantage of use IT2-FS here, because the initial tuning would be based in some knowledge of the control process.

As it is proposed in (Leottau & Melgarejo, 2010b), it is used the middle of universe as center of antecedent sets location and the 10% of universe size as a measure for uncertainty. As universe size of $E\varphi$ and ΔE is 2π, the FOU wide of this IT2-FSs is initially defined as $2\pi \cdot 10\% = 0.2\pi$ and center of sets are located at zero. Please, see Figure 7.a for more clarity.

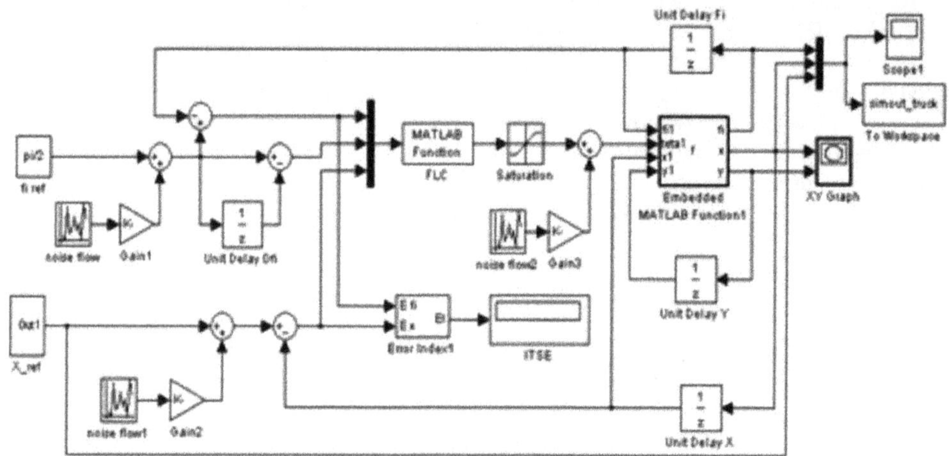

Fig. 3. Simulink model for the truck system.

The whole system is simulated using Matlab®, fuzzy Toolbox and Simulink. In addition, the IT2-FLC is designed and simulated with the IT2 Fuzzy Toolbox by the Tijuana Institute of Technology and Baja California Autonomous University (Castro et al., 2007). The Simulink model for the truck system is depicted in Figure 3.

As a figure of merit, the integral time square error (ITSE) is used (Dorf & Bishop, 1998). The objective of this example application is not to achieve an optimal performance in the controller. Thus, tuning parameters is carried out using the prove-error method, looking that the mobile robot follows a defined trajectory (see Figure 6).

3.4 The reduction of dimensionality

The reduction of dimensionality consists of converting the IT2-FLC into a T1-FLC. It is possible compressing the FOU in the IT2-FSs up to transform it in a T1-FS as is shown in Figure 4. The main objective of this procedure is making easier the tuning of fuzzy sets parameters (Leottau & Melgarejo, 2010b). The amount of parameters used for modelling a T2-FS is greater than for a T1-FS. In this way, tuning these parameters in a T1-FLC is easier and faster (Mendel, 2001).

In this application, the embedded T1-FS that are located in the middle of each FOU is used as a tuning start point. Once the IT2-FSs are converted to T1-FSs, the MF parameters' tuning is carried out using Microsoft Excel® solver tool that uses the Generalized Reduced Gradient Method as non linear optimization algorithm (Frontline Systems, 2010). A screen of used Excel® spreadsheet is shown in Figure 5.

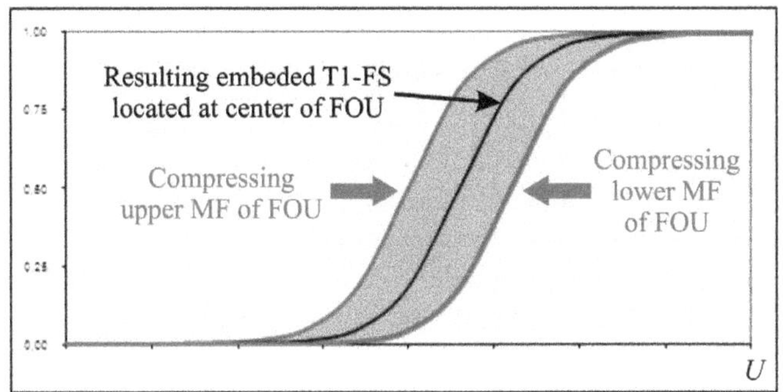

Fig. 4. Reducing the dimensionality of an IT2-FS.

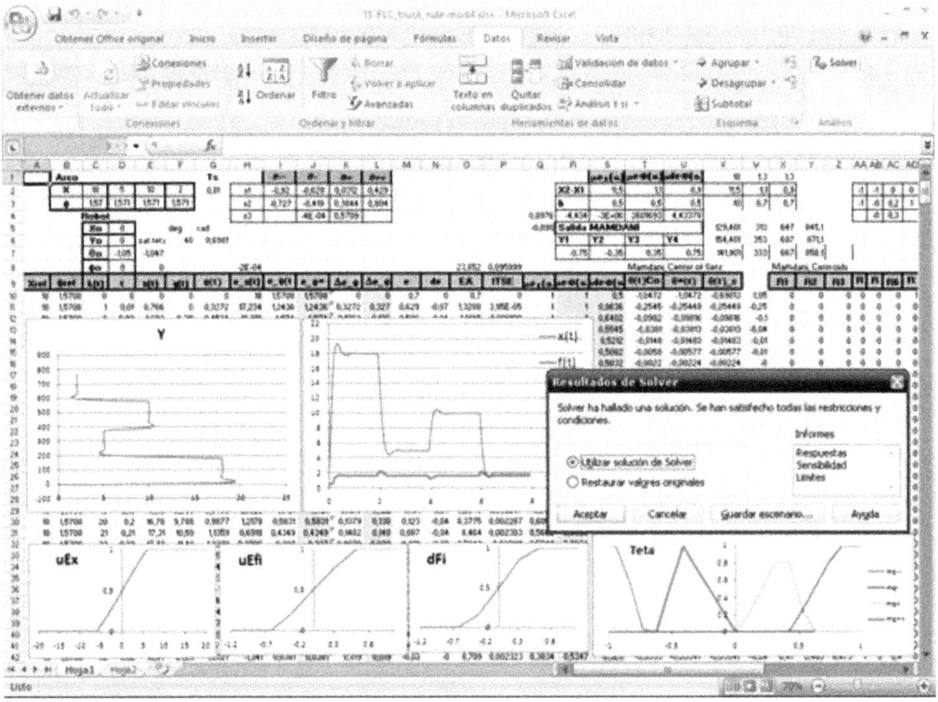

Fig. 5. A screen of used Excel® spreadsheet for MF parameters' tuning.

3.5 Returning to Type-2

Once the T1-FLC parameters are tuned, returning to T2-FLS scheme using the same uncertainties defined in step 3.3 is proposed in (Leottau & Melgarejo, 2010b), but in this case, locating the FOUs over their embedded T1-FSs obtained in previously tuning. So, By using the tuned T1-FSs obtained in step 3.4 as the central embedded T1-FS as in Figure 4, the same FOUs wide resulting by step 3.3 are included, so the FLC returns to its initial interval type-2 nature. Then a fine tuning is carried out again by using the prove-error method taking into account the system response obtained by simulation.

4. Hardware implementation of the IT2-FLC for the truck

Designed IT2-FLC showing in section 3 has the following characteristics: three inputs ($N=3$), two interval type-2 fuzzy sets (IT2-FSs) by input (M=2), six antecedent sets (M_A=$M \cdot N$=6), eight rules (R=8), four IT2-FSs in the consequent (M_C=4). It must be taken into account that negative values in antecedent sets are handled as the complement of their respective positive values, so in practical terms it is defined one IT2-FSs by input ($M^\dagger = 1$ and $M_A^\dagger = 3$). It is applicable for example for calculating the data memory usages. Universe of discourse' sizes have been defined in section 3.2. Its discretization levels in antecedents (D_A) and consequent (D_C) must be defined according computational and hardware resources available on the embedded platform to this implementation.

4.1 Available embedded platform for implementing the IT2-FLC

The available embedded platform for implementing the IT2-FLC is the Digital Signal Controller DSC-56F8013 evaluation board (Freescale Semiconductor, 2010). Next, some hardware resources of DSC-56F8013 are listed in Table 1.

Word length	16-bit
Core frequency	Up to 32 MIPS at 32MHz
Program Flash Memory	8KB (4KWord)
RAM	2KB (1KWord)
PWM module	One 6-channel
ADC	Two 3-channel 12-bit
Communication Interfaces	One Serial Communication Interface (SCI) with LIN slave functionality. One Serial Peripheral Interface (SPI). One Inter-Integrated Circuit (I2C) Port.
Quad Timer	One 16-bit Quad Timer.
On-Chip Emulation	JTAG/Enhanced On-Chip Emulation (OnCE™) for unobtrusive, real-time debugging.

Table 1. DSP-56F8013 basic features.

4.2 Determining computational model for implementing the IT2-FLC for the truck system

A summary of computational models for implementing the embedded IT2-FLC is shown in Table 2. The rest of this section explains why these computational models are chosen and how can be determined some of its features.

A methodological proposal for implementing interval type-2 fuzzy processors (IT2-FP) over DSC technology is reported in (Leottau & Melgarejo, 2010a). There, several computational

	IT2-FLC with centroid TR	IT2-FLC with CoS TR
Fuzzyfier	FCA	MBA
U_A =D_A	250	250
Inference Engine	FCA	-
U_C	1000	1000
D_C	100	4
RAM Consumption	456 Bytes (22.8%)	44 Bytes (0.28%)
DATA Mem. Consumption	4600 Bytes (57.5%)	136 Bytes (0.017%)

Table 2. Computational models features.

models have been characterized and tested over an IT2-FLS with the following characteristics: two inputs ($N=2$), three IT2-FSs by input ($M=3$), six antecedent sets ($M_A=M{\cdot}N=6$), nine rules ($R=9$), nine IT2-FSs in the consequent ($M_C=9$). Since characteristics of that IT2-FLS are similar and even more complex than the IT2-FLC designed for this mobile robot application, following considerations are based in that methodological proposal.

4.2.1 Processing time

According inference times reported in (Leottau & Melgarejo, 2010a) for a DSC-56F8013, a complete inference is carried out between 500uS and 60mS for the fastest (MBA, Dc=10) and the slowest (FCA, Dc=1000) computational model respectively. That times can be used as references for determine the IT2-FLC sample time (*Ts*) for the truck system. Among this time, the embedded FLC samples the inputs variables for computing a new crisp output.
Since θ output is limited to $\pm 40^\circ$, it is necessary to define a sampling time such that θ can rotate at least 80°. If it is used a Hitec servo motor (Hitec Servo Motors, 2010) that handling operating speed from 0.05 sec/60°, defining a sampling time of $Ts=100mS$ is enough, maintaining a convenient extra range.

4.2.2 Computational model with centroid type-reduction

Computational model chosen for the IT2-FLC with centroid TR is implemented with FCA for fuzzyfier and inference engine. Although (Leottau & Melgarejo, 2010a) evidences that processors implemented with FCA are slower than those implemented with MBA, this is not a problem taking into account that selected sampling time is enough. On the other hand, available DSC56F8013 offers limited memory resources, with that respect (Leottau & Melgarejo, 2010a) evidence that FCA is the computational model with less data memory consumption. In this way, we consider that cost-benefit trade-off of this computational model is convenient for this application.

4.2.3 Computational model with CoS type-reduction

Computational models with CoS TR offer small data memory consumption but lower accuracy regarding methods as Centroid (Mendel, 2001). On the other hand, computational models based on MBA offer the highest memory consumptions and the highest accuracy (Leottau & Melgarejo, 2010a). Thus, by combining CoS TR with a fuzzyfier based on MBA can be a convenient trade-off between accuracy and data memory usage. Thus, computational model chosen for the IT2-FLS with CoS TR is implemented with MBA for the fuzzyfier.

4.2.4 Interconnecting the emulated mobile robot plant with the embedded IT2-FLC

The truck system is emulated by using Matlab® over Simulink, thus it is necessary interconnect the DSC56F8013 with the computer in order to communicate the embedded IT2-FLC with the truck system. Since 56F8013 evaluation board include a RS-232 interface, for easy connection to a host processor the serial port is used. Since emulation is carried out over a general propose processor which serial port baud-rate is limited to 115200, it is selected a sped of 57600 bauds in order to not to force the platform.

As it is mentioned in section 3, in a context of simulation, a block of Matlab® fuzzy toolbox or it2fuzzy toolbox is included in the Simulink model for simulating the T1-FLC and the IT2-FLC respectively (Figure 3). In order to interconnect the truck system plant with the IT2-FLC embedded in the DSC56F8013, these blocks must be replaced by a Matlab function that carries out the RS-232 serial port communication.

4.2.5 Universe sizes and discretización levels regarding RS-232 communication

Universes of discourse in the process are defined as: $Ex \in [-20, 20]$ ($U_{Plant\ X} = 40$), $E\varphi$ and $\Delta\varphi \in [-\pi, \pi]$($U_{(Plant\ \varphi \text{ and } \Delta\varphi)} = 2\pi$) and $\theta \in [-\pi/3, \pi/3]$ ($U_{(Plant\ \theta)} = 2\pi/3$).

RS-232 interface has been chosen for connecting the embedded IT2-FLC with the truck system emulated by using Matlab® over Simulink. Since RS-232 interface can send and receive up to one Byte per time (2^8=256 levels), a universe of discourse between [0, 249] is defined in order to transmit just one byte per input, so U_A=D_A=250. Resolution of input universes is determined as: Input_res=U_{Plant}/D_A. Thus, Ex_res=40/250=0.16, $E\phi$_res=$2\pi/250 \simeq 0.025$ and $\Delta\phi$_res=$2\pi/250 \simeq 0.025$.

Output universe is defined between [-π/3, π/3]. If U_C is the output universe size in the embedded IT2-FLC, calculating output resolution by using D_A=U_C, θ_res= $2\pi/(3\cdot 250)$=0.00837=0.48°. Since computational models chosen particularly offer a low accuracy and output is limited to ±40°, we consider that 0.48° is a low resolution. Thus, a universe of discourse between [0, 999] is defined, so U_C=1000 and θ_res= 0.00209=0.12°.

It is necessary to transmit from the PC serial port to the IT2-FLC embedded on the DSC56F8013 three bytes, one per input of IT2-FLC. Then, it is necessary to receive one data as its output. Since the output universe of the embedded IT2-FLC has been to a thousand points. In order to receive the output of the IT2-FLC from DSC56F8013, it is necessary to divide this result. So, units, tens and cents are transmitted as three independent bytes those must be concatenated by a Matlab RS-232 function before to be injected to the truck plant.

Universes of discourse for the embedded IT2-FLC are defined in the previous paragraphs as and. So, it is necessary a scalization as:

$$V_{FLC} = V_{Plant} \cdot U_{FLC}/U_{Plant} + U_{FLC}/2 \tag{5}$$

Where V_{FLC} and U_{FLC} are respectively the scaled value and the universe size that handles the embedded FLC and U_{Plant} and V_{Plant} are respectively the universe size and the un-scaled value that handles the truck plant.

4.2.6 Discretization levels in the consequent regarding available memory resources

According (Leottau & Melgarejo, 2010a), the RAM usage expressed in Words for the inference engine with centroid type-reduction is $2\cdot D_C+2\cdot R$ and for all discussed fuzzyfiers is $2\cdot M_A$. Since DSC56F8013 offers 1KWord in RAM, the maximum discretization levels in the consequent regarding RAM resources can be calculated as:

$$D_{C(max)} = (RAM_{available} - 2 \cdot M_A - 2 \cdot R)/2 = 486 \quad (6)$$

On the other hand, the DATA memory usage for the inference engine with centroid TR and MBA is $2 \cdot D_C \cdot M_C$ and $2 \cdot M_A \cdot D_A$ for the fuzzyfier. Since DSC56F8013 offers 4KWord in Flash memory, the maximum discretization levels in the consequent regarding DATA memory resources can be calculated as:

$$D_{C(max)} = (Flash_{available} - 2 \cdot M_A^{\dagger} \cdot D_A)/(2 \cdot M_C) = 312 \quad (7)$$

Thus, with centroid TR and MBA for fuzzyfier and inference engine, Dc must be less than 312. On the other hand, RAM usage for inference engine with CoS TR is $2 \cdot R = 16$ *Words*. Equation 7 evidences that Dc must be than three hundred twelve points. So, Dc is defined as one hundred taking into account the study carried out in (Leottau & Melgarejo, 2010a). This study evidence that by using $Dc=100$, results maintaining a trade-off between accuracy, memory consumption and processing time.

5. Tests, results and discussion

5.1 Obtained IT2-FLC for the truck backer-upper

The obtained IT2-FLC for tracking the trajectory of the truck Backer-Upper is presented in this sub-section. The IT2-FLC has been designed following the procedure described throughout section 3. Preliminary response of x and φ obtained in step 3.3 is presented in Figure 6. This response has been obtained with the IT2-FSs resulting in the initialization step. These IT2-FSs and resulting T1-FSs tuned in step 3.4 are shown in Figure 7.a. The final IT2-FSs obtained after step 3.5 are shown in Figure 7.b.

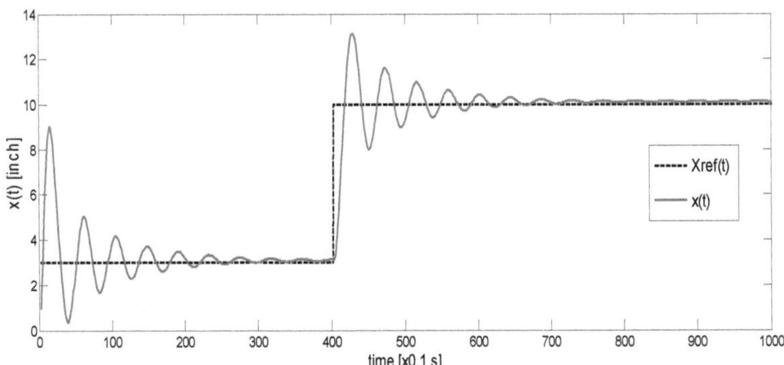

Fig. 6. IT2-FLC design procedure. Preliminary response of x and obtained in step 3.3.

5.2 Tests

Figure 8 resumes the approach outlined throughout this chapter applied for designing and implementing an IT2-FLC for the truck system. After modelling and defining parameters steps in 3.1 and 3.2, initialization and the first design in step 3.3 is carried out by simulation with Matlab® over Simulink. Subsequent to reduction of dimensionality in step 3.4, the tuning of the T1-FLC is carried out by using Excel®. Then, tuned T1-FLC is tested over simulation and emulation. Later than returning to Type-2 and the final fine tuning in step

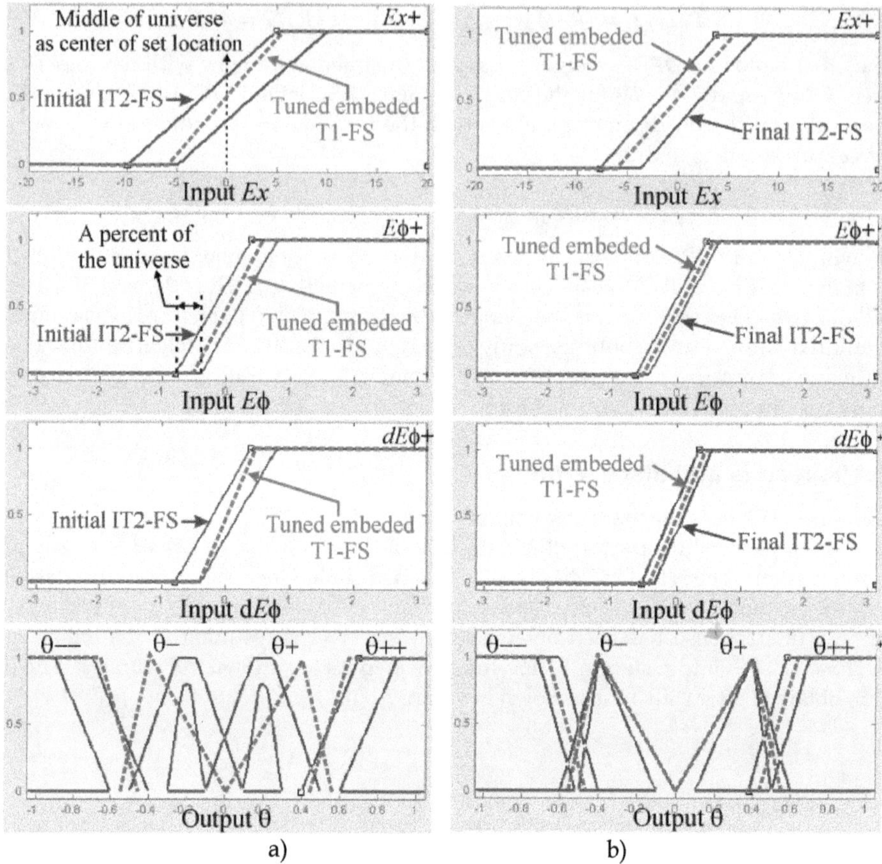

Fig. 7. IT2-FLC design procedure. a) IT2-FSs first design in step 3.3 (Blue lines) and tuned T1-FSs resulting in step 3.4. (Red dotted lines) b) IT2-FSs final implementation after step 3.5 (Blue lines) and newly tuned T1-FSs resulting in step 3.4. (Red dotted lines).

3.5, the resulting IT2-FLC is tested over simulation and emulation too. As it is mentioned before, the simulations are carried out by using fuzzy and it2-fuzzy toolboxes and emulations are carried out over the 56F8013 board whose programming is debugged and loaded by using the Codewarrior® suite for DSC56800E v8.2.3.

Different tests are achieved for evaluating the final performance of designed T1-FLC and IT2-FLC. A combined trajectory in x is used as reference form emulating different operating points to the FLC. In addition, random noise generators are inserted in the inputs and outputs of FLC in order to emulate some sources of uncertainty such as sensors measurements and changes of actuators characteristics. The FLCs are tested under three different noise environments: (1) without noise, (2) moderate noise with amplitude as 0.2% of maximum value that respective variable can take and (3) high noise with amplitudes as 1%. E.g. Ex input can take values within $|0, 20|$, then the amplitude of moderate noise is: $A_{mN}(Ex)$ =20 x 0.2% = 0.04. In order to emulate a stochastic nature of uncertainty, every noise environment is tested five times changing the initial seed of the random noise generator.

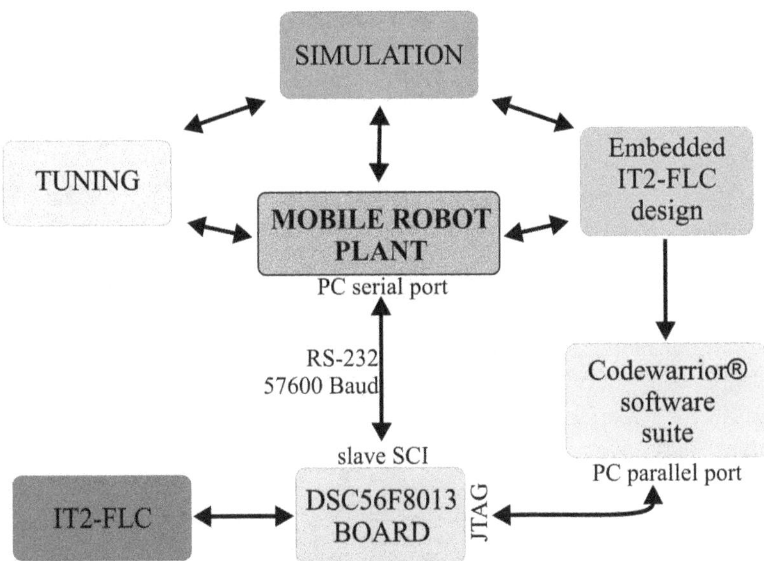

Fig. 8. Conceptual diagram of proposed approach for implementing the embedded IT2-FLC for the truck system.

As it is mentioned in section 2, there are several type reducer and defuzzyfier methods. In order to determine the impact of that in the final performance of designed T1-FLC and IT2-FLC, a test is carried out changing the TR or defuzzyfier method in each FLC. Under simulation, some of available methods in its respective Matlab® toolbox (*fuzzy* for type-1 and *it2fuzzy* for type-2) are tested: centroid and center of sets (CoS) by the T2-FLC and centroid, bisector and mean of maximum (Mom) by T1-FLC. Under emulation, considered methods in section 2 are tested: Centroid and CoS.
The tests procedure is described as follows:

1. Set a TR or defuzzification method for the FLC.
2. Set a noise environment.
3. Run one complete simulation or emulation and register ITSE.
4. Repeat the same simulation (steps 2-3) five times with a different initial noise seed in order to obtain the average and standard deviation of ITSE.
5. Repeat steeps 2-4 for the three noise environments considered.
6. Repeat steeps 1-5 for the TR and defuzzification methods considered.
7. Repeat the procedure for the T1-FLC and the IT2-FLC over simulation and over emulation.

5.3 Results

Results for x and φ are presented in Figure 9 to Figure 12. Figure 9 and 10 show the response of FLCs tested without noise, under simulation and emulation respectively. Figure 11 shows the response of simulated FLCs and Figure 12 shows the response of hardware implemented FLCs, both cases tested under the worst noise condition, emulating a high uncertainty environment. Since all FLCs are tested under three different noise environments and five different initial seeds for each noise environment, obtained results for ITSE as performance indices are presented as an average and its standard deviation. It is shown in tables 3 and 4.

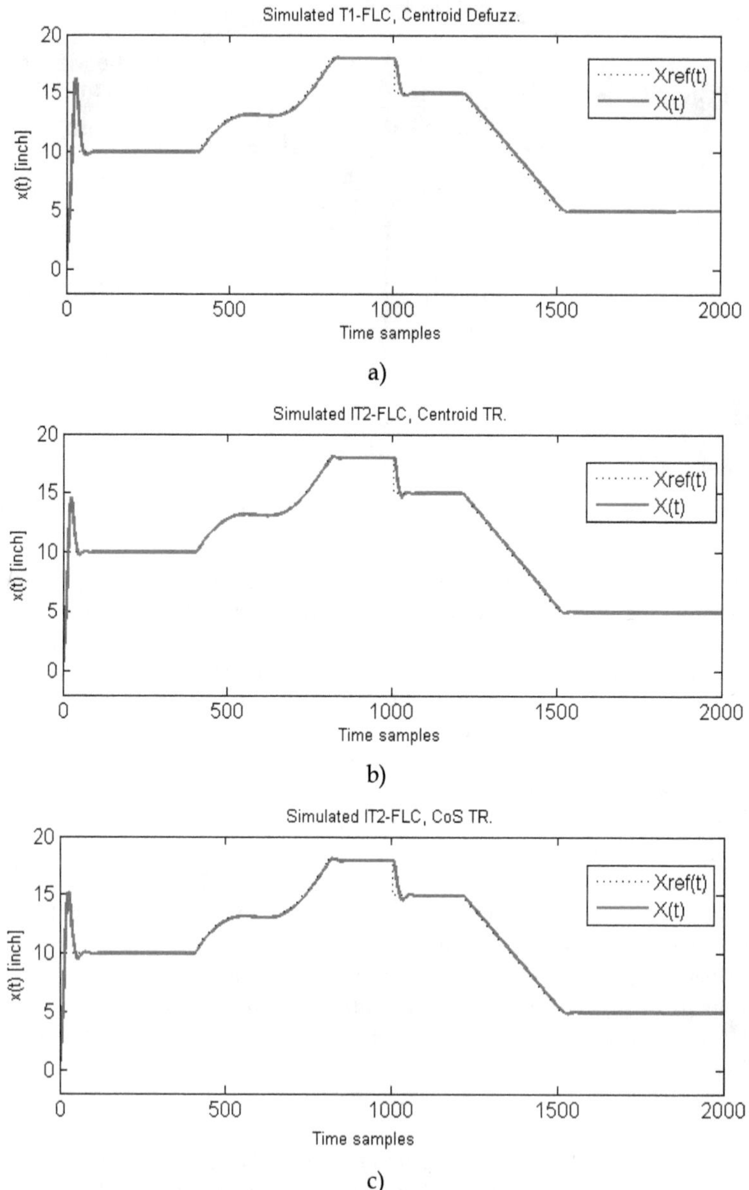

Fig. 9. Without noise and under simulation, trajectory for x of: *a)* T1-FLC with Centroid defuzzification. *b)* IT2-FLC with Centroid type-reduction. *c)* IT2-FLC with CoS type-reduction. (*time=Time_samples/Ts*, where*Ts=0.1s*)

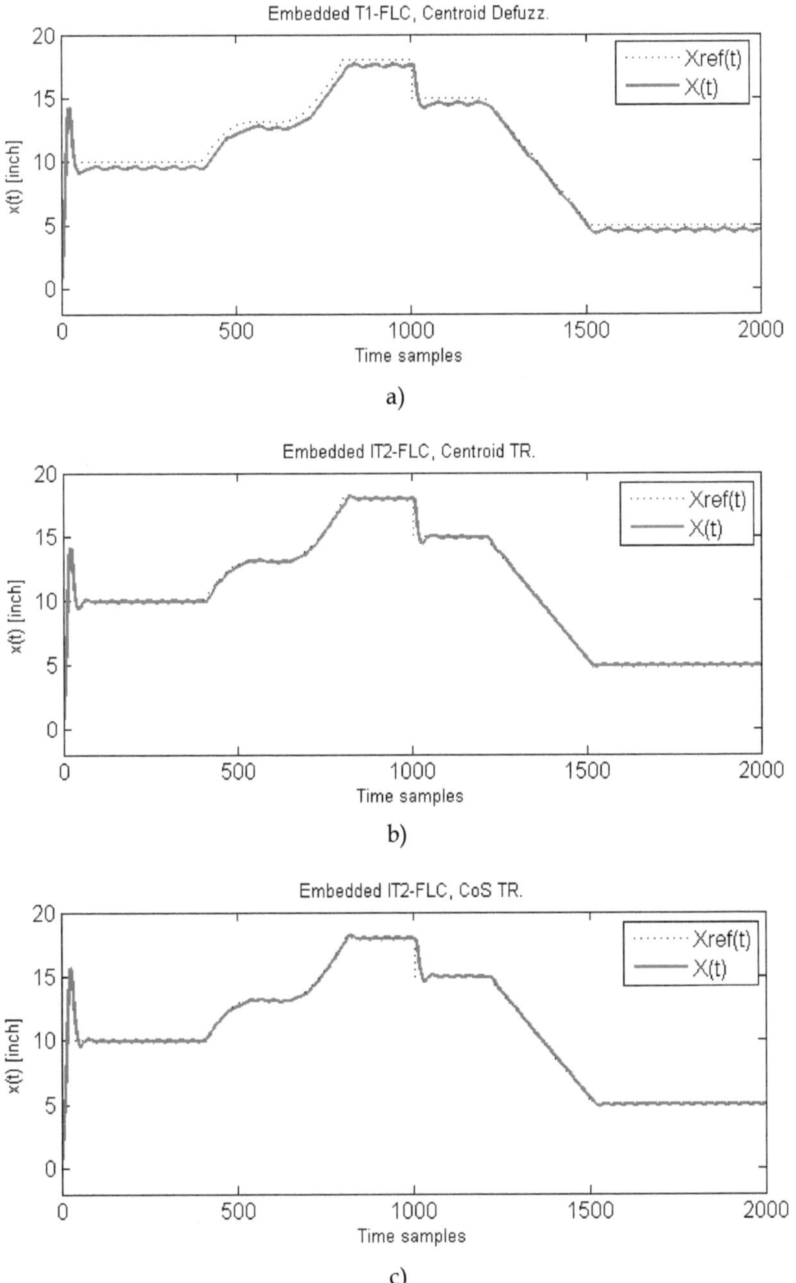

Fig. 10. Without noise, trajectory for x of hardware implemented: a*)* T1-FLC with centroid defuzzification. b*)* IT2-FLC with centroid type-reduction. c*)* IT2-FLC with CoS type-reduction. (*time=Time_samples/Ts*, where*Ts=0.1s*)

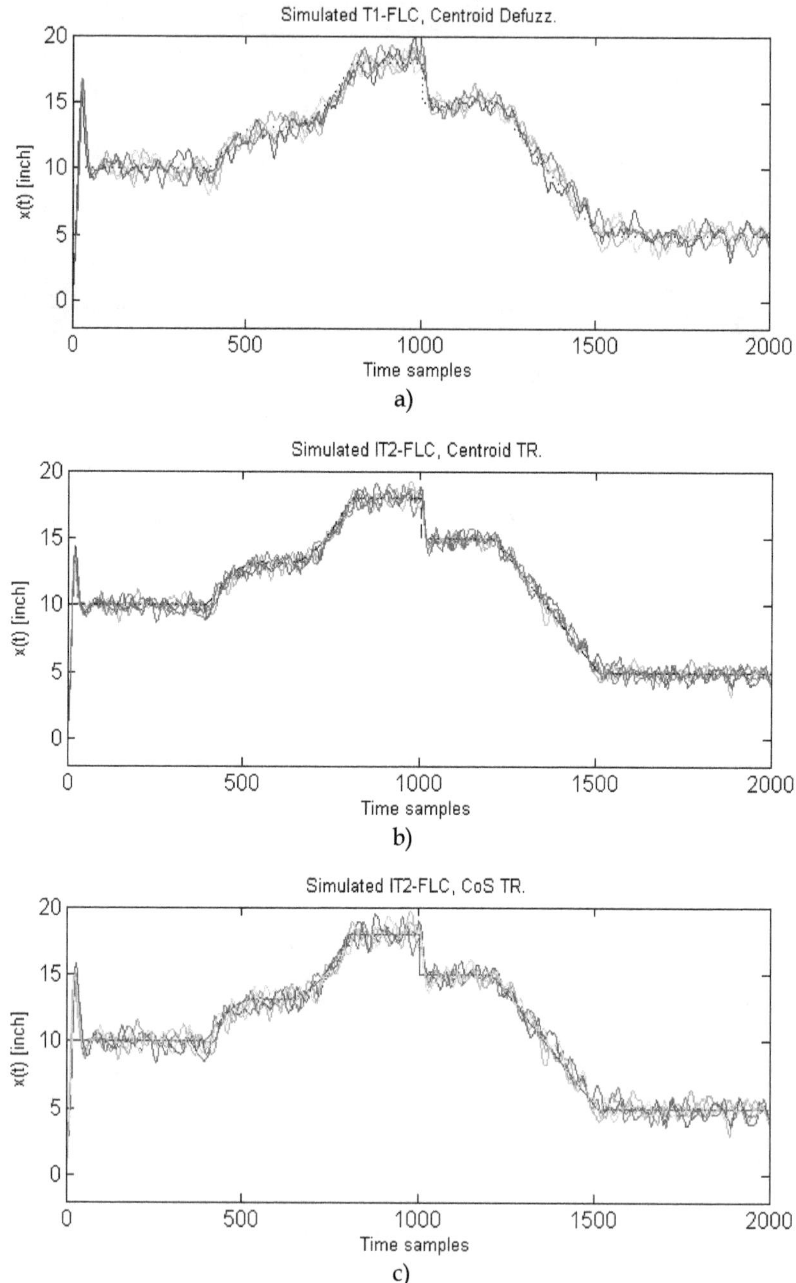

Fig. 11. Under the worst noise condition and simulation, trajectory for x of: *a)* T1-FLC with centroid defuzzification. *b)* IT2-FLC with centroid type-reduction. *c)* IT2-FLC with CoS type-reduction. (*time=Time_samples/Ts*, where*Ts=0.1s*)

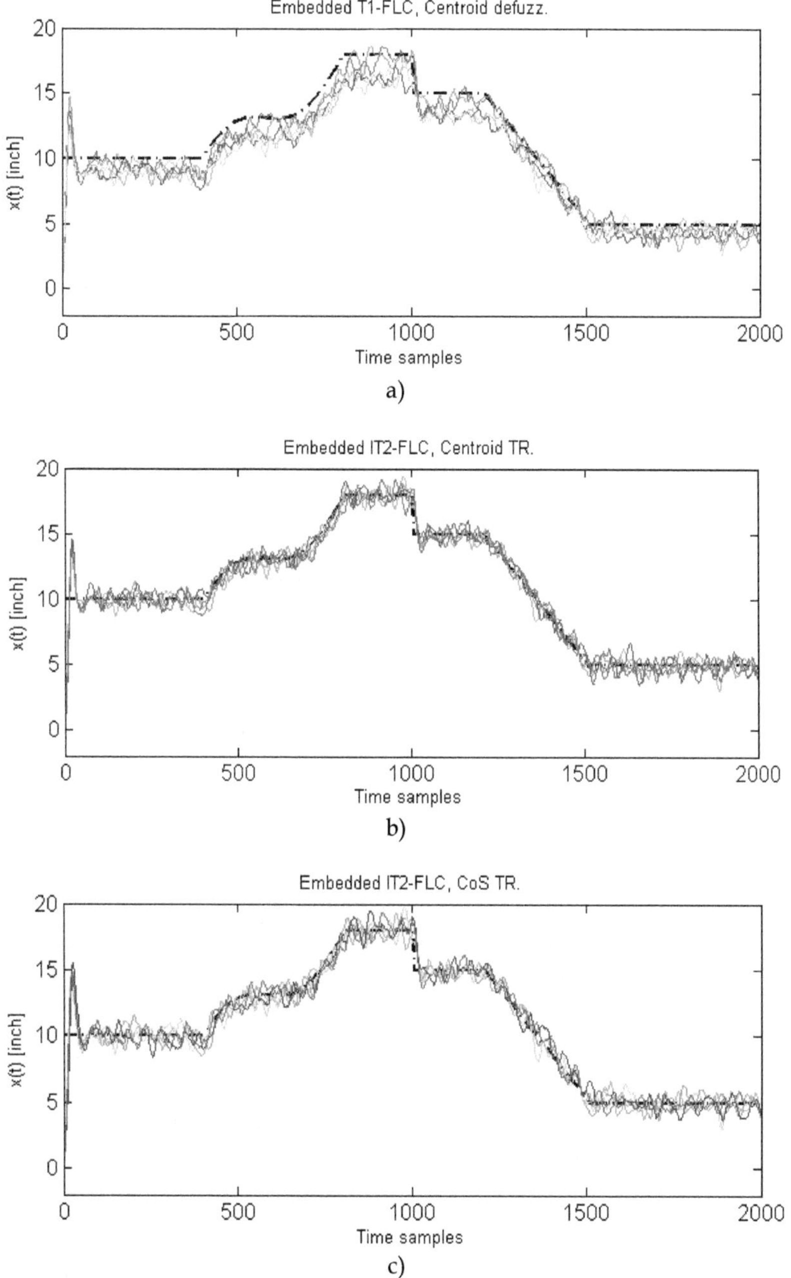

Fig. 12. Under the worst noise condition, trajectory for x of Hardware implemented: a*)* T1-FLC with Centroid defuzzification. b*)* IT2-FLC with centroid type-reduction. c*)* IT2-FLC with CoS type-reduction. (*time=Time_samples/Ts*, where*Ts=0.1s*)

Noise level	Cent.		Bisector		MoM		Cent. Hardware		CoS Hardware	
	ITSE	*Desv.*	*ITSE*	*Desv.*	*ITSE*	*Desv.*	*ITSE*	*Desv.*	*ITSE*	*Desv.*
Zero	18.48	-	†	-	††	-	68.025	-	†††	-
Mod.	22.27	2.10	†	-	††	-	84.84	5.625	†††	-
High	98.93	14.71	†	-	††	-	137.38	13.94	†††	-

† Does not reach the reference.
†† Response turns unstable.
††† Does not reach the reference and presents oscillations.
Since tests have evidenced that the T1-FLCs with MoM and Bisector defuzzyfier and the embedded T1-FLC with CoS defuzzyfier do not present a successfully performance, response for these FLCs have not shown in Figs 9 to 12.

Table 3. Results of ITSE for the T1-FLC.

Noise level	Cent.		CoS		Cent. Hardware		CoS Hardware	
	ITSE	*Desv.*	*ITSE*	*Desv.*	*ITSE*	*Desv.*	*ITSE*	*Desv.*
Zero	11.23	-	13.73	-	11.7	-	13.77	-
Mod.	13.78	0.93	18.02	1.70	13.84	0.85	18.37	0.87
High	44.57	4.81	77.41	10.49	62.122	6.45	82.92	8.76

Table 4. Results of ITSE for the IT2-FLC.

5.4 Discussion

Without noise and using centroid TR, Simulated IT2-FLC ITSE is about 39.2% better than simulated T1-FLC ITSE and embedded IT2-FLC ITSE is about 82.8% better than embedded T1-FLC ITSE. Under moderate noise and using centroid TR it can be said that: Simulated T2-FLC ITSE is about 39.4% better than simulated T1-FLC ones and embedded T2-FLC ITSE is about 83.69% better than embedded T1-FLC. Under high noise and using centroid TR Simulated T2-FLC ITSE is about 54.9% better than simulated T1-FLC ones and embeddedT2-FLC ITSE is about 54.78% better than embedded T1-FLC ones.

ITSE of simulated IT2-FLC with centroid type reduction is between 18% (without noise) and 42% (with high noise) better than the IT2-FLC with CoS. Besides, ITSE of hardware implemented IT2-FLC with centroid type reduction is between 15% (without noise) and 25% (with high noise) better than the IT2-FLC with CoS.

The ITSE of the simulated IT2-FLC with Cent type reducer is between 0.4% (with moderate noise) and 28% (with high noise) better than the embedded IT2-FLC whit the same type reducer. Besides, the ITSE of the simulated IT2-FLC with CoS type reducer is between 0.3% (without noise) and 6.7% (with high noise) better than the embedded IT2-FLC whit the same type reducer. This difference is caused by the integer numeric format that handles the hardware platform which causes lost in the accuracy of embedded FLCs regarding their models tested over simulation.

Taking into account the comparison presented above, it is possible to consider that the IT2-FLC outperforms the T1-FLC in all cases. In addition, by looking Figure 11 and 12, it can be observed that the T2-FLC presents fewer oscillations around the reference, being the IT2-FLC implemented with centroid type reduction, those with better performance.

T2-FLC shows a good immunity to type reducer changes, which is a great advantage taking into account the reduction of computational complexity and inference time when strategies as CoS or heights are implemented.

6. Conclusions

An IT2-FLS for controlling the trajectory of a truck backer-upper mobile robot application and its embedded implementation has been presented in this chapter. The IT2-FLC has been designed based on the approach proposed in (Leottau & Melgarejo, 2010b) and implemented following methodological considerations reported in (Leottau & Melgarejo, 2010a) to the hardware implementation of IT2-FLS over DSC technology.

Two computational models have been selected taking into account the available and demanded hardware and computational resources. Several tests have been carried out in order to evaluate and to compare the performance of developed T2-FLCs and a T1-FLC. Simulated and emulated results evidence that the IT2-FLC is robust to type reducer and defuzzyfier changes and exhibits better performance than a T1-FLC when noise is added to inputs and outputs emulating some sources of uncertainty.

As future work, it is considered to test the developed IT2-FLC with a real mobile robot platform in order to carry out a comparative study of simulation, emulation and real platform performance. By this way, it is possible to extend the design and implementation methodology to other applications, involving in a more formal way a modelling of uncertainty.

7. References

Baturone, I.; Barriga, A.; Sanchez, S.; Jimenez, C.J. & Lopez, D.R. (2000). *Microelectronics Design of Fuzzy Logic-Based Systems*, CRC Press LLC, London.

Castro, J.; Castillo, O. & Melin, P. (2007). An Interval Type-2 Fuzzy Logic Toolbox for Control Applications, *Proceedings of the 2007 IEEE International Conference on Fuzzy Systems, FUZZ-IEEE 2007,* London, UK.

Dorf, R. & Bishop, R. (1998). *Modern Control Systems*, Addison-Wesley.

Figueroa, J.; Posada, J.; Soriano, J.; Melgarejo, M. & Rojas, S. (2005). A Type-2 Fuzzy Logic Controller For Tracking Mobile Objects In The Context Of Robotic Soccer Game, *Proceedings of the 2005 IEEE International Conference on Fuzzy Systems, FUZZ-IEEE 2005.*

Freescale Semiconductor. (January 2010). *Digital Signal Controller 56800/E Reference.* 29.01.2010, Available from: Http://www.Freescale.Com/Dsc

Frontline Systems. (March 2010). *Smooth Nonlinear Optimization.* 17.03.2010, Available from: http://www.solver.com/technology4.htm

Hagras, H. (2004). A Type-2 Fuzzy Logic Controller for Autonomous Mobile Robots, *Proceedings of the 2004 IEEE International Conference on Fuzzy Systems, FUZZ-IEEE 2004*, Budapest, Hungary.

Hagras, H. (2007). Type-2 FLC's: a New Generation of Fuzzy Controllers, *IEEE Computational Intelligence Magazine*, February 2007, pp. 30-43.

Hagras, H. (2008). *Developing a type-2 FLC through embedded type-1 FLCs,* Proceedings of the 2008 IEEE International Conference on Fuzzy Systems, FUZZ-IEEE 2008, Hong Kong.

Hitec Servo-Motors, *Servo-motors catalogue*. 05.08.2010, Available from: http://www.hitecrcd.com/products/servos/index.html

John, R. & Coupland, S. (2007). Type-2 Fuzzy Logic A Historical View, *IEEE Computational Intelligence Magazine*, 2,1, pp. 57-62.

Karnik, N. & Mendel, J. (2001). Centroid of a Type-2 Fuzzy Set, *Information Sciences*, 132, pp. 195-220.

Kuo, B. & Golnaraghi, F. (1996). *Automatic Control Systems*, Prentice Hall.

Leottau, L. & Melgarejo, M. (2010a). A Methodological Proposal for Implementing Interval Type-2 Fuzzy Processors over Digital Signal Controllers, *Journal of Applied Computer Science Methods*, v.2-1, June 2010, pp.61-81.

Leottau, L. & Melgarejo, M. (2010b). A Simple Approach for Designing a Type-2 Fuzzy Controller for a Mobile Robot Application, *Proceedings of the North American Fuzzy Information Processing Society's NAFIPS 2010*, Toronto, Canada, July 2010.

Nguyen, D. & Widrow, B. (1989). The Truck Backer-Upper: An Example of self-learning in Neural Networks. *Proceedings of the International Joint Conference in Neural Networks*, pp. II-357-363, June 1989.

Martinez, R.; Castillo, O. & Aguilar, L. (2008). Optimization with Genetic Algorithms of Interval Type-2 Fuzzy Logic controllers for an autonomous wheeled mobile robot: A comparison under different kinds of perturbations, *Proceedings of the 2008 IEEE International Conference on Fuzzy Systems, FUZZ-IEEE 2008.*

Melgarejo, M.; Garcia, A. & Pena-Reyes, C. (2004). Computational Model and architectural proposal for a hardware Type-2 Fuzzy System, *Proceedings of the 2nd IASTED conference on Neural Networks and Computational Intelligence*, Grindewald, Switzerland.

Melgarejo, M. & Pena-Reyes, C. A. (2007). Implementing Interval Type-2 Fuzzy Processors, *IEEE Computational Intelligence Magazine*, 2,1, pp. 63-71.

Melin, P. & Castillo, O. (2003). A new method for adaptive model-based control of non-linear plants using type-2 fuzzy logic and neural Networks, *Proceedings of the 2003 IEEE International Conference on Fuzzy Systems, FUZZ-IEEE 2003*, St. Louis, MO.

Mendel, J. (2001). Uncertain Rule-Based Fuzzy Logic Systems: Introduction and New Directions, Prentice Hall, New Jersey.

Mendel, J. (2007). Advances In Type-2 Fuzzy Sets and Systems, *Information Sciences*, 177,1, pp. 84-110.

Torres, P. & Saez, D. (2008). Type-2 fuzzy logic identification applied to the modelling of a robot hand, *Proceedings of the 2008 IEEE International Conference on Fuzzy Systems, FUZZ-IEEE 2008.*

Wang, L. X. (1997). *A Course in Fuzzy Systems and Control*, Prentice Hall, New Jersey.

Wu, D. & Mendel, J. (2009). Enhanced Karnik-Mendel Algorithms, *IEEE Transactions on Fuzzy Systems*, 17, pp. 923-934.

Wu, D. & Tan, W. (2004). A type-2 fuzzy logic controller for the liquid-level process, Proceedings of the 2004 IEEE International Conference on Fuzzy Systems, FUZZ-IEEE 2004.

7

Statistical Video Based Control of Mobile Robots

Krzysztof Okarma and Piotr Lech
West Pomeranian University of Technology, Szczecin
Poland

1. Introduction

A typical approach to the control of mobile robots is based on the analysis of signals from various kinds of sensors (e.g. infra-red or ultrasound). Another type of input data used for motion control of robots can be video signals acquired by the cameras. In such case many algorithms can be utilised e.g. SLAM (Simultaneous Localization and Mapping) and its modifications, being still an active field of research (Kalyan et al., 2010), which require usually the full analysis of the images (or video frames) being the input data (Se et al., 2001; 2002), in some applications using also additional operations such as Principal Component Analysis (Tamimi & Zell, 2004). Similar problems may also occur in the robot localisation and map building processes (Hähnel et al., 2003).
An interesting alternative can be the application of the fast image analysis techniques based on the statistical experiments. Presented statistical approach is related to the first step of such control systems, which can be considered as the pre-processing technique, reducing the amount of data for further analysis with similar control accuracy. Discussed methods can be utilised for many control algorithms based on the image and video data e.g. the proportional steering of line following robots (Cowan et al., 2003) considered as a representative example in further discussions. Nevertheless, some other applications of the presented approach are also possible e.g. fast object classification based on the shape analysis (Okarma & Lech, 2009) or motion detection, as well as fast reduced-reference image quality estimation (Okarma & Lech, 2008b).

2. Reduction of power consumption by mobile robots

Considering the autonomous mobile robots, one of their most important parameters, corresponding directly to their working properties, is the power consumption related to the maximum possible working time. Limited energy resources are often the main element reducing the practical applicability of many such constructions regardless of many modern energy sources which can be used such as e.g. solar based solutions. The limited capacity of the installed batteries reduces the range and capabilities of these devices and each of the possible attempts to optimise energy consumption usually affects the mobility and efficiency of the robots.
It is possible to reduce the consumed energy by finding the shortest path to the target (robot's drive) or conducting the optimisation of the amount of processed data by the control algorithm. Since the computational cost of mapping algorithms, searching for the optimal path, finding robot's own location, visual SLAM algorithms etc., based on the image analysis is strongly dependent on the resolution and representation of the analysed image, a significant "smart" decrease of the amount of processed data seems to be an interesting direction of

research in this area. Reducing the computational load of the processor the overall demand for the electrical energy decreases, so it can be assumed that limiting the amount of processed data in the system, energy savings can be obtained, despite using a non-optimal (longer) motion path. For the video based control algorithms such restriction can be achieved simply by reducing the image resolution or reducing the frame rate of the acquired image sequence used for he robot's motion control. Nevertheless, such decrease of the amount of analysed data may lead to some steering errors, as illustrated in Fig. 1, where the optimal shortest path cannot be used due to the presence of some obstacles. Each of the indicated alternative paths has a different length related to the different energy necessary for the robot to reach the target point.

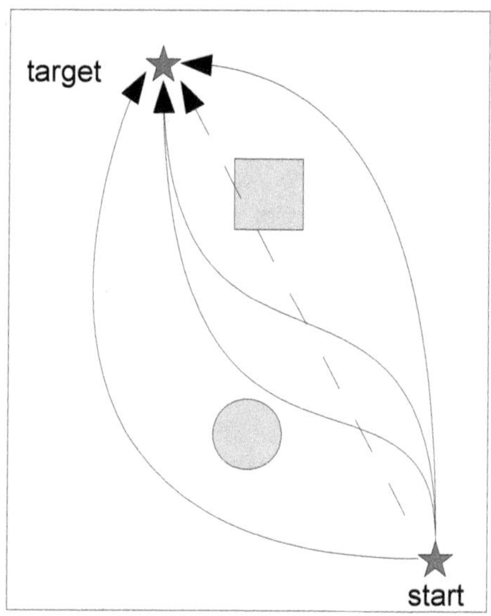

Fig. 1. Illustration of the problem of the optimal path searching in the presence of obstacles

The optimal robot's path in the known environment can be chosen using the machine vision approach with motion detection and robot's motion parameters estimation, using the general algorithm presented in Fig. 2 such that the total energy consumption, related both to the robot's drive and the control part dependent mainly on the processing of the vision data, is minimum. Conducting the experimental verification of the idea for the mobile robot based on a simple netbook platform using the joystick port for controlling the drives, for the testing area of 6 square meters illustrated in Fig. 1, the maximum power consumption is about 5.2 A (42.5% by the drive part and 57.5% by the control part of the system). The length of the optimum straight path, assuming no obstacles on the scene, is equal to 2 meters so all the alternative trajectories are longer.

The experiments conducted in the environment described above has been related to the decrease of the resolution of the image (obtained results are presented in Table 1) and the frame rate for the fixed resolution of 320 × 240 pixels (results shown in Table 2). For the reduction to 8 frames per second the vision based control algorithm could not find any appropriate path.

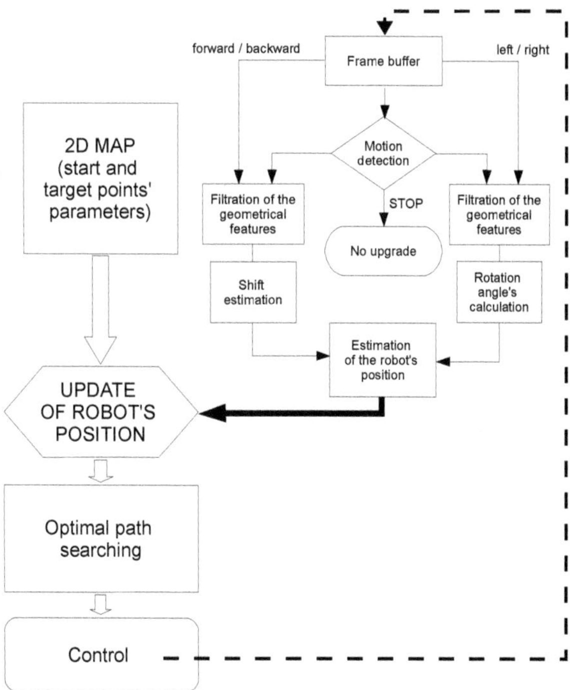

Fig. 2. Illustration of the general vision based control algorithm of the mobile robot

Resolution in pixels	Length of the path [m]	Total energy consumption [kJ]
640 × 480	2.15	5.022
352 × 288	2.39	4.519
320 × 240	2.69	4.268
176 × 144	2.80	4.168
160 × 120	2.95	4.369

Table 1. Path length and power consumption of the mobile robot used in the experiments dependent on the image resolution

Frame rate [fps]	Length of the path [m]	Total energy consumption [kJ]
25	2.39	4.519
21	2.44	4.432
16	2.66	4.332
12	3.23	4.668
10	3.33	4.798

Table 2. Path length and power consumption of the mobile robot used in the experiments dependent on the frame rate of the analysed video sequence

Analysing the results presented in Tables 1 and 2 a significant influence of the amount of the analysed video data on the obtained results and the power consumption can be noticed. The decrease of the resolution as well as the reduction of the video frame rate causes the increase of the path's length but due to the reduction of the processor's computational load the total energy consumption may decrease. Nevertheless, the decrease of the amount of the processed data should be balanced with the control correctness in order to prevent an excessive extension of the robot's path.
Since the simple decrease of the resolution may cause significant control errors as discussed above, a "smart" decrease of the number of analysed pixels, which does not cause large control errors should be used for such purpose much more efficiently. In the proposed statistical approach the main decrease of the amount of data is related to the random draw of a specified number of pixels for further analysis using the modified Monte Carlo method. In a typical version of this method (as often used for the fast object's area estimation) the pixels are chosen randomly by an independent drawing of two coordinates (horizontal and vertical), what complicates the error estimation. For the simplification, the additional mapping of pixels to the one-dimensional vector can be used, what is equivalent to the data transmission e.g. as an uncompressed stream. The additional reduction takes place during the binarization performed for the drawn pixels.

3. The Monte Carlo method for the fast estimation of the objects' geometrical properties

3.1 The area estimation

Conducting a statistical experiment based on the Monte Carlo method, an efficient method of estimation of the number of pixels in the image fulfilling a specified condition can be proposed, which can be used for the motion detection purposes as well as for the area estimation of objects together with their basic geometrical features. Instead of counting all the pixels from a high resolution image, the reduction of the number of pixels used for calculations can be achieved by using a pseudo-random generator of uniform distribution. Due to that the performance of the algorithm can be significantly increased. The logical condition which has to be fulfilled can be defined for a greyscale image as well as for the colour one, similarly to the chroma keying commonly used in television. leading to the binary image, which can be further processed. In order to prevent using two independent pseudo-random generators it is assumed that the numbered samples taken from the binary image are stored in the one-dimensional vector, where "1" denotes the black pixels and "0" stands for the white ones representing the background (or reversely). Then a single element from the vector is drawn, which is returned, so each random choice is conducted independently. For simplifying the further theoretical analysis the presence of a single moving dark object (Ob) in the scene (Sc) with a light background can be assumed.
In order to determine the object's geometrical features as well as some motion parameters, such as direction and velocity, the random choice of pixels cannot be performed using the whole image due to the integrating properties of the Monte Carlo method. For preventing possible errors caused by such integration the scene can be divided into smaller and smaller squares, so the K_N squares from N elements of the scene would represent the object. In such case the probability of choosing the point on the object's surface (assuming the generator's uniform distribution) is equal to:

$$p = \frac{K_N}{N} \tag{1}$$

For the infinite number of samples the reduction of the sampling distance takes place and the probability of choosing the point representing the object on the image can be expressed as:

$$p = \lim_{N \to \infty} \frac{K_N}{N} = \frac{A_{Ob}}{A_{Sc}} \tag{2}$$

so the estimated area of the object is:

$$A_{Ob} \approx K_N \cdot \frac{A_{Sc}}{N} \tag{3}$$

Since the total area of the scene is equal to $A_{Sc} = N \cdot k_x \cdot k_y$, the estimated object's area is equal to $A_{Ob} \approx K_N \cdot k_x \cdot k_y$, where k_x and k_y are the scale factors for horizontal and vertical coordinates respectively, equivalent to the number of samples per unit.

Considering the above analysis, the probability of choosing the point belonging to the object can be used in the proposed algorithm with a reduced number of analysed samples instead of the full image analysis using a statistical experiment based on the Monte Carlo method. This method originates directly from the law of large numbers, because the sequence of successive approximations of the estimated value is convergent to the sought solution and the distance of the actual value after performing the specified number of statistical tests to the solution can be determined using the central limit theorem.

After the binarization the luminance (or colour) samples represented as "ones" or "zeros" corresponding to the allowed values of the specified logical condition, which are stored in one-dimensional vector are chosen randomly. For a single draw a random variable X_i of the two-way distribution is obtained:

$$X_i = \begin{cases} 1 & \text{for black samples} \\ 0 & \text{for white samples,} \end{cases} \tag{4}$$

leading to the following probability expressions:

$$P(X_i = 1) = p \qquad P(X_i = 0) = q \tag{5}$$

where $p + q = 1$, $E(X_i) = p$, $V(X_i) = p \cdot q$.

An important element of the method is the proper choice of the logical condition for the binarization allowing a proper separation of the object from the background. Such choice depends on the specific application, for example for the light based vehicles' tracking purposes the chroma keying based on the CIELAB colour model can be appropriate (Mazurek & Okarma, 2006).

For n independent draws the variable Y_n is obtained:

$$Y_n = \frac{1}{n} \cdot \sum_{i=1}^{n} X_i \tag{6}$$

According to the Lindberg-Levy's theorem, the distribution of Y_n tends to the normal distribution $N(m_y, \sigma_y)$ if $n \to \infty$. Since the expected value and variance of Y_n are equal to $E(Y_n) = p$ and $V(Y_n) = \frac{p \cdot q}{n}$ respectively, the distrubution of the random value Y_n is normal with the following parameters:

$$m_y = p \tag{7}$$

$$\sigma_y = \sqrt{\frac{p \cdot q}{n}} \tag{8}$$

Considering the asymptotic normal distribution $N(p, \sqrt{p \cdot q/n})$ it can be stated that the central limit theorem is fulfilled for the variable Y_n.
Substituting:

$$U_n = \frac{Y_n - m_y}{\sigma_y} \tag{9}$$

obtained normal distribution can be standardized towards the standard normal distribution $N(0,1)$.
In the interval estimation method the following formula is used:

$$p(|U_n| \leq \alpha) = 1 - \alpha \tag{10}$$

Assuming the interval:

$$|U_n| \leq u_\alpha \tag{11}$$

considering also the formulas (3), (7), (8) and (9), the following expression can be achieved:

$$\left| Y_n - \frac{K_N}{N} \right| \leq \varepsilon_\alpha \tag{12}$$

where

$$\varepsilon_\alpha = \frac{u_\alpha}{\sqrt{n}} \cdot \sqrt{\frac{K_N}{N} \cdot \left(1 - \frac{K_N}{N}\right)} \tag{13}$$

The probability estimator p (eq. 1), for k elements from n draws, fulfilling the specified logical condition used for the definition of X_i (eq. 4) and representing the number of the drawn pixels representing the object, can be expressed as:

$$\hat{p} = \frac{k}{n} = \frac{1}{n} \cdot \sum_{i=1}^{n} X_i = Y_n \tag{14}$$

and the object's area estimator as:

$$\hat{A}_{Ob} = \hat{p} \cdot A_{Sc} = \frac{k}{n} \cdot A_{Sc} \tag{15}$$

Using the equations (12) and (15) the obtained formula describing the interval estimation for the object's area is:

$$\left| \frac{\hat{A}_{Ob}}{A_{Sc}} - \frac{K_N}{N} \right| \leq \varepsilon_\alpha \tag{16}$$

where ε_α is specified by the equation (13).
It is worth to notice that all the above considerations are correct only for a random number generator with the uniform distribution, which should have as good statistical properties as possible. The discussed algorithm are identical to the method of area estimation of the 2-D object's (expressed in pixels). Nevertheless, the applicability of such approach is limited by the integrating character of the method so an additional modification based on the block approach is necessary, as mentioned earlier.

Such prepared array, considered as a reduced resolution "greyscale" image, can be directly utilised in some typical control algorithms. Further decrease of the computational cost can be obtained by the additional binarization of this image leading to the ultra-low resolution binary image considered in further discussions, where each of the blocks can be classified as a representative of an object or the background. For the line following robot control purposes the objects is equivalent to the path. The control accuracy of the mobile robots corresponds to the quality loss of the data present in this image. Such quality can be treated as the quality of the binary image assuming the knowledge of the reference image (without any distortions). Unfortunately, most of the image quality assessment methods, even the most recent ones, can be successfully applied only for greyscale or colour images and the specific character of the binary images is not respected by them. For this reason only the methods designed exclusively for the binary images can be used. Such binary image quality assessment methods can be used as the optimisation criteria for the proposed algorithm. In the result the proper choice of the block size, binarization threshold and the relative number of randomly drawn pixels can be chosen.

Depending on the logical condition used for the statistical analysis (the construction of the 1-D binary vector), the algorithm can be implemented using various colour spaces (or only some chosen channels), with the possibility of utilising independent logical conditions for each channel (chroma keying).

3.2 The Monte Carlo based motion detection

The real-time navigation of an autonomous mobile robot based on the machine vision algorithms requires a fast processing of images acquired by the camera (or cameras) mounted on the robot. In some more sophisticated systems the data fusion based on the additional informations acquired by some other sensors, such as PIR sensors, ultrasound detectors, a radar technique equipment or optical barriers, is also possible. Most of such sensors are responsible for the motion detection, especially if the presence of some moving obstacles in the robot's surrounding is assumed. The vision based sensors can be divided into three main groups: analogue (i.e. comparing the luminance with a given threshold value), digital (obtained by the sampling and quantization of acquired analogue signal) and the digital ones acquired directly from digital video cameras. There are also some restrictions of typical alternative solutions, e.g. passive infrared detectors are sensitive to temperature changes and they are useless for the motion detection of objects with the same temperature as the background. Some more robust solutions, such as ultrasound and radar detectors, are usually active (energy emitting) and require an additional power supply. The application of a vision based motion detection procedure utilising the Monte Carlo method for the mobile robots control purposes is caused mainly by the fact that no additional hardware is necessary, since the images used by the motion detector are acquired by the same camera as in the classification procedure.

Assuming the scene with the constant light conditions without any moving objects, it can be characterised by a constant value defined as the number of black (or white) pixels of the respective binary image, which usually represent the object visible on the image obtained from the camera. The algorithm of the motion detection utilises the rapid changes of the number of such pixels (treated as "ones") when an object moves respectively to the light sources. Such change is caused by the dependence of the image pixel's luminance (and colour) on the angle between normal vector to its surface and the direction of the light ray passing the pixel, as well as the influence of the surface's reflection coefficient, shape, roughness etc. Real objects usually have a heterogeneous structure of their surface so the number of analysed points changes dynamically (assuming constant threshold or chroma keying range). For slow

changes of the light conditions, especially for outside environment, the binarization threshold can be updated. It it worth to notice that in the proposed approach the necessity of the storage of the single value only (the number of pixels corresponding to the "ones") allows the reduction of the required system's operating memory size.
A typical well known video based algorithm of motion detection is based on the comparison of two neighbouring frames of the video signal but it requires the storage of the images used during the comparison so relatively large amount of memory is needed. Some more robust algorithms of background estimation, directly related to motion detection, usually based on moving average or median filter, are also more computationally demanding and require the analysis of several video frames (Cucchiara et al., 2003; Piccardi, 2004). Even a comparison of all the pixels from two video frames can be treated as a time consuming operation for an autonomous mobile robot with limited computational performance and relatively small amount of operating memory, so in the real-time applications some high performance processing units would be required. Instead of them the statistical Monte Carlo approach described above can be used for the reduction of the computational complexity, where only some randomly chosen pixels are taken into account. Therefore the comparison of full frames is reduced to the comparison of only two values representing the number of "ones" in each frame. Any rapid change of that value is interpreted as the presence of a moving object.
A slightly modified method based on the additional division of the image into smaller blocks can be applied for a more robust detection. In the presence of moving objects the changes can be detected only in some of obtained blocks so the integrating character of the Monte Carlo method is eliminated.
The fast motion detection based on the comparison of two values representing the estimated areas of the object in the two consecutive binary frames is based on the following formula:

$$\hat{A}_{Ob}(i+\delta) - \hat{A}_{Ob}(i) > threshold \tag{17}$$

where $threshold >> \varepsilon_{\alpha}$ and $\delta \geq 1$ denotes the shift between both analysed frames.
In the proposed block based approach the application of the formula (17) for each of the $r \times r$ pixels blocks allows proper motion detection, using the array of the Monte Carlo motion detectors, also in the situation when the motion of the mobile robot causes the relative motion of the objects observed by the integrated camera. For the perspectively looking camera, typical for such applications, even if the objects do not change their size on the image, the changes can be observed among the blocks. The objects moving towards the camera always cause the increase of their estimated area in the image used by the motion detector. The estimation of the geometrical parameters of the objects, often necessary for the proper navigation of the controlled robot, can be performed for the frame with the maximum value of the estimated area occupied by the object on the image with eliminated background, assuming that robot is not moving directly towards the object. If the size of the object increases in consecutive frames, such estimation should be performed immediately preventing possible collision with the object.

3.3 The estimation of geometrical parameters

The extraction of some geometrical parameters, such as perimeters or diameters, using the discussed approach is also possible. For this purpose the analysed binary image has to be divided into $T \times S$ blocks of $r \times r$ pixels each using a square grid. Then the area of each object's fragment in the elementary square elements (blocks) is calculated and the results of such estimation can be stored in the array P containing $T \times S$ elements.

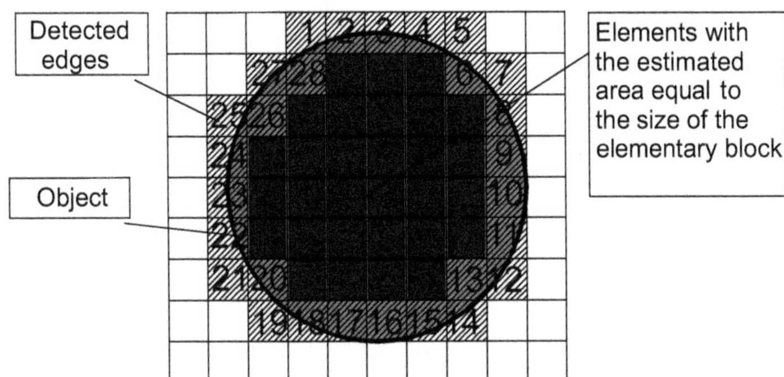

Fig. 3. The idea of the block based edge estimation

The simplest approach to the edge detection, which is necessary for further processing, is based on the array P. For this purpose the second array K with the same size is created and its elements have the following values:

- zero if the corresponding element's value in the array P is equal to zero (background),
- zero if the corresponding element's value in the array P is equal to the size of the elementary block (all pixels represent the inner part of the object) and none of its neighbouring blocks (using the 8-directional neighbourhood) has the zero value,
- one for the others (representing the edge).

The idea of the edge detection is illustrated in Fig. 3
The obtained array K is the projection of edges detected from the source image, so the number of its non-zero elements represents the estimated value of object's perimeter expressed in squares of $r \times r$ pixels. For a better estimation the number of square elements can be increased (smaller values of the parameter r) and then the previously discussed operation should be repeated. In such case the analysis of the blocks in the array P with the zero values (representing the inner parts of the object) is not necessary, so the further analysis can be conducted only for the significantly reduced number of indexed elements classified as the edge in the previous step. Such alternative edge detection method based on the two-step algorithm with the same initial step of the algorithm and another pass of the previous algorithm conducted as the second step with the edge detection calculated using the array K (the edge obtained in the first step is treated as the new object without fill) allows more accurate estimation of the edge with slight increase of the computational complexity. All the pixels with the zero values should be replaced from the array K and using such corrected image the two curves are obtained as the result (assuming the scale factor of the block greater than $t = 2$ - at least 3×3 elements). As the estimate of the edge the middle line between the two obtained curves can be used, providing slightly lower accuracy but also a reduced computational cost. The estimate of the total object's area can also be determined with similarly increased accuracy as the sum of the values in the array P. The limit accuracy of the algorithm is determined by the size of the elementary block equal to 1 pixel what is equivalent to the well-known convolution edge detection filters.
Geometrical parameters of the objects, which can be estimated using the modified Monte Carlo method, can be divided into the two major groups: local (such as mean diameter or the

average area) and global (e.g. the number of objects in the specified area). The estimation of the relative global parameters is not necessary if only a single object is present on the analysed fragment of the image. The most useful parameters used for image analysis, which can be also used for robot's control purposes, are insensitive to image deformations which can occur during the acquisition, as well as to some typical geometrical transformations. In such sense the usefulness of the simplest parameters such as area and perimeter is limited. Nevertheless, the majority of more advanced parameters (such as moments) can be usually determined using the previously described ones, similarly as some motion parameters e.g. direction or velocity (Okarma & Lech, 2008a).

4. Application for the line following robots

One of the most important tasks of the control systems designed for mobile robots is line tracking. Its typical optical implementation is based on a line of sensors receiving information about the position of the traced line, usually located underneath the robot. Such line sensor is built from a specified number of cells limiting the resolution of the optical system. Another important parameter of such systems is the distance from the sensors to the centre of steering, responsible for the maximum possible speed of the properly controlled robot. The smoothness of the robot's motion is also dependent on the spacing between the cells forming the line sensor.

A significant disadvantage of such optical systems is relatively low resolution of the tracking system, which can be increases using some other vision based systems with wide possibilities of analysing data acquired from the front Dupuis & Parizeau (2006); Rahman et al. (2005). Nevertheless the analysis of the full acquired image even with low resolution is usually computationally demanding and time consuming, especially in the presence of some obstacles, some line intersections etc., assuming also varying lighting conditions. Another serious drawback of using line of sensors is the short time for the reaction limited by the distance between the sensors and the steering centre (or wheels). The proposed approach, based on the fast image analysis using the Monte Carlo method, preserves the main advantages of the vision systems allowing the reduction of the amount of processed data. Considering its application for the control of the mobile robots, the camera moves relatively to the static scene, differently than in the primary version of the algorithm, but the working properties of the method are similar.

In order to filter the undesired data, usually related to some contaminations which should not be used for the robot control purposes, the binarization threshold should be properly set. In the conducted experiments the additional "cut-off" value has been set as minimum 20% black pixels possibly representing the object within the block in order to avoid the influence of the small artifacts, especially close to the region of interest. Finally, the simplified binary representation of the image is obtained where such artifacts (the elements which in fact do not represent the followed line) have been removed during the "cut-off" operation illustrated in Fig. 4 where the original binary image and the intermediate result obtained by the Monte Carlo method are also presented.

4.1 A simple proportional control algorithm for controlling a line following robot

The mobile robot control process can be based on the popular proportional control approach. It is typically used for controlling the robots with sensors based on the infra-red receivers grouped into the line containing a specified number of the infra-red cells. In the proposed method the differential steering signals for the motors are obtained using the first (the lowest)

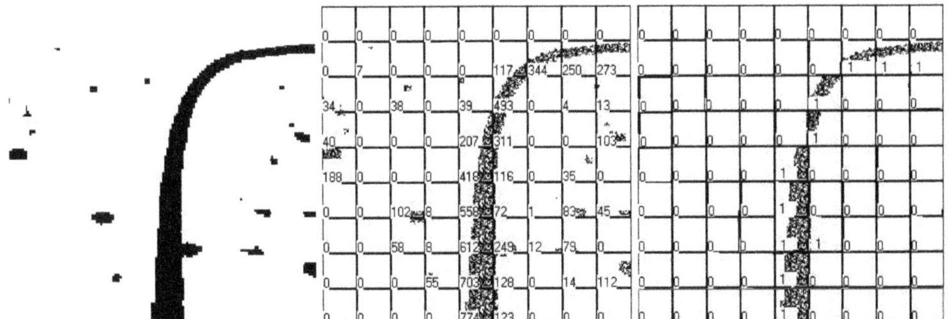

Fig. 4. The idea of the "cut-off" operation

row of the obtained simplified binary image which is equivalent to the typical line of the infra-red sensors.

Nevertheless, the machine vision algorithm can be disturbed by some artifacts caused by the errors during the preprocessing step (e.g. insufficient filtration of some undesired elements on the image). An optimal situation would take place for a single "one" in the lowest row of the image representing the tracked line.

In the simplified version of the algorithm no line crossings can be assumed. The reduction of the possible disturbances can be achieved using the following additional operations:

- limitation of the line thickness by the proper choice of the binarization threshold (lines containing more than a single "one" can be ignored assuming no line crossings),
- orphan blocks removal (applicable for all blocks with the value "one" without any neighbouring "ones"),
- ignoring the empty lines (without any "ones").

4.2 Extension of the robot control by vision based prediction

The control algorithm of the robot should point towards such a position that only a single block of the bottom image row will be filled (representing the line). In the case when two neighbouring blocks represent the line the choice of the block should be based on the simple analysis of the current trend, assuming that the turn angles of the line match the motion and control possibilities of the controlled robot.

The control algorithm for a single frame can be described as follows (Okarma & Lech, 2010):

- binarization of the current frame,
- orphan blocks removal,
- filling the gaps in the detected line using the approximation methods,
- control operation:
  - the detection of the crossing lines
    - if not turning: moving forward with the maximum speed for the symmetrical line-crossing or using the speed control
    - else: turning and using the speed control if the maximum value is not in the middle block
  - speed control: velocity should be proportional to the sum of the values in the middle blocks of each horizontal line (if zero, the minimum speed is set before the turning and the control flag is set to 0).

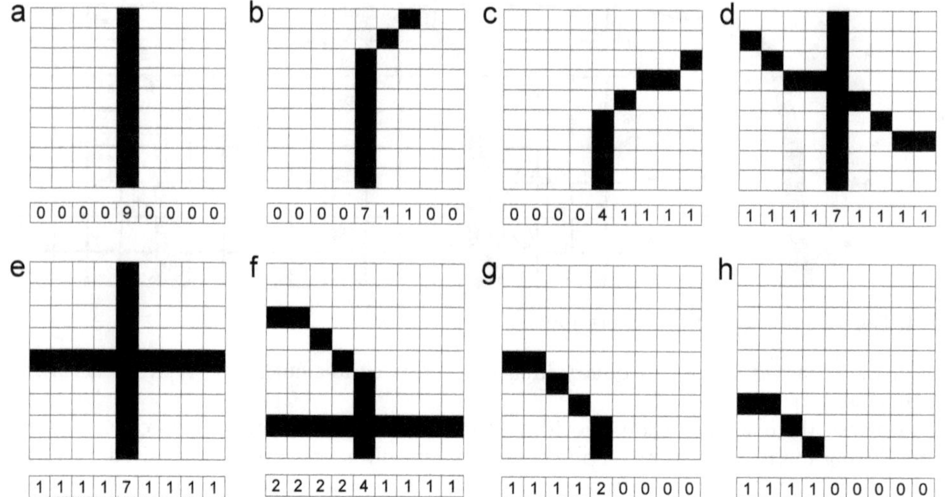

Fig. 5. The illustration of the various configurations of the followed line

The main element of the modified control system is the variable (flag), determined using the previous frames, informing the control system about the current state (e.g. 0 for moving forward or turning, 1 for turning with detected crossing lines).

In the case of detected crossing line the control depends on the symmetry. For the symmetrical line-crossing the robot should move forward with the maximum speed or the additional speed control. If the maximum value is not in the middle block the line-crossing is not symmetrical and the robot should turn left or right and decrease it speed. The speed of the robot should be proportional to the sum of the values in the middle blocks of each horizontal line. If that sum is equal to zero, the minimum speed should be set before the turning. In that case the mentioned above flag should be set to 1. The illustration of the various scenarios is presented in Fig. 5.

One of the typical features of the proportionally controlled systems is the oscillating character of the path for the robot reaching the specified line, causing also the characteristic oscillations of the images acquired from the camera. Nevertheless, the conducted experiments have verified the usefulness of the fast image analysis based on the Monte Carlo method for the efficient controlling of the line following robot in a simplified but very efficient way, especially useful for the real-time applications. The application of the proposed solution for the control of the mobile robot, despite of the simplicity of its control algorithm, is comparable with some typical ones based on the classical approach (proportional steering based on the infra-red sensors).

The main advantage of the proposed method is the possibility of using some prediction algorithms allowing the increase of the robot's dynamic properties and fluent changes of its speed depending on the shape of the line in front of the robot. Another relevant feature is its low computational complexity causing the relatively high processing speed, especially important in the real-time embedded systems.

4.3 Application of the Centre of Gravity

If the robot is located directly over the straight line its central point should be exactly between the two edges obtained using the two-step algorithm of edge detection discussed above. In such situation the sum of the pixel values on its left side should be identical as on its right side. Any significant disturbance can be interpreted as the beginning of a line curve with the assumption that no additional elements, except two line edges, are visible on the image, so the mobile robot should begin turning. An efficient and simple method for controlling the robot which can be used for such purpose is based on the image's Centre of Gravity (Centre of Mass). It can be determined for a grayscale image as:

$$x_c = \frac{\sum\limits_{i,j} i \cdot X_{i,j}}{\sum\limits_{i,j} X_{i,j}} \tag{18}$$

$$y_c = \frac{\sum\limits_{i,j} j \cdot X_{i,j}}{\sum\limits_{i,j} X_{i,j}} \tag{19}$$

where $X_{i,j}$ represents the luminance of the pixel located at position (i, j).
Calculating the coordinates of the Centre of Gravity (CoG), they can be used by the control algorithm as the current target point for the mobile robot. The horizontal coordinate x_c can be used for controlling the turning angle and the vertical y_c for the speed control, assuming the previous filtration of any artifacts from the image.
The approach discussed above can also be implemented using the block based Monte Carlo method using the "cut-off" operation for decrease of the influence of noise, leading to the hybrid control method. For each block the local area estimator $\hat{A}$ is used as the input for the "cut-off" operation leading to the simplified binary representation utilised for steering using the method discussed above. In comparison to the classical Center of Gravity method its implementation using the Monte Carlo approach causes about 8 times reduction of the computation time as has been verified in the conducted experiments.
One of the main modifications proposed in the hybrid method is the switch between the CoG and Monte Carlo based control in order to increase the robustness of the control. For this purpose the sum of each column of the simplified binary image is calculated and stored in an additional vector. If the vector containing the sums of the values, obtained by counted blocks with "ones", has more than one local minimum (e.g. 2 4 0 0 1 6 0 0 1), some additional objects are present on the image. In such situation the calculation of the Centre of Gravity coordinates would be disturbed by a neighbouring object located close to the line, so the motion control should be based on the previous value. Otherwise, the control values determined by the Monte Carlo method can be additionally verified by the CoG calculations.
A more complicated situation may take place for the presence of artifacts together with some discontinuities of the followed line. In this situation the vector containing the sum of the column in the simplified block representation of the image has more than one local maximum. In such case the additional elimination of the blocks which are not directly connected to the followed line (detected as represented by the maximum value in the vector) should be conducted e.g. using the morphological erosion, before the calculations of the Centre of Gravity used for determining the current trend (direction of the line).

5. Simulations in the Simbad environment

In order to verify the working properties of the proposed methods, regardless of some tests conducted using a netbook based prototype, a test application has been created using Java programming language in NetBeans environment allowing the integration of many libraries developed for Java e.g JARToolKit (Geiger et al., 2002) allowing to access the Augmented Reality Toolkit (ARToolKit) functionality via Java and the use of different rendering libraries allows high and low level access. The environment of the mobile robot as well as the robot itself have been modelled using the Simbad simulator (Hugues & Bredeche, 2006), which is available for this platform.

A significant advantage of the simulator is its 3D environment, due to the utilisation of Sun Java3D technology. Nevertheless, an important restrictions of Simbad, in opposite to some other similar products, are the limited working properties for the controlling the real robots die to its internal structure related to the modelling of physical phenomena. However, this results in a simplified and transparent programming source code, so the implementation of the presented algorithms does not cause any considerable problems. The only restriction in the initialisation part of the robot controller is the directing the camera towards the "ceiling", caused by the inability to put the followed line on the ground. Instead of it the line has been located over the robot and the lines in the front of the robot are shown in the top part of the image. The only negative consequence is the reverse transformation of the left and right directions between the robot and the camera. The proportional controlled differential drive robots used in the experiments are typically implemented in Simbad and conducted simulations have confirmed the working properties of the presented steering methods for all control modes with different simulation speed values. The test application used for the simulations is illustrated in Fig . 6.

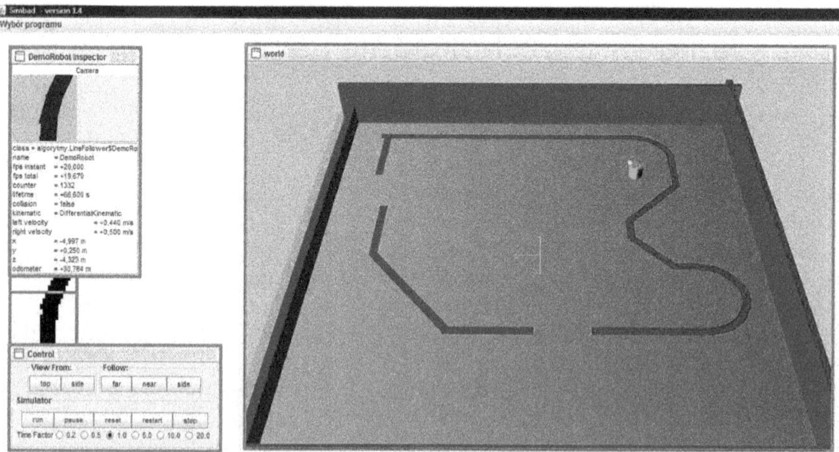

Fig. 6. Illustration of the test application

6. Conclusions and future work

Obtained results allowed the full control of the line following robot (speed, direction and "virtual" line reconstruction). Due to the ability of path prediction, in comparison to the typical line following robots based on the line of sensors mounted under the robot, the "virtual"

line reconstruction can be treated as one of the most relevant advantages of the image based motion control. Such reconstruction can also be performed using the ultra-low resolution binary image as the input data. Another possibility, being analysed in our experiments, is the adaptation of the Centre of Gravity method for the real time steering based on the binary image analysis.
The additional robustness of the method can be achieved by the adaptive change of the block size used in the modified Monte Carlo method according to the quality assessment of the binary image and the additional analysis of the line continuity.
Positive results of the simulations performed in Java 3D based Simbad environment allow the proper implementation of the proposed soft computing approach in the experimental video controlled line following robot. Good simulation results, as weel as the low energy consumption by the experimental netbook-based robot, have been also obtained considering the synthetic reduced resolution "grayscale" image constructed using the Monte Carlo method, where the relative areas of the objects in each block have been used instead of the luminance level (after normalisation).
The planned future work is related to the physical application of the proposed approach for some other types of robot control algorithms based on video analysis, e.g. SLAM. Another direction of the future research should be the development of a dedicated binary image quality assessment method, which could be used for the optimisation of the proposed algorithm.
Another interesting direction of the future work is the extension of the proposed method towards direct operations using the colour images. For this purpose a verification of the usefulness of the typical colour models is necessary by means of the chroma keying used for the thresholding in the Monte Carlo method.

7. References

Cowan, N., Shakernia, O., Vidal, R. & Sastry, S. (2003). Vision-based follow-the-leader, *Proceedings of IEEE/RSJ International Conference on Intelligent Robots and Systems*, Vol. 2, Las Vegas, Nevada, pp. 1796–1801.

Cucchiara, R., Grana, C., Piccardi, M. & Prati, A. (2003). Detecting moving objects, ghosts, and shadows in video streams, *IEEE Transactions on Pattern Analysis and Machine Intelligence* 25(10): 1337–1342.

Dupuis, J.-F. & Parizeau, M. (2006). Evolving a vision-based line-following robot controller, *Proceedings of the 3rd Canadian Conference on Computer and Robot Vision*, Quebec City, Canada, p. 75.

Geiger, C., Reimann, C., Stocklein, J. & Paelke, V. (2002). JARToolkit - a Java binding for ARToolkit, *Proceedings of the The First IEEE International Workshop on Augmented Reality Toolkit*, Darmstadt, Germany, p. 5.

Hähnel, D., Triebel, R., Burgard, W. & Thrun, S. (2003). Map building with mobile robots in dynamic environments, *International Conference on Robotics and Automation*, Vol. 2, Taipei, Taiwan, pp. 1557–1563.

Hugues, L. & Bredeche, N. (2006). Simbad: An autonomous robot simulation package for education and research, *in* S. Nolfi, G. Baldassarre, R. Calabretta, J. C. T. Hallam, D. Marocco, J.-A. Meyer, O. Miglino & D. Parisi (eds), *SAB*, Vol. 4095 of *Lecture Notes in Computer Science*, Springer, pp. 831–842.

Kalyan, B., Lee, K. & Wijesoma, W. (2010). FISST-SLAM: Finite set statistical approach to simultaneous localization and mapping, *International Journal of Robotics Research* 29(10): 1251–1262.

Mazurek, P. & Okarma, K. (2006). Car-by-light tracking approach based on log-likelihood track-before-detect algorithm, *Proceedings of the 10th International Conference "Computer Systems Aided Science Industry and Transport" TRANSCOMP*, Vol. 2, Zakopane, Poland, pp. 15–20.

Okarma, K. & Lech, P. (2008a). Monte Carlo based algorithm for fast preliminary video analysis, *in* M. Bubak, G. D. van Albada, J. Dongarra & P. M. A. Sloot (eds), *ICCS (1)*, Vol. 5101 of *Lecture Notes in Computer Science*, Springer, pp. 790–799.

Okarma, K. & Lech, P. (2008b). A statistical reduced-reference approach to digital image quality assessment, *in* L. Bolc, J. L. Kulikowski & K. W. Wojciechowski (eds), *ICCVG*, Vol. 5337 of *Lecture Notes in Computer Science*, Springer, pp. 43–54.

Okarma, K. & Lech, P. (2009). Application of Monte Carlo preliminary image analysis and classification method for automatic reservation of parking space, *Machine Graphics and Vision: International Journal* 18(4): 439–452.

Okarma, K. & Lech, P. (2010). A fast image analysis technique for the line tracking robots, *in* L. Rutkowski, R. Scherer, R. Tadeusiewicz, L. A. Zadeh & J. M. Zurada (eds), *ICAISC (2)*, Vol. 6114 of *Lecture Notes in Computer Science*, Springer, pp. 329–336.

Piccardi, M. (2004). Background subtraction techniques: a review, *Proceedings of the IEEE International Conference on Systems, Man & Cybernetics*, Vol. 4, The Hague, Netherlands, pp. 3099–3104.

Rahman, M., Rahman, M., Haque, A. & Islam, M. (2005). Architecture of the vision system of a line following mobile robot operating in static environment, *Proceedings of the 9th International Multitopic Conference, IEEE INMIC 2005*, Karachi, pp. 1–8.

Se, S., Lowe, D. & Little, J. (2001). Local and global localization for mobile robots using visual landmarks, *Proceedings of the IEEE/RSJ International Conference on Intelligent Robots and Systems*, Vol. 1, Maui, Hawaii, pp. 414–420.

Se, S., Lowe, D. & Little, J. (2002). Mobile robot localization and mapping with uncertainty using scale-invariant visual landmarks, *International Journal of Robotics Research* 21: 735–758.

Tamimi, H. & Zell, A. (2004). Vision based localization of mobile robots using kernel approaches, *Proceedings of the 2004 IEEE/RSJ International Conference on Intelligent Robots and Systems*, Sendai, Japan, pp. 1896–1901.

8

LQR Control Methods for Trajectory Execution in Omnidirectional Mobile Robots

Luis F. Lupián and Josué R. Rabadán-Martin
Mobile Robotics & Automated Systems Lab, Universidad La Salle
Mexico

1. Introduction

Omnidirectional mobile robots present the advantage of being able to move in any direction without having to rotate around the vertical axis first. While simple straight-line paths are relatively easy to achieve on this kind of robots, in many highly dynamic applications straight-line paths might just not be a feasible solution. This may be the case because of two main reasons: (1) there may be static and moving obstacles between the initial and desired final position of the robot, and (2) the dynamic effects of the inertia of the robot may force it to execute a curved path. This chapter will address these two situations and present a segment-wise optimal solution for the path execution problem which is based on a Linear-Quadratic Regulator.

It must be emphasized that, rather than attempting to perform an exact path tracking, the approach presented here deals with the problem of visiting a sequence of target circular regions without specifying the path that will connect them. The freedom given to the connecting paths brings the opportunity for optimization. In fact, the path that the robot will take from one circular region to the next will emerge as the solution of an optimal control problem, hence the term *segment-wise optimal* solution.

Omnidirectional wheels have the property of sliding laterally with almost zero force while providing full traction in the rolling direction. This effect is achieved by adding a set of smaller wheels around the periphery of the main wheel, as depicted in Fig. 1.

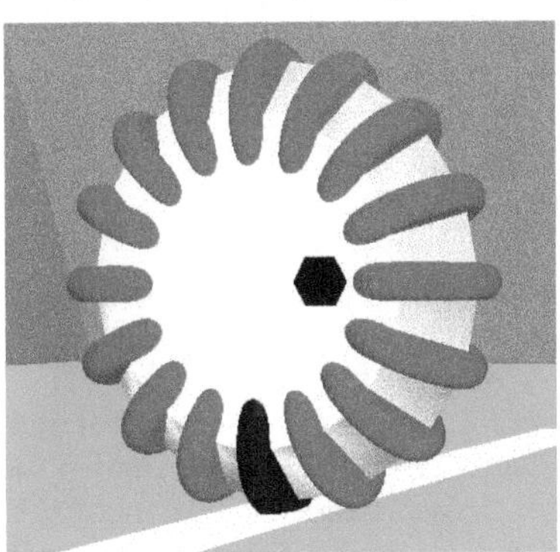

Fig. 1. Omnidirectional wheel

By using several omnidirectional wheels distributed around the periphery of a cylindrical robot one can achieve the effect of driving the robot in a direction dependent on a vector sum of forces. This idea is illustrated in Fig. 2. The forces applied at the wheels by the motors give translational and rotational movement to the mobile robot.

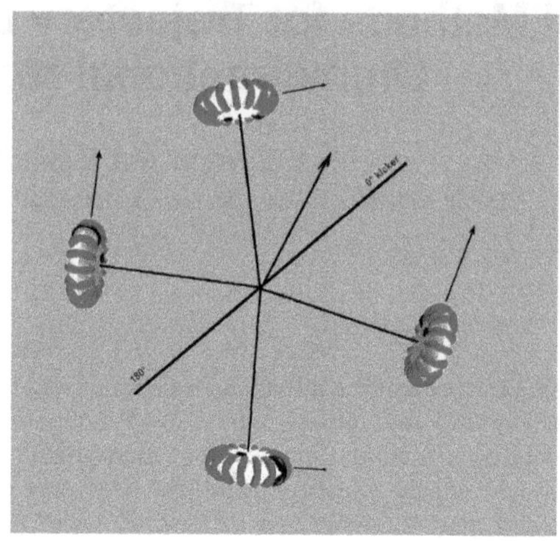

Fig. 2. Omnidirectional drive

Omnidirectional movement on a mobile robot has many applications, and thus has received attention from the scientific community over several years. Borenstein and Evans developed a control strategy that allows trajectory control on a mobile robot that uses conventional non-omnidirectional wheels and thus requires the drive wheels to rotate with respect to the vertical axis (Borenstein & Evans, 1997). With regard to omnidirectional mobile robots based on omnidirectional wheels there are several works that deserve attention. Ashmore and Barnes presented a detailed analysis of the kinematics of this kind of mobile robots and show that under certain circumstances a curved path may be faster than the straight line between two points (Ashmore & Barnes, 2002). Balkcom et al. characterized time-optimal trajectories for omnidirectional mobile robots (Balkcom et al., 2006a;b). Kalmár-Nagy et al. developed a control method to generate minimum-time trajectories for omnidirectional mobile robots (Kalmár-Nagy et al., 2004; 2002), then Purwin and D'Andrea presented the results of applying this method to a RoboCup F180 omnidirectional mobile robot (Purwin & D'Andrea, 2005).
Section 2 deals with the first necessary step towards applying LQR control for trajectory execution, which is formulating a state-space model for the dynamics of the omnidirectional mobile robot (Lupián & Rabadán-Martin, 2009).
This state-space model is non linear with respect to the control due to the fact that the robot rotates around the vertical axis and the pan angle is one of the state variables. In Sec. 3 of this chapter we show how to overcome this problem to successfully apply a Linear Quadratic Regulator for the case of three, four and five-wheeled omnidirectional mobile robots. In Sec. 4 we present a method to generate a segment-wise optimal path by solving an LQR control problem for each segment between the initial state through a sequence of target regions that ends at the desired final state of the omnidirectional mobile robot. Finally, Sec. 5 presents the results of several simulation experiments that apply the methods described in this chapter.

2. State-space dynamic model

The analysis presented pertains to a specific class of omnidirectional mobile robots that are of cylindrical shape with n omnidirectional wheels distributed around the periphery of the body of the robot, with the axes of the wheels intersecting at the geometrical vertical axis of the robot. Figure 3 shows an instance of this class of omnidirectional mobile robots for the case of $n = 5$ omnidirectional wheels. It is not necessary to have a uniform distribution of the wheels around the periphery of the robot. That is, the angle that separates the axis of one wheel to the next does not need to be the same. However, there is an obvious restriction that must be met in order to maintain stability, which is that the projection of the center of mass of the robot onto the ground must be contained within the convex hull of the set of contact points of the n wheels with the ground. For simplicity of the analysis, the mass of the robot is assumed to be distributed uniformly, so the center of mass is contained within the geometrical vertical axis of the robot.

Fig. 3. Omnidirectional robot

In the literature, there are several proposals for the dynamic model of an omnidirectional mobile robot. One of the main problems with these models is that they do not provide a complete state-space representation, so it is not possible to perform state-space control by using one of them. Most of these proposals are based on the force coupling matrix (Gloye & Rojas, 2006), which provides a direct relationship between the torques applied by the driving motors and the accelerations in the x, y and angular directions.
In compact form, these equations may be written as follows

$$\begin{aligned} a &= \frac{1}{M}(F_1 + F_2 + \cdots + F_n) \\ \dot{\omega} &= \frac{R}{I}(f_1 + f_2 + \cdots + f_n) \end{aligned} \tag{1}$$

where a is the acceleration of the robot with respect to the inertial reference, $\dot{\omega}$ is the angular acceleration, n is the number of omnidirectional wheels around the periphery of the robot, F_i is the vector force applied by motor i, f_i is the signed scalar value of F_i (positive for counter clock-wise rotation), M is the mass of the robot, R is its radius and I is its moment of inertia.

Taking into account the driving motor distribution shown in Fig. 4, Eq. 1 may be re-written as

$$\begin{aligned} Ma_x &= -f_1 \sin\theta_1 - f_2 \sin\theta_2 - \cdots - f_n \sin\theta_n \\ Ma_y &= f_1 \cos\theta_1 + f_2 \cos\theta_2 + \cdots + f_n \cos\theta_n \\ MR\dot{\omega} &= \frac{1}{\alpha}(f_1 + f_2 + \cdots + f_n) \end{aligned} \tag{2}$$

where x and y are the axes of the inertial frame of reference, a_x is the x component of the acceleration, a_y is the y component of the acceleration, α is such that $I = \alpha MR^2$, and θ_i is the angular position of driving motor i with respect to the robot's referential frame as shown in Fig. 4. Equation 2 is expressed in matrix form as

$$\begin{bmatrix} a_x \\ a_y \\ R\dot{\omega} \end{bmatrix} = \frac{1}{M} \begin{bmatrix} -\sin\theta_1 & -\sin\theta_2 & \cdots & -\sin\theta_n \\ \cos\theta_1 & \cos\theta_2 & \cdots & \cos\theta_n \\ \frac{MR}{I} & \frac{MR}{I} & \cdots & \frac{MR}{I} \end{bmatrix} \begin{bmatrix} f_1 \\ f_2 \\ \vdots \\ f_n \end{bmatrix} \tag{3}$$

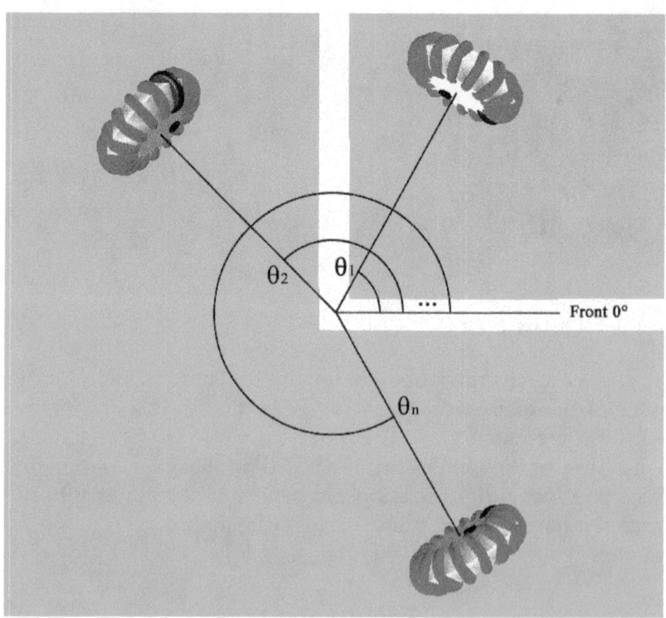

Fig. 4. Driving motor distribution

Equation 3 gives a general idea about the dynamic behavior of the mobile robot. However, it can not be considered a complete state-space model since it does not deal with the inertial effects of the mass of the robot, so we define a state vector that is complete enough to describe the dynamic behavior of the mobile robot (Lupián & Rabadán-Martin, 2009). We define the state vector z as

$$z = \begin{bmatrix} x & y & \beta & \dot{x} & \dot{y} & \dot{\beta} & \mu_1 & \cdots & \mu_n \end{bmatrix}^{\mathrm{T}} \tag{4}$$

where (x, y) is the position of the robot with respect to the inertial reference, $(\dot{x}, \dot{y})$ is the vector velocity of the robot with respect to the field plane, β is the angular position and $\dot{\beta}$ is

the angular velocity. Those are the six state variables we are interested in controlling, so it would suffice to use them to describe the state of the robot.
However, for simulation purposes we introduce variables μ_i, which correspond to the angular positions of each of the omnidirectional wheels of the robot. The decision to include these state variables was initially motivated by the fact that the dynamic response of the robot would be simulated on a virtual environment developed in OpenGL, and the motion of the wheels would make this simulation more realistic. On the physical robot, however, these state-variables will have a more important role because, unlike the other six state variables, the angular positions of the wheels can be measured directly through motor encoders, so one can implement a state observer to estimate the remaining variables.
In order to express the state equations in a compact and clear form, the state vector will be partitioned as follows:

$$\begin{aligned} z_1 &= \begin{bmatrix} x & y & \beta & \dot{x} & \dot{y} & \dot{\beta} \end{bmatrix}^{\mathrm{T}} \\ z_{11} &= \begin{bmatrix} x & y & \beta \end{bmatrix}^{\mathrm{T}} \\ z_{12} &= \begin{bmatrix} \dot{x} & \dot{y} & \dot{\beta} \end{bmatrix}^{\mathrm{T}} \\ z_2 &= \begin{bmatrix} \mu_1 & \dots & \mu_n \end{bmatrix}^{\mathrm{T}} \end{aligned} \tag{5}$$

In terms of z_1 y z_2 the state vector z can be expressed as

$$z = \begin{bmatrix} z_1 \\ z_2 \end{bmatrix} \tag{6}$$

and the state-space model becomes

$$\begin{bmatrix} \dot{z}_1 \\ \dot{z}_2 \end{bmatrix} = \begin{bmatrix} A_{11} & A_{12} \\ A_{21} & A_{22} \end{bmatrix} \begin{bmatrix} z_1 \\ z_2 \end{bmatrix} + \begin{bmatrix} B_1 \\ B_2 \end{bmatrix} u \tag{7}$$

where u is the control vector composed of the scalar forces f_i divided by the mass of the robot

$$u = \frac{1}{M} \begin{bmatrix} f_1 \\ f_2 \\ \vdots \\ f_n \end{bmatrix}, \tag{8}$$

and the matrices are defined as follows. Matrix A_{11} simply expresses the relationships among the first six state variables, and is given by

$$A_{11} = \begin{bmatrix} 0_{3\times 3} & I_{3\times 3} \\ 0_{3\times 3} & 0_{3\times 3} \end{bmatrix} \tag{9}$$

Matrix B_1 is obtained from Eq. 3 and becomes

$$B_1 = \begin{bmatrix} 0 & 0 & \cdots & 0 \\ 0 & 0 & \cdots & 0 \\ 0 & 0 & \cdots & 0 \\ -\sin\theta_1 & -\sin\theta_2 & \cdots & -\sin\theta_n \\ \cos\theta_1 & \cos\theta_2 & \cdots & \cos\theta_n \\ \frac{MR}{I} & \frac{MR}{I} & \cdots & \frac{MR}{I} \end{bmatrix} \tag{10}$$

Matrix B_1 expresses the correct influence of each force over the respective acceleration only for the case in which the angular position β of the robot is zero. However, in order to take into account the fact that the robot will rotate as time goes by this matrix should also depend on β as follows

$$B_1(\beta) = \begin{bmatrix} 0 & 0 & \cdots & 0 \\ 0 & 0 & \cdots & 0 \\ 0 & 0 & \cdots & 0 \\ -\sin(\theta_1+\beta) & -\sin(\theta_2+\beta) & \cdots & -\sin(\theta_n+\beta) \\ \cos(\theta_1+\beta) & \cos(\theta_2+\beta) & \cdots & \cos(\theta_n+\beta) \\ \frac{MR}{I} & \frac{MR}{I} & \cdots & \frac{MR}{I} \end{bmatrix}$$
$$B_1(\beta) = \begin{bmatrix} 0_{3\times n} \\ B_{12}(\beta) \end{bmatrix} \tag{11}$$

A_{12} and A_{22} are zero matrices of size $6 \times n$ and $n \times n$ respectively. Matrix A_{21} expresses the angular motion of the wheels with respect to the motion of the mobile robot and, like B_1, it is dependent on the angular position β of the robot:

$$A_{21}(\beta) = \frac{1}{r}\begin{bmatrix} 0_{n\times 3} & A_{212}(\beta) \end{bmatrix}$$
$$A_{212}(\beta) = \frac{1}{r}\begin{bmatrix} \sin(\theta_1+\beta) & -\cos(\theta_1+\beta) & -R \\ \sin(\theta_2+\beta) & -\cos(\theta_2+\beta) & -R \\ \vdots & \vdots & \vdots \\ \sin(\theta_n+\beta) & -\cos(\theta_n+\beta) & -R \end{bmatrix} \tag{12}$$

where r is the radius of the omnidirectional wheels.
Taking into account that both A_{21} and B_1 depend on β the state model in Eq. 7 may be reformulated as

$$\begin{bmatrix} \dot{z}_1 \\ \dot{z}_2 \end{bmatrix} = \begin{bmatrix} A_{11} & 0_{6\times n} \\ A_{21}(\beta) & 0_{n\times n} \end{bmatrix}\begin{bmatrix} z_1 \\ z_2 \end{bmatrix} + \begin{bmatrix} B_1(\beta) \\ 0_{n\times n} \end{bmatrix} u \tag{13}$$

3. Global linearization of the state-space dynamic model

Since the angular position β is one of the state variables this implies that the model in Eq. 13 is non-linear, and that represents an important difficulty for the purpose of controlling the robot. This is why it became necessary to find a way to linearize the model.
The solution to this problem required a shift in perspective in relation to the model. The only reason why the angle β has a non-linear effect on the dynamics of state variables in z_1 is because as the robot rotates also the driving forces, which are the control variables, rotate.
Let $F = F_1 + F_2 + \ldots + F_n$ be the resulting vector force applied by the omnidirectional wheels to the robot when it is at the angular position β and the control vector is u. Let $\check{u}$ be the control vector that will produce the same resulting vector force F when the robot is at the angular position $\beta = 0$. This idea is explained by Fig. 5.
The control vectors u and $\check{u}$ are then related by

$$\begin{aligned} \begin{bmatrix} \sin(\theta_1) & \cdots & \sin(\theta_n) \end{bmatrix} \check{u} &= \begin{bmatrix} \sin(\theta_1+\beta) & \cdots & \sin(\theta_n+\beta) \end{bmatrix} u \\ \begin{bmatrix} \cos(\theta_1) & \cdots & \cos(\theta_n) \end{bmatrix} \check{u} &= \begin{bmatrix} \cos(\theta_1+\beta) & \cdots & \cos(\theta_n+\beta) \end{bmatrix} u \end{aligned} \tag{14}$$

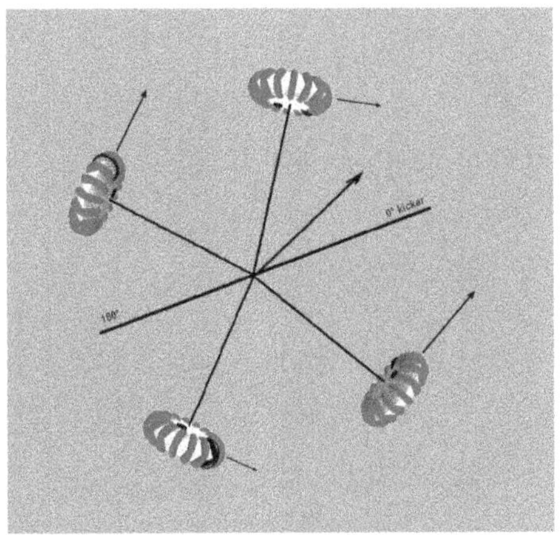

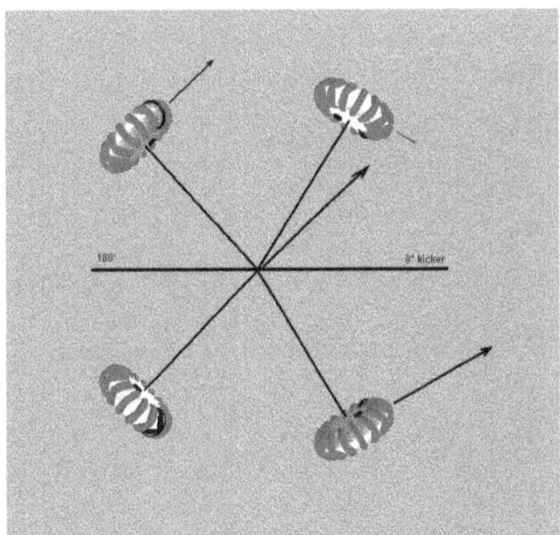

Fig. 5. Change of variable in control vector

Moreover, in order to ensure that the angular acceleration will remain unchanged it is necessary that the resulting scalar force $f = f_1 + f_2 + \ldots + f_n$ is the same in both cases, which translates to

$$\sum_{i=1}^{n} \check{u}_i = \sum_{i=1}^{n} u_i \tag{15}$$

What we need from Eqs. 14 and 15 is a one-to-one mapping that can take us from u to $\check{u}$ back and forth. Since u is n-dimensional and we have three equations the system is under-determined whenever $n > 3$. Although the pseudo-inverse would provide a linear transformation from one domain to the other, this transformation would be rank-deficient and thus would not be one-to-one. Our solution requires then to add $n - 3$ *complementary equations*. In order to avoid unnecessary numerical problems, these equations should be chosen so that the linear transformation is well-conditioned for any value of the angle β.

3.1 Transformation matrix for $n = 3$

The simplest case comes when the number of omnidirectional wheels of the robot is $n = 3$, see Fig. 6. For this particular case the number of equations provided by Eqs. 14 and 15 is equal to the number of control variables, so there is no need for additional equations to complete the transformation.

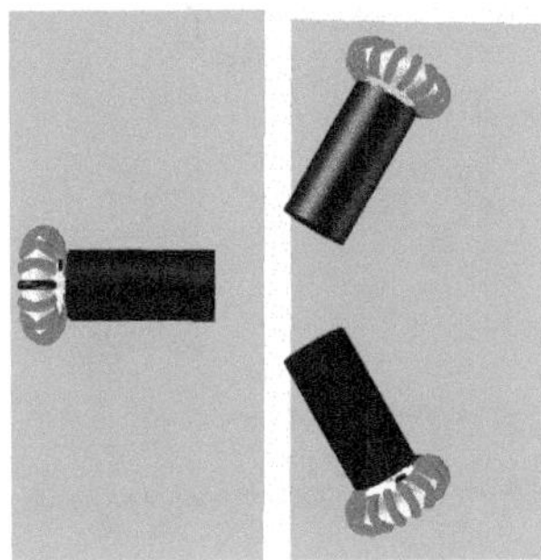

Fig. 6. Omnidirectional drive for $n = 3$

Equations 14 and 15 can be written in matrix form as follows

$$\begin{bmatrix} \sin(\theta_1) & \sin(\theta_2) & \sin(\theta_3) \\ \cos(\theta_1) & \cos(\theta_2) & \cos(\theta_3) \\ 1 & 1 & 1 \end{bmatrix} \check{u} = \begin{bmatrix} \sin(\theta_1+\beta) & \sin(\theta_2+\beta) & \sin(\theta_3+\beta) \\ \cos(\theta_1+\beta) & \cos(\theta_2+\beta) & \cos(\theta_3+\beta) \\ 1 & 1 & 1 \end{bmatrix} u \quad (16)$$

We can then define the transformation matrix $\Omega_3(\beta)$ according to

$$\Omega_3(\beta) = \begin{bmatrix} \sin(\theta_1) & \sin(\theta_2) & \sin(\theta_3) \\ \cos(\theta_1) & \cos(\theta_2) & \cos(\theta_3) \\ 1 & 1 & 1 \end{bmatrix}^{-1} \begin{bmatrix} \sin(\theta_1+\beta) & \sin(\theta_2+\beta) & \sin(\theta_3+\beta) \\ \cos(\theta_1+\beta) & \cos(\theta_2+\beta) & \cos(\theta_3+\beta) \\ 1 & 1 & 1 \end{bmatrix} \quad (17)$$

3.2 Transformation matrix for $n = 4$

For the particular case of $n = 4$ (Fig. 7), it is easy to see that Eq. 18 satisfies the requirement of completing a well-conditioned 4×4 transformation since it is orthogonal to two of the other equations and still sufficiently linearly independent from the third one.

$$\begin{bmatrix} 1 & -1 & 1 & -1 \end{bmatrix} \check{u} = \begin{bmatrix} 1 & -1 & 1 & -1 \end{bmatrix} u \quad (18)$$

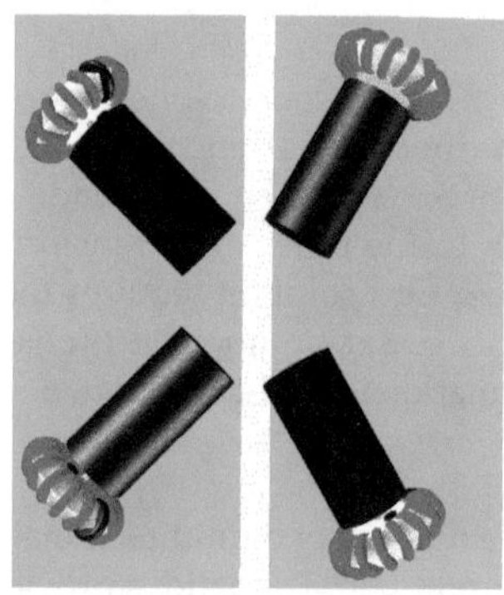

Fig. 7. Omnidirectional drive for $n = 4$

Equations 14, 15 and 18 can be put in matrix form as follows

$$\begin{bmatrix} \sin(\theta_1) & \sin(\theta_2) & \sin(\theta_3) & \sin(\theta_4) \\ \cos(\theta_1) & \cos(\theta_2) & \cos(\theta_3) & \cos(\theta_4) \\ 1 & 1 & 1 & 1 \\ 1 & -1 & 1 & -1 \end{bmatrix} \check{u} =$$

$$\begin{bmatrix} \sin(\theta_1+\beta) & \sin(\theta_2+\beta) & \sin(\theta_3+\beta) & \sin(\theta_4+\beta) \\ \cos(\theta_1+\beta) & \cos(\theta_2+\beta) & \cos(\theta_3+\beta) & \cos(\theta_4+\beta) \\ 1 & 1 & 1 & 1 \\ 1 & -1 & 1 & -1 \end{bmatrix} u \quad (19)$$

We can then define the transformation matrix $\Omega_4(\beta)$ according to

$$\Omega_4(\beta) = \begin{bmatrix} \sin(\theta_1) & \sin(\theta_2) & \sin(\theta_3) & \sin(\theta_4) \\ \cos(\theta_1) & \cos(\theta_2) & \cos(\theta_3) & \cos(\theta_4) \\ 1 & 1 & 1 & 1 \\ 1 & -1 & 1 & -1 \end{bmatrix}^{-1} \begin{bmatrix} \sin(\theta_1+\beta) & \sin(\theta_2+\beta) & \sin(\theta_3+\beta) & \sin(\theta_4+\beta) \\ \cos(\theta_1+\beta) & \cos(\theta_2+\beta) & \cos(\theta_3+\beta) & \cos(\theta_4+\beta) \\ 1 & 1 & 1 & 1 \\ 1 & -1 & 1 & -1 \end{bmatrix} \quad (20)$$

3.3 Transformation matrix for $n = 5$

For the particular case of $n = 5$, shown in Fig. 8, an adequate choice for the complementary equations is shown in Eqs. 21.

$$\begin{aligned} \begin{bmatrix} 1 & -1 & 0 & 1 & -1 \end{bmatrix} \check{u} &= \begin{bmatrix} 1 & -1 & 0 & 1 & -1 \end{bmatrix} u \\ \begin{bmatrix} 1 & -1 & 0 & -1 & 1 \end{bmatrix} \check{u} &= \begin{bmatrix} 1 & -1 & 0 & -1 & 1 \end{bmatrix} u \end{aligned} \quad (21)$$

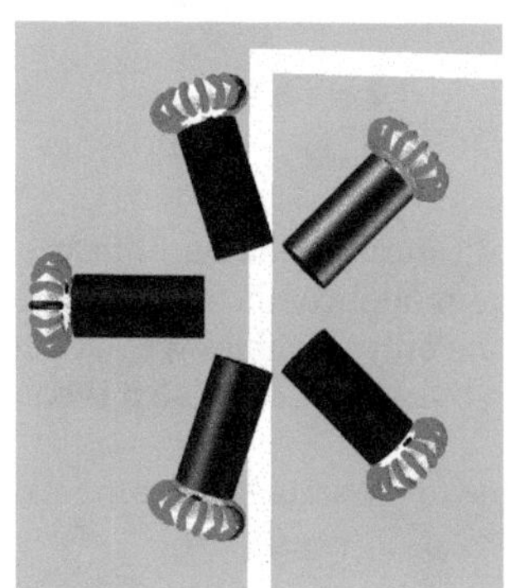

Fig. 8. Omnidirectional drive for $n = 5$

Equations 14, 15 and 21 can be put in matrix form as follows

$$\begin{bmatrix} \sin(\theta_1) & \sin(\theta_2) & \sin(\theta_3) & \sin(\theta_4) & \sin(\theta_5) \\ \cos(\theta_1) & \cos(\theta_2) & \cos(\theta_3) & \cos(\theta_4) & \cos(\theta_5) \\ 1 & 1 & 1 & 1 & 1 \\ 1 & -1 & 0 & 1 & -1 \\ 1 & -1 & 0 & -1 & 1 \end{bmatrix} \check{u} = \begin{bmatrix} \sin(\theta_1+\beta) & \sin(\theta_2+\beta) & \sin(\theta_3+\beta) & \sin(\theta_4+\beta) & \sin(\theta_5+\beta) \\ \cos(\theta_1+\beta) & \cos(\theta_2+\beta) & \cos(\theta_3+\beta) & \cos(\theta_4+\beta) & \cos(\theta_5+\beta) \\ 1 & 1 & 1 & 1 & 1 \\ 1 & -1 & 0 & 1 & -1 \\ 1 & -1 & 0 & -1 & 1 \end{bmatrix} u \quad (22)$$

We can then define the transformation matrix $\Omega_5(\beta)$ according to

$$\Omega_5(\beta) = \begin{bmatrix} \sin(\theta_1) & \sin(\theta_2) & \sin(\theta_3) & \sin(\theta_4) & \sin(\theta_5) \\ \cos(\theta_1) & \cos(\theta_2) & \cos(\theta_3) & \cos(\theta_4) & \cos(\theta_5) \\ 1 & 1 & 1 & 1 & 1 \\ 1 & -1 & 0 & 1 & -1 \\ 1 & -1 & 0 & -1 & 1 \end{bmatrix}^{-1} \begin{bmatrix} \sin(\theta_1+\beta) & \sin(\theta_2+\beta) & \sin(\theta_3+\beta) & \sin(\theta_4+\beta) & \sin(\theta_5+\beta) \\ \cos(\theta_1+\beta) & \cos(\theta_2+\beta) & \cos(\theta_3+\beta) & \cos(\theta_4+\beta) & \cos(\theta_5+\beta) \\ 1 & 1 & 1 & 1 & 1 \\ 1 & -1 & 0 & 1 & -1 \\ 1 & -1 & 0 & -1 & 1 \end{bmatrix} \quad (23)$$

Further generalization is possible by taking into account that the set of $n-3$ complementary equations must be chosen such that the transformation matrix is full rank and of minimum condition number.

3.4 Global linearization using the transformation matrix $\Omega_n(\beta)$

The change of variable from u to $\breve{u}$ for any number of wheels n can now be expressed as

$$\begin{aligned} \breve{u} &= \Omega_n(\beta)u \\ u &= \Omega_n(\beta)^{-1}\breve{u} \end{aligned} \quad (24)$$

Matrix $\Omega_n(\beta)$ has the property of canceling the non-linear effect of β on matrix $B_1(\beta)$ since $B_1(0) = B_1(\beta)\Omega_n(\beta)^{-1}$ hence the model can be linearized at least with respect to the state variables in z_1, which are in fact the only variables we need to control. It is important to note that this is not a local linearization but a global one, so it is equally accurate in all of the control space.

By applying this change of variable to the state-space model of Eq. 13 we obtain the following model

$$\begin{aligned} \begin{bmatrix} \dot{z}_1 \\ \dot{z}_2 \end{bmatrix} &= \begin{bmatrix} A_{11} & 0_{6\times n} \\ A_{21}(\beta) & 0_{n\times n} \end{bmatrix} \begin{bmatrix} z_1 \\ z_2 \end{bmatrix} + \begin{bmatrix} B_1(\beta) \\ 0_{n\times n} \end{bmatrix} \Omega_n(\beta)^{-1}\breve{u} \\ \begin{bmatrix} \dot{z}_1 \\ \dot{z}_2 \end{bmatrix} &= \begin{bmatrix} A_{11} & 0_{6\times n} \\ A_{21}(\beta) & 0_{n\times n} \end{bmatrix} \begin{bmatrix} z_1 \\ z_2 \end{bmatrix} + \begin{bmatrix} B_1(\beta)\Omega_n(\beta)^{-1} \\ 0_{n\times n}\Omega_n(\beta)^{-1} \end{bmatrix} \breve{u} \\ \begin{bmatrix} \dot{z}_1 \\ \dot{z}_2 \end{bmatrix} &= \begin{bmatrix} A_{11} & 0_{6\times n} \\ A_{21}(\beta) & 0_{n\times n} \end{bmatrix} \begin{bmatrix} z_1 \\ z_2 \end{bmatrix} + \begin{bmatrix} B_1(0) \\ 0_{n\times n} \end{bmatrix} \breve{u} \end{aligned} \quad (25)$$

If we separate this model according to the state vector partition proposed in Eq. 5 it is easy to see that the transformed model is linear with respect to state variables in z_1 and control variables $\breve{u}$

$$\begin{aligned} \dot{z}_1 &= A_{11}z_1 + B_1(0)\breve{u} \\ \dot{z}_2 &= A_{21}(\beta)z_1 \end{aligned} \quad (26)$$

From the state-space model in Eq. 26 it is possible to formulate a wide variety of linear state-space controllers for state variables in z_1.

4. Segment-wise optimal trajectory control

Our solution to the trajectory execution control problem requires the specification of the desired trajectory as a sequence of target regions as shown in Fig. 9. Each of the target regions is depicted as a red circle. This figure shows the robot in its initial state near the top left corner. The robot will have to visit each of the target regions in sequence until it reaches the final target which contains the red ball near the bottom right corner. As soon as the center of the robot enters the current target region it will start moving towards the next one. In this way, the trajectory is segmented by the target regions. The desired precision of the trajectory is determined by the size of the next target region. A small target requires higher precision than a larger one. Each target region makes additional specifications for the desired final state of the robot as it reaches the target. These specifications include the desired final scalar speed and the desired final heading direction. This last specification is useful for highly dynamic applications, such as playing soccer, since it allows the robot to rotate as it moves along the trajectory in order to align its manipulating device with the location of the target (a ball in the case of soccer).

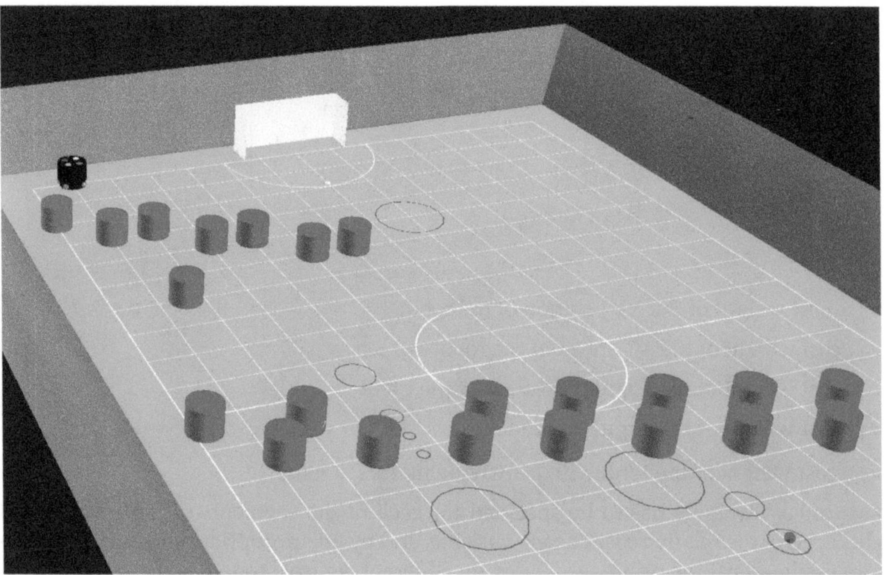

Fig. 9. Sequence of target regions

Each of the segments of the trajectory can be treated as a different control problem in which the final state of the previous segment is the initial state of the current one. Each segment has its own desired final state and in this way the mobile robot is forced to loosely follow the specified trajectory.

Once the trajectory has been segmented by this sequence of target regions the problem becomes how to force the robot to move from one target to the next. Our proposal is to solve an infinite-horizon LQR control problem for each of the segments. Although this approach may provide a sub-optimal solution rather than the optimal solution, that would be obtained from the corresponding finite-horizon LQR problem formulation, it has the advantage of requiring the solution of an algebraic Riccati equation rather than the more computationally demanding

differential Riccati equation. Computational efficiency is of course an important concern in this application since the solution will have to be obtained in real time in a high speed mobile robot environment.
The infinite-horizon LQR control problem consists on finding the state-feedback matrix K such that $\check{u} = -Kz_1$ minimizes the performance index J given by

$$J = \int_0^\infty ({z_1}^{\mathrm{T}} Q z_1 + \check{u}^{\mathrm{T}} R \check{u}) dt \tag{27}$$

taking into account that z_1 and $\check{u}$ are restricted by the dynamics of the mobile robot given by Eq. 26.
The performance index J specifies the total cost of the control strategy, which depends on an integral quadratic measure of the state z_1 and control $\check{u}$. Q and R represent positive definite matrices that give a weighted measure of the cost of each state variable and control variable, respectively.
For simplicity, in our solution Q and R are defined to be diagonal matrices

$$Q = \begin{bmatrix} w_{xy} & 0 & 0 & 0 & 0 & 0 \\ 0 & w_{xy} & 0 & 0 & 0 & 0 \\ 0 & 0 & w_\beta & 0 & 0 & 0 \\ 0 & 0 & 0 & w_v & 0 & 0 \\ 0 & 0 & 0 & 0 & w_v & 0 \\ 0 & 0 & 0 & 0 & 0 & w_{\dot{\beta}} \end{bmatrix} \tag{28}$$

$$R = w_m I_{n \times n}$$

where

- w_{xy} : cost weight of the XY position of the robot
- w_v : cost weight of the XY speed of the robot
- w_β : cost weight of the angular position of the robot
- $w_{\dot{\beta}}$: cost weight of the angular speed of the robot
- w_m : cost weight of the torque of the driving motors

This set of cost weights has to be specified for each of the segments of the trajectory. The weights specify the relative cost of each variable, and by an appropriate choice of their values one can easily adjust the optimality index for different control strategies. For example, if w_m is very large in comparison to the other weights then our strategy will be to save energy, if w_{xy} is large in comparison to the rest then the strategy dictates that the robot should reach the target region as soon as possible without regard to a specific target final speed, angle β or energy consumption. It would make sense to keep w_β very low during most of the trajectory except for those segments in which the robot is approaching the final target region or, in the case of soccer applications, orienting itself to receive a pass from a teammate robot.
The LQR approach is much more natural from the point of view of the designer as compared to other linear state-feedback controller design techniques such as *pole placement*. Although pole placement may help in specifying a desired time-domain transient behavior, the resulting feedback control may turn out to be higher than what the actuators can actually achieve. In LQR, on the other hand, the solution strikes a balance between the transient behavior of the state variables and the energy consumed by the actuators. In the case of LQR the resulting

poles are implicitly determined by the choice of matrices Q and R, rather than being explicitly specified.
Our segment-wise optimal solution to the trajectory execution control problem allows then to formulate control strategies that can easily adapt to the circumstances of the application. So, for example, the robot may be forced to move quickly in segments of the trajectory that do not require precision, and it may be left free to rotate in those segments in which its orientation is not of concern. In this way, its limited energy is consumed only for those objectives that are important at the moment.

5. Experimental results

A benchmark sequence of target regions was used for the purpose of testing our proposed solution. This sequence is shown in Fig. 9, and has a total of nine segments. Several obstacles were placed on the testing field to give more realism to the 3D OpenGL simulation of the resulting trajectory. A four wheel omnidirectional robot was used for this simulation with wheels distributed around its periphery according to Table 1. Wheels 1 and 4 are the frontal wheels. They are intentionally separated wider apart than the rest in order to allow for a manipulating device to be installed between them.

Table 1. Omnidirectional wheels distribution

θ_1	θ_2	θ_3	θ_4
60°	135°	−135°	−60°

The specific optimality-index parameters used for each of the segments of the trajectory are shown in Table 2. Throughout the whole trajectory, parameter w_m is kept at a moderate value in order to conserve energy without degrading the time-domain performance too much. Parameter w_β is kept at a low value for all segments except for the final two. This saves the energy that would otherwise be required to maintain the orientation of the robot towards a specific angle in the first seven segments where that would not be of interest, but then forces the robot to align itself with the location of the final target region in order to capture the object of interest located at the center of such region. Parameter $w_{\dot{\beta}}$ is kept at a low value throughout the whole trajectory. Parameter w_{xy} is set to a high value in segments 2, 3, 4, 5 and 9. The first four of these segments correspond to the part of the trajectory that takes the robot through the narrow opening in the obstacle barrier, where high precision is required in order to avoid a collision. High precision is also desirable in the last segment to ensure a successful capture of the object of interest. Finally, parameter w_v is given a moderate value throughout the first seven segments of the trajectory and then is lowered in the last two segments in order to allow for the energy to be used for the more important objective of orienting the robot towards the final target region. In this analysis what matters is the relative value of the weights. However, in order to give a more precise idea, in this analysis *moderate* means 1, *low* means 0.1 and *high* means 5.
The resulting trajectory is shown in Fig. 10. The continuous black line shows the path of the robot, while the smaller black straight lines show the orientation of the robot at each point. This gives an idea of how the robot rotates as it moves along the trajectory. One surprising result of this simulation experiment is that the robot started the second segment almost in reverse direction and slowly rotated to approach a forward heading motion. The final section of the trajectory shows how the object of interest gets captured by the robot. Figure 11 shows

Par.	1	2	3	4	5	6	7	8	9
w_{xy}	1	5	5	5	5	1	1	1	5
w_v	1	1	1	1	1	1	1	0.1	0.1
w_β	0.1	0.1	0.1	0.1	0.1	0.1	1	5	5
$w_{\dot{\beta}}$	1	0.1	0.1	0.1	0.1	1	0.1	0.1	0.1
w_m	1	1	1	1	1	1	1	1	1

Table 2. Optimality index parameters for each segment

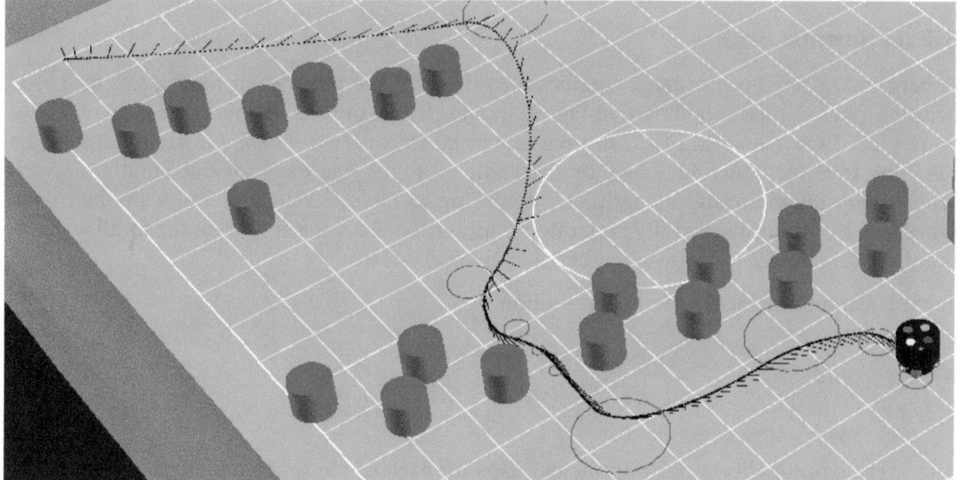

Fig. 10. Simulation of the segment-wise optimal trajectory execution

how the speeds and torques of each driving motor evolve against time. From these graphs we can interpret that there are only two required high energy transients. The first one comes at the initial time, when the robot must accelerate from a motionless state. The second high energy transient comes at the transition from the first to the second segment, where the robot must abruptly change direction while moving at a relatively high speed.

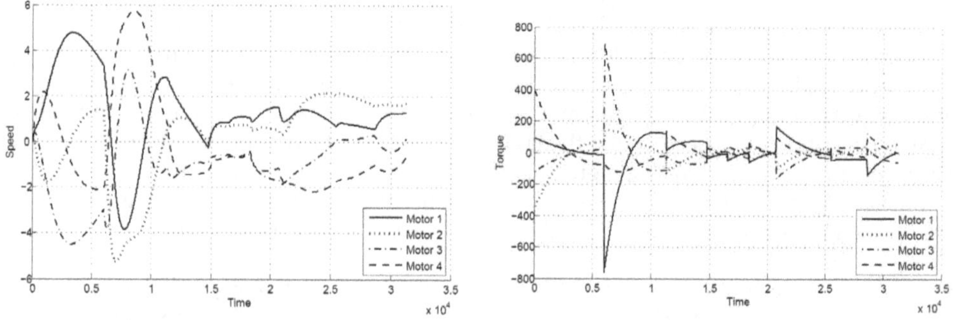

Fig. 11. Speed and torque vs. time for each wheel

A second experiment that we performed to test the segment-wise optimal solution to the trajectory execution problem is that of performing an 8-shaped trajectory, which was successfully executed as shown in Fig. 12. Although the resulting 8 is not perfectly smooth the advantage here is that the path is achieved without the need for a computationally intensive path tracking algorithm but rather only loosely specified by a small set of target regions.

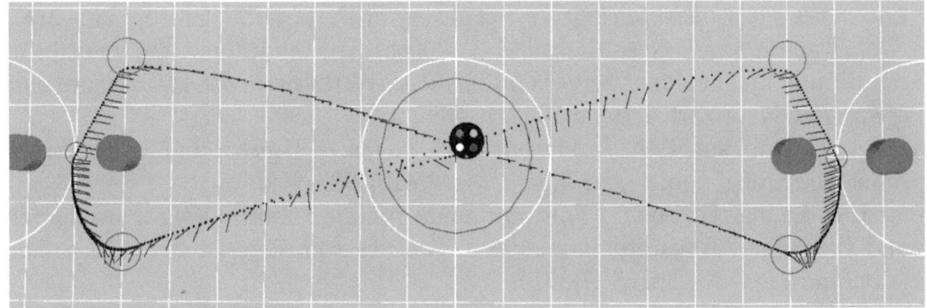

Fig. 12. Simulation of the segment-wise optimal 8-shaped trajectory

6. Conclusion

This work presents the formulation of a complete state-space model for the omnidirectional mobile robot. Although the model is nonlinear with respect to the control vector, it is shown how it can be globally linearized by a change of variables on the control vector. It has been shown how to apply this linearization for the case of $n = 3$, $n = 4$ and $n = 5$ omnidirectional wheels. The distinction among number of wheels is far from trivial since there is the need of introducing *complementary equations* that will make the proposed change of variable invertible.

The model used for analysis in this paper assumes an idealized omnidirectional robot with uniform density. Nonetheless, the same fundamental ideas that were tested under simulation in this work would still apply for a real robot.

There are still two major considerations to take into account before our proposed control method can be applied in the real world. First, our solution assumes availability of all state variables in z_1. However, in reality these state variables would have to be estimated from the available measurements using an approach similar to that presented in (Lupián & Avila, 2008). These measurements would come from the encoders of the driving motors and from the global vision system. The global vision system provides a measurement of variables x, y and β, but this measurement comes with a significant delay due to the computational overhead, so in actuality these variables would have to be predicted (Gloye et al., 2003). The second major consideration is that this solution will be implemented on a digital processor so, rather than a continuous-time model, the problem should be formulated in discrete-time.

The benchmark sequence of target regions used in the "Simulation results" section assumes the environment is static. In a real highly dynamic application the sequence of target regions will have to be updated as the trajectory execution progresses. This dynamic path-planning problem would be the task of a higher-level artificial intelligence algorithm which is left open for future research.

The first of the two simulation experiments can be better appreciated on the 3D OpenGL animation developed for this purpose. The reader can access a video of this animation online at (Lupián, 2009).

7. References

Ashmore, M. & Barnes, N. (2002). Omni-drive robot motion on curved paths: The fastest path between two points is not a straight-line, *AI '02: Proceedings of the 15th Australian Joint Conference on Artificial Intelligence*, Springer-Verlag, London, UK, pp. 225–236.

Balkcom, D. J., Kavathekar, P. A. & Mason, M. T. (2006a). The minimum-time trajectories for an omni-directional vehicle, *Seventh International Workshop on the Algorithmic Foundations of Robotics*, New York City.

Balkcom, D. J., Kavathekar, P. A. & Mason, M. T. (2006b). Time-optimal trajectories for an omni-directional vehicle, *Int. J. Rob. Res.* 25(10): 985–999.

Borenstein, J. & Evans, J. (1997). The omnimate mobile robot: Design, implementation, and experimental results, *International Conference on Robotics and Automation, IEEE*, Albuquerque, NM, pp. 3505–3510.

Gloye, A. & Rojas, R. (2006). *Künstliche Intelligenz*, Springer, chapter Holonomic Control of a Robot with an Omnidirectional Drive.

Gloye, A., Simon, M., Egorova, A., Wiesel, F., Tenchio, O., Schreiber, M., Behnke, S. & Rojas, R. (2003). Predicting away robot control latency, *Technical report B-08-03*, FU-Berlin.

Kalmár-Nagy, T., D'Andrea, R. & Ganguly, P. (2004). Near-optimal dynamic trajectory generation and control of an omnidirectional vehicle, *Robotics and Autonomous Systems* 46: 47–64.

Kalmár-Nagy, T., Ganguly, P. & D'Andrea, R. (2002). Real-time trajectory generation for omnidirectional vehicles, *American Control Conference*, Anchorage, AK, pp. 286–291.

Lupián, L. F. (2009). Simulación de robot F180 en AIR-GL, World Wide Web electronic publication.
URL: *http://www.youtube.com/watch?v=-rBP3VwDPAs*

Lupián, L. F. & Avila, R. (2008). Stabilization of a wheeled inverted pendulum by a continuous-time infinite-horizon LQG optimal controller, *Robotic Symposium, IEEE Latin American*, Bahia, Brazil, pp. 65–70.

Lupián, L. F. & Rabadán-Martin, J. R. (2009). Segment-wise optimal trajectory execution control for four-wheeled omnidirectional mobile robots, *Robotics Symposium (LARS), 2009 6th Latin American*, Valparaíso, Chile, pp. 1 – 6.

Purwin, O. & D'Andrea, R. (2005). Trajectory generation for four wheeled omnidirectional vehicles, *American Control Conference* pp. 4979–4984.

9

Control Architecture Design and Localization for a Gas Cutting Robot

KiSung You, HwangRyol Ryu and Chintae Choi
Research Institute of Industrial Science and Technology
Korea

1. Introduction

Conventional control architecture which has been employed in mobile robot control software is divided into two categories such as knowledge-based and behavior-based control architecture. Early implementation of control architecture was mainly focused on building for sensing the environment, modeling it, planning based on this perceived model and executing the planned action to achieve a certain task. This design approach is called sense-model-plan-act (SMPA) or knowledge-based. A mobile robot making use of a knowledge-based controller tries to achieve its goal by following closely the sense-model-plan-act procedure. SMPA controller also needs initial knowledge, required to model its task environment prior for the robots to executing the planned task. Hence, if the initial knowledge is suited to its working environment, the resulting tasks guarantee success. Although the resulting overall behavior is predictable, the controller often suffers from being slow and becomes complex as it deals with a dynamic environment because most of the controller processing time is consumed in building a model, doing general perception and planning. Therefore, it is suitable controller for robots to require high-level intelligence and work in static and predictable environment.

Brooks proposed a radically different approach in the design of mobile robot control architecture to address the drawback of knowledge-based control architecture. This control architecture functions a horizontal computation scheme so that each behavior is a fixed action pattern with respect to the sensory information. When the mobile robot is confronted by a dynamic environment, a behavior-based robot can react fast because of the direct coupling between its behaviors and the sensed states. The robot controller can be built incrementally, thus making it highly modular and easy to construct. However, this reactive approach still suffers from planning the productive and efficient actions in an unstructured environment because they are only confined to reactions to sensors and the changing states of other modules.

In this chapter, we address the functional safety problems potentially embedded in the control system of the developed mobile robot and introduce a concept of Autonomous Poor Strip Cutting Robot (APSCR) control architecture with a focus on safety in order to design the walkthrough procedure for each behavior. In section 2, we explain the working environment where the robot will be manually or autonomously operated. Section 3 explains the control architecture of APSCR. In section 4, we explain the localization system of APSCR. Finally, section 5 shows some experimental result and in section 6 conclusions will be lastly addressed.

2. Working environment explanation

This section makes an explanation of the robot's working environment in Hot Rolling facility, Gwangyang steel work and describes the challenges and difficulties which the robot shall confront based on the results of CCR (customer critical requirements) and TCR (technical critical requirements) are.

2.1 Robot's working environment

Hot Rolling, used mainly to produce sheet metal or simple cross sections from billets is the most efficient process of primary forming used for the mass production of steel. The first operation of hot rolling process is to heat the stock to the proper deformation temperature. During heating, a scale forms on the stock surface which must be systematically removed and thus descaling is performed mechanically by crushing during a hot forming operation or by spraying with water under high pressure.

Fig. 1. Picture of poor strip produced during hot rolling.

During rolling, deformation of material occurs in between in the forms of rotation, driven rolls. The transporting force during rolling is the friction between rolls and a processed material. Due to this tremendous force applied to the slabs into wrong direction, the poor strips are produced as shown in Figure 1. Due to these poor strips, the whole facility has to be stopped until those poor strips are to be taken away from the roller conveyor by cutting them into the right size and weight so that the heavy-weight crane can hold and carry them out. In order for the workers to cut the poor strips (generally the poor strip is 1.5m x 60m), they have to put on the fire resistant work wear and do the oxygen cutting on the surface of the poor strip whose temperature is over 200 °C.

During inspecting the working environment, we gathered the strong demands and voices about the development of new type of automation which can provide the safety and the cutting operation. We carefully determined the requirements in the respects to the workers or technical barrier. We concluded that the APSCR must be equipped with all the safety requirements and be designed to respond to the severe faults that might be done by operators or malfunctioned by the mechanical faults.

2.2 Technical requirement

We must take into consideration robot's functionalities and safety with respect to all cutting operations because the working environment is extremely bad to the robot for itself. For the purpose of fulfilling all the tasks, we analyzed the survey results in technically critical requirements in the engineer's point of view and took into consideration task achievements in the customer's point of view as well. Below is a list of important information required for the robot design in the customer's point of view.

- Robot must cut a poor strip at single cutting operation, which means not to move back to the points where cutting is done.
- When a flame is not detected due to the malfunction of flame detection sensor or clogged by the heavy dust, robot must stop all the operations.
- Poor strips should be cut within 10 ~15 minutes regardless of the thickness of the poor strips in the width direction.
- Poor strip (approximately 1.5 m x 60 m) cutting operation should be done within 1 hour.
- Robot should be equipped with a backfire protection device.
- We take into consideration repulsiveness of torching at the moment of emitting the gas and oxygen.
- Maximum thickness of the poor strips is 150 mm.
- Electrical spark must be sealed off because gas tank and oxygen tank are loaded on the robot.

We also take the design of the robot into consideration in the engineer's point of views as follows:

- Torch tip should be designed to rotate automatically.
- It should be an automatic ignition system.
- Robot should move along the guide fence installed at the edge of the conveyor stand.
- Backfire protection device should be mounted in the pipe of gas.
- Workers should control a flow of gas and oxygen manually.
- Robot body should sustain a fully stretching arm (full reach: 3.8 m) so that links of arm segment should not hang downwards with no firmness.
- Cutting operation is divided by two steps: pre-heating and blowing. Each step must be accurately controlled by a system or worker.
- For the safety protection against gas explosion, a flame detection sensor should be interrupted by a hardware and software.
- Object detection mechanism, implemented by high sensitive ultrasonic sensor must be considered in order to avoid collision on moving next cutting point.
- Weight balance mechanism should be taken into consideration since robot's total weight is approximately 700 kg.
- Intensity of flame shouldn't be strong enough to cut the bottom plate under a poor strip.

We had collected many opinions and suggestions regarding to robot's functionalities and behaviors. Above lists are a part of surveys, which workers thought of the fundamental tasks and operations which the robot developer should consider.

Fig. 2. APSCR stretches its arm up to 3.8 m.

3. System configuration for APSCR

When we develop industrial robots under an unpredictable environment such as steelworks, there are many critical requirements in the sense that hardware parts would comprise of the supplementary functionalities with a help of software parts. On the contrary, software and hardware are separately developed and combined without considering control architecture. The proposed control architecture mainly makes use of the advantages of conventional reactive control architecture. In this section we focus on the software architecture assuming that the hardware architecture is well designed and describe how the safety walkthrough procedure is sufficiently cooperated within the reactive controller by utilizing hardware and software interrupts as well. Figures 3 through 4 are pictures illustrating the proposed gas cutting robot as mentioned in section 2.

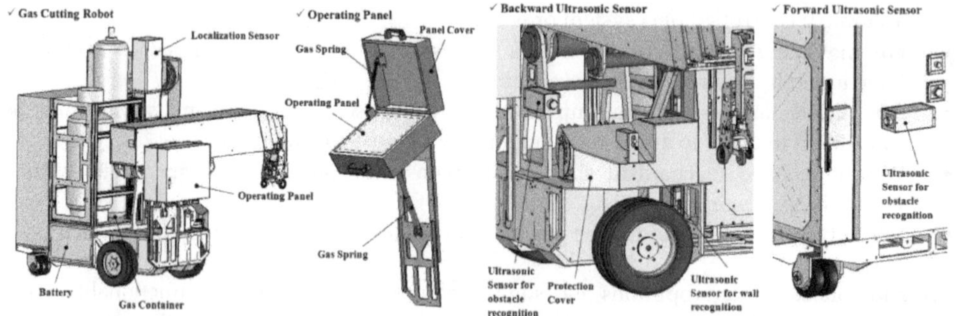

Fig. 3. Layout of a gas cutting robot.

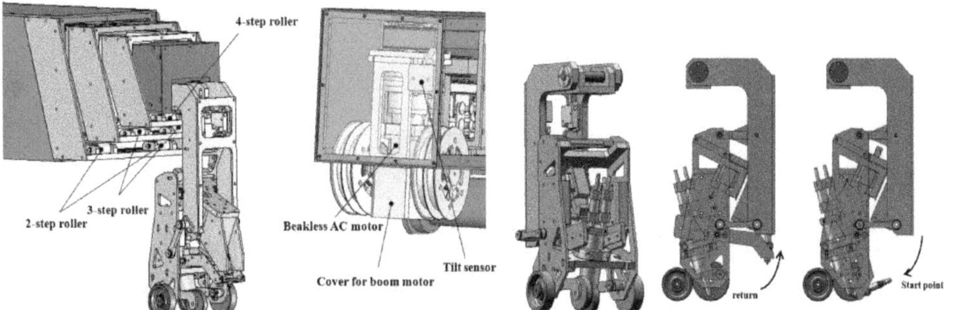

Fig. 4. Structure of boom(robot arm) and touch holder.

3.1 Design of safety interaction for software interrupts

In the context of developing safety-oriented software, safety protection functionalities in control architecture become a key to making a robot function as was designed with a focus on the safety in this project. The safety protection functionalities are embodied within the task-achieving behaviors as follows.

- *BEHAVIOR_SelfWalkThrough*
- *BEHAVIOR_CheckDistance10cm*
- *BEHAVIOR_InitialAlignmentWithLane*
- *BEHAVIOR_MeasureDepthOfPoorStrip*
- *BEHAVIOR_SetStartingPosition4Cutting*
- *BEHAVIOR_StartFiring*
- *BEHAVIOR_SetFinishingPosition4Cutting*
- *BEHAVIOR_MovingNextCuttingPosition*
- *BEHAVIOR_CheckSensorStatus4Navigation*
- *BEHAVIOR_CheckSensorStatus4Cutting*
- *BEHAVIOR_Preheating4Cutting*
- *BEHAVIOR_SettingCuttingParameter*

The task-achieving behaviors, which are described above provide the functional mechanism for APSCR to provide the poor strip cutting procedure autonomously. The following Figure 5 shows how behaviors modules depend on each other and are grouped according to their functions.

To be able to operate the task-achieving behaviors sufficiently as defined in Figure 6, the APSCR starts with checking its system by *BEHAVIOR_SelfWalkThrough* and do the alignment job in order to previously position cutting procedure by calling the following functions *CheckDistance10cm* and *InitialAlignemtnwithStand* in Before Cutting. After initial setup is done, the robot starts to measure the thickness of the poor strip (*MeasureDepthOfPoorStrip*), finds the starting position in which the edge of the poor strip is detected by the roller mounted on the torch holder (*SetStartingPostion4Cutting*), cuts the poor strip with controlling the speed of Boom and Derrick and maintaining the gap between the torch tip and surface of the strip (*StartFiring, Preheating4Cutting, SettingCuttingParameter*), and finally detects the finishing edge of the strip (*SetfinishigPosition4cutting*). After finishing the cutting procedure, while monitoring the alignment between the robot and the poor strip stand and making obstacle avoidance modules running (*CheckSensorStatus4Navigation*), the robot moves to the next cutting position

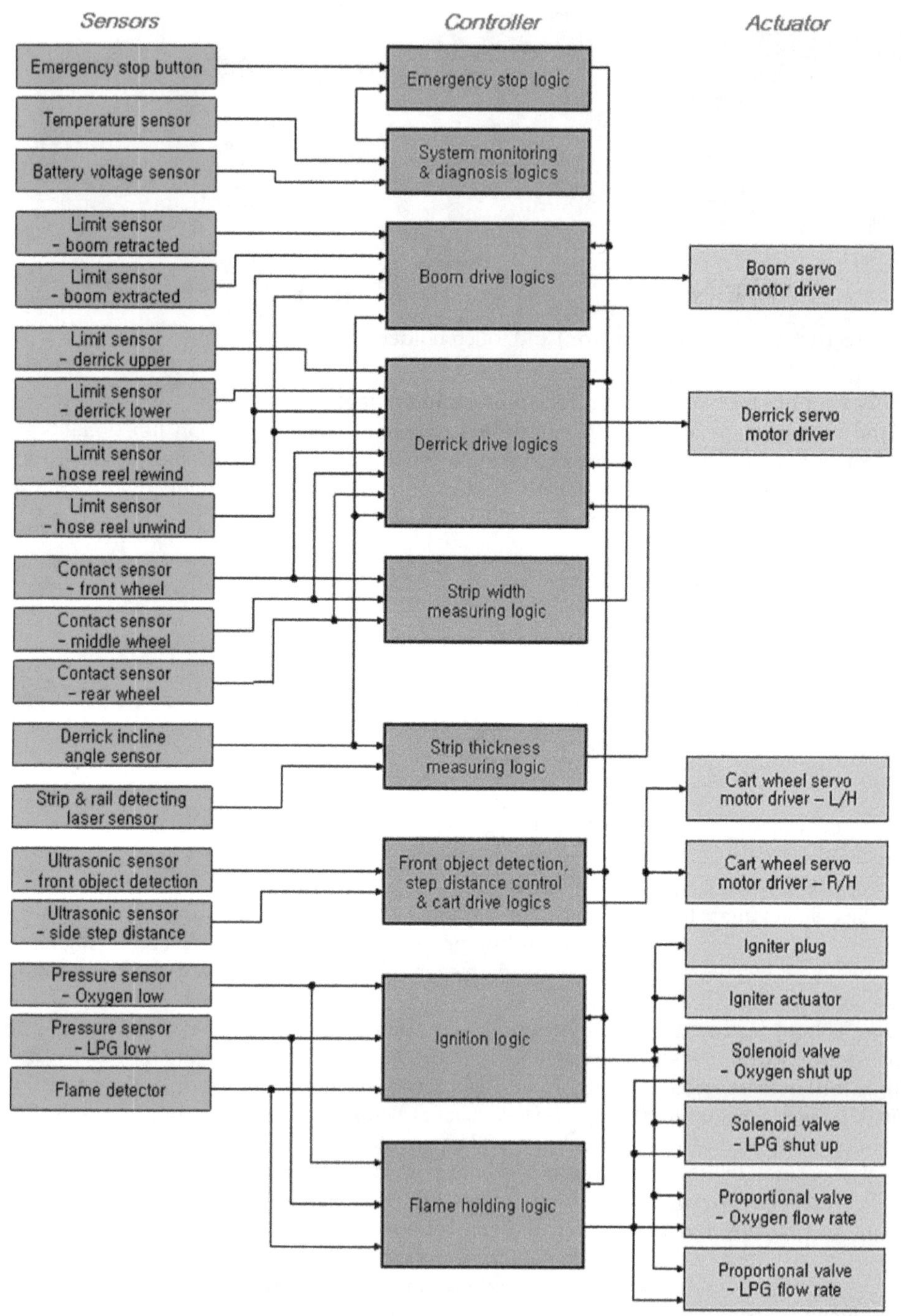

Fig. 5. Interaction between sensors and actuators.

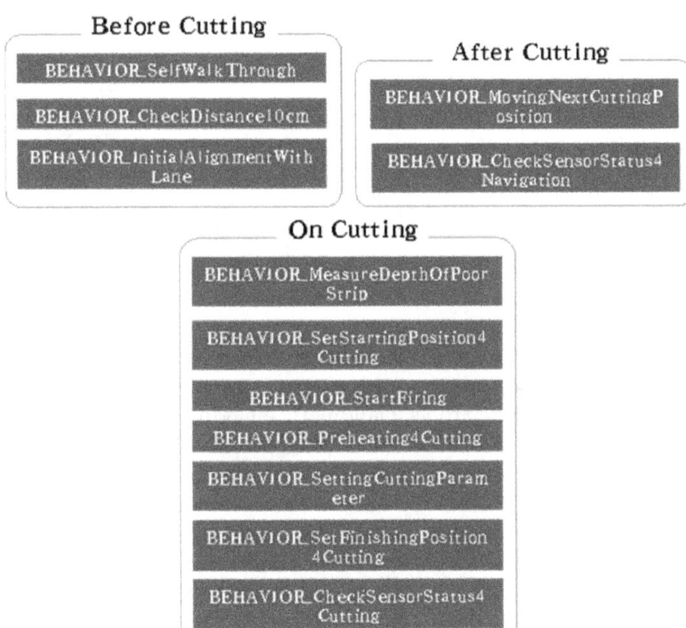

Fig. 6. Autonomous poor strip cutting procedure.

(*MovingNextCuttingPosition*). Each of these task-achieving behaviors comprises of the primitive action modules to support the functional mechanism of the robot and sort out the interactions between the behaviors. Every primitive action module has its own timer or interrupts because each behavior or primitive action should be considerably taken for the safety.

3.2 Design of motion controller

As shown in Figure 6, every primitive action modules are implemented in the manner that each sensor considers a different property and I/O interface. Controller integrates all modules discussed previously using the safety-oriented reactive control architecture. One of the significant advantages for encapsulating the behavior modules is that when programming, we can hide a system level architecture such as sensor calibration, hardware or software interrupts, and so on. More importantly, we do not have to program the detailed safety modules of relevance. The motion controller for a gas cutting robot is composed of a microprocessor and peripheral circuits with a digital input and output (DIO), an analog to digital converter (ADC), a digital to analog converter (DAC), a serial communication circuit, and timers. The microprocessor of the controller is the Atmega128 made in Atmel Co., Ltd.

4. Localization for a gas cutting robot

Localization system for a gas cutting robot is a unique sensor system for indoor localization of industrial mobile robots. It analyzes infrared ray image which is reflected from a passive landmark with an independent ID. The output of position and heading angle of a robot is given with very precise resolution and high speed. It is seldom affected by surroundings such as an infrared ray, a fluorescent light and sunshine. It is composed of an IR Projector part and an image processing unit. It can calculate high resolution and high speed localization of position and heading angle. Also, landmark is used by being attached on ceiling. This localization system doesn't require any device for any synchronization or communication between a robot and a landmark. The area that localization system covers is extended by only adding landmarks to ceiling. Then each section can be distinguished easily each other by using landmarks with different IDs. This system automatically measure and calibrate distance between landmarks and ceiling height. The greatest advantage of this system is that it is nearly not affected in environment such as lamp and sunlight and works excellent localization function at night as well as in the daytime.

4.1 Background art

In general, to control an indoor mobile robot, it is required to recognize a position of the robot. There are two self-localization calculation methods performed by a robot itself using a camera. First, there is a method of using an artificial landmark. a landmark having a certain meaning is installed on a ceiling or a wall, the landmark is photographed by a CMOS image sensor, the landmark is extracted from an image, and coordinates on a screen are allowed to be identical with coordinates of a mobile robot, thereby calculating a self-localization of the mobile robot by itself On the other hand, the landmark is installed on the top of the mobile robot and the CMOS image sensor is installed on the ceiling. Second, there is a method of using a natural landmark. A ceiling is photographed by a camera; information of structures such as lightings installed on the ceiling and straight lines and edges of interfaces between the ceiling and walls is extracted, thereby calculating a self-localization of a mobile robot by itself using the information. However, when using the artificial landmark, the artificial landmark may be affected by lightings and color information of the landmark may be distorted by sunlight. Also, when using the natural landmark, since the natural landmark is much affected by brightness of an ambient light and there is required odometer information or another robot position reader when recording a position of a feature of the landmark, a large memory is required and an additional device is essential. Particularly, when there is no illumination, it is very difficult to use the natural landmark. Accordingly, it is required a new self-localization recognition method of a mobile robot that is capable of be disaffected by lightings and reduces a calculation time of the image processing. Also, when using the two conventional methods described above, since coordinates of a camera and coordinates of a landmark attached to a ceiling calculate position information of a mobile robot while assuming that there is no rotation of in directions excluding the direction of gravity, the robot position information calculated using an image obtained by the camera may have many errors when the mobile robot goes over a small mound or is inclined by an external force or an inertial force of rapid acceleration or deceleration. On the other hand, though there may be an initial correction for an inclination occurring when attaching one of a CMOS or CCD sensors used for a camera device to a robot, the initial correction is merely for an error occurring when it is initially installing in the robot but not for an error caused by an inclination occurring while the robot actually is driving.

An aspect of the present localization system provides a landmark for recognizing a position of a mobile robot, the landmark capable of allowing one of position and area information of the mobile robot to be detected from a landmark image photographed by a camera and being recognized regardless of indoor lightings. Also it provides an apparatus and method for recognizing a position of a mobile robot, in which an image of the landmark is obtained by an infrared camera and the position and area information of a gas cutting robot can be obtained without particular preprocessing the obtained image. This proposed method provides an apparatus and method for recognizing a position of a mobile robot, the apparatus having an inclination correction function to precisely recognizing the position of the mobile robot by correcting position information of an image of a landmark photographed by an infrared camera by measuring inclination information of the mobile robot, such as a roll angle and a pitch angle, by using two axis inclinometer.

4.2 Technical solution

According to aspect of the present localization system, the landmark is used as recognizing coordinates and azimuth information of a mobile robot. The landmark including a position recognition part is formed of a mark in any position and at least two marks on an X axis and Y axis centered on the landmark. The landmark may further include an area recognition part formed of a combination of a plurality of marks to distinguish an individual landmark from others. According to another aspect of the present localization system, there is an apparatus provided for recognizing a position of a mobile robot. The apparatus including: an infrared lighting unit irradiating an infrared ray to a landmark formed of a plurality of marks reflecting the infrared; an infrared camera photographing the landmark and obtaining a binary image; a mark detector labeling a partial image included in the binary image and detecting the mark by using a number and/or dispersion of labeled pixels for each the partial image; and a position detector detecting coordinates and an azimuth of the mobile robot by using centric coordinates of the detected mark. The landmark may include a position recognition part formed of a mark in any position and at least two marks located on an X axis and Y axis centered on the mark. Also, it is provided a method of recognizing a position of a mobile robot having an inclination correction function, the method including: (a) obtaining a binary image by irradiating an infrared ray to a landmark including a position recognition part formed of a mark in any position and at least two marks located on an X axis and Y axis centered on the mark to reflect the infrared ray and photographing the landmark; (b) detecting two-axis inclination information of the mobile robot to the ground and obtaining a binary image again when the detected two-axis inclination information is more than a predetermined threshold; (c) labeling a partial image included in the binary image and detecting the mark by using a number and/or dispersion of labeled pixels for each the partial image; and (d) detecting coordinates and an azimuth of the mobile robot by using centric coordinates of the detected mark. Coordinates and an azimuth of the mobile robot can be detected by correcting the centric coordinates of the detected mark by using a coordinate transformation matrix according to the two axis inclination information. Artificial landmark to enhance reflective gain is shown in Figure 7.

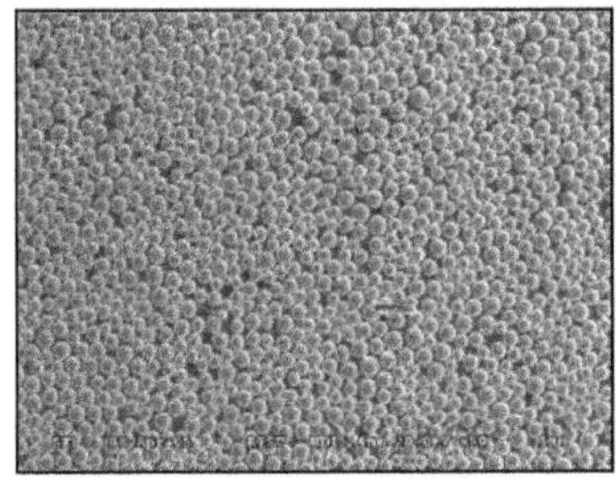

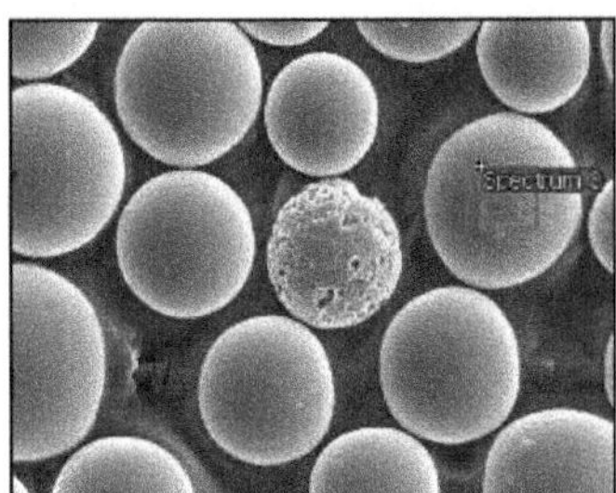

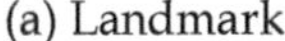

(a) Landmark (b) enlarged landmark

Fig. 7. Retro high gain reflective landmarks

4.3 Camera calibration

A pinhole model generally uses cameral model when describing an image process. Figure 8 is a configuration diagram illustrating the pinhole model. Referring to Figure 8, a point m_r of a point M_c on a three dimensional space corresponds to a point at which a straight line connecting the point M_c to a point C meets with a plane r. In this case, the point C is designated as an optical center, and the plane r is designated as a retinal plane. A straight line passing the point C and vertical to the plane r may exist, which is designated as an optical axis. Generally, the point C is allowed to be an origin point of camera coordinates, and the optical axis is allowed to be identical with Z axis of an orthogonal coordinate system. After the camera model is determined, a structure of the camera may be expressed with various parameters. The parameters may be divided into two kinds of parameters used for describing a camera, intrinsic parameters and extrinsic parameters.

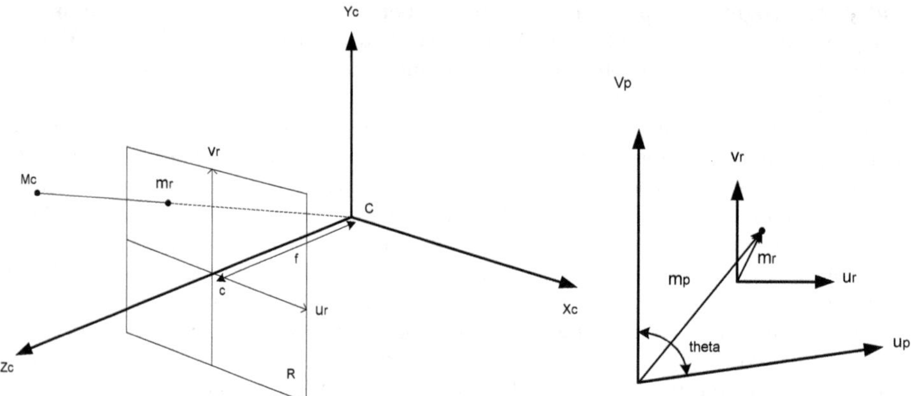

Fig. 8. Camera Model & In case: out of square in CCD array.

The intrinsic parameters describe corresponding relationships between points on the camera coordinates, which is expressed within three dimensional coordinates and the retinal plane with two dimensional coordinates, where the points are projected. The extrinsic parameters describe a transform relationship between the camera coordinates and world coordinates. Hereinafter, the intrinsic parameters will be described.

Referring to Figure 8, it may be known that a relationship between points $M_C = [X_C \; Y_C \; Z_C]^T$ on the camera coordinates and a point $m_r = [u_r \; v_r]^T$ on the corresponding retinal plane is provided as follows.

$$\begin{cases} u_r = f\dfrac{X_C}{Z_C} \\ v_r = f\dfrac{Y_C}{Z_C} \end{cases} \tag{1}$$

wherein f indicates a focal length that is a distance between the optical center C and a point c at which the optical axis meets the retinal plane. The point c indicates a principal point.

A phase formed on the retinal plane is sampled by the CCD or CMOS array, converted into a video signal, outputted from the camera, and stored in a frame buffer. Accordingly, a finally obtained coordinate value is not a coordinate value of the retinal plane but a pixel coordinate value. When pixel coordinates corresponding to m_r is $m_p = [u_p \ v_p]^T$, a transform relationship between the two coordinates is provided as follows.

$$\begin{cases} u_p = k_u u_r + u_0 \\ v_p = k_v v_r + v_0 \end{cases} \tag{2}$$

wherein α and β are values indicating a scale transformation between the two coordinates and u_0 and v_0 are pixel coordinates values of the principal point c.

The relationship given by Eq. (2) is effective when an array of the CCD array is formed by a perfect right angle. However, since it is actually difficult to form a perfect right angle, it is required to obtain a relationship equation considering the difficulty. As shown in Figure 8, when an angle formed by both axes forming the pixel coordinates is designated as θ, there is a relationship between coordinates on the retinal plane and the pixel coordinates, as follows.

$$\begin{cases} u_p = k_u u_r - k_u \cot\theta v_r + u_0 \\ v_p = k_v \csc\theta v_r + v_0 \end{cases} \tag{3}$$

When applying Eq. (1) to Eq. (3), a relationship equation between the three dimensional coordinates on the camera coordinates and the pixel coordinates is finally obtained as follows.

$$\begin{cases} u_p = \alpha_u \dfrac{X_C}{Z_C} - \alpha_u \cot\theta \dfrac{Y_C}{Z_C} + u_0 = \alpha \dfrac{X_C}{Z_C} + \gamma \dfrac{Y_C}{Z_C} + u_0 \\ v_p = \alpha_v \csc\theta \dfrac{Y_C}{Z_C} + v_0 = \beta \dfrac{Y_C}{Z_C} + v_0 \end{cases} \tag{4}$$

As described above, the intrinsic parameters are formed of five such as α, β, γ, u_0, and v_0. Hereinafter, the extrinsic parameters will be described.

Generally, points on a three dimensional space are described in different coordinates from the camera coordinates, the coordinates generally designated as world coordinates. Accordingly, a transformation equation from the world coordinates to the camera coordinates is required, the transformation equation capable of being shown by a displacement vector indicating a relative position between origin points of respective coordinates and a rotation matrix showing a rotation amount of each coordinate axis. When a point shown in world coordinates is $M_w = [X_w \ Y_w \ Z_w]^T$ and is converted into M_c in the camera coordinates, a relationship equation between Mw and Mc is shown as follows.

$$M_C = R(M_w + t) \tag{5}$$

wherein R indicates the rotation matrix and t indicates the displacement vector. Since R includes three independent parameters and t also includes three independent parameters, the number of extrinsic parameters is six. Hereinafter, it will be described to show a camera mode using projective geometry.

A pinhole model of a camera may be linearly shown by using a concept of homogeneous coordinates. When a point on two dimensional pixel coordinates is defined as $m = [u \ v]^T$ and the coordinates on three dimensional world coordinates corresponding to the point, are

defined as $M=[X\,Y\,Z]^T$, homogeneous coordinates formed by adding 1 to the last term of the coordinates become $\tilde{m}=[u\,v\,1]^T$, and $\tilde{M}=[X\,Y\,Z\,1]^T$.
A relationship equation between the three dimensional point M and m that is formed by projecting the point M is expressed using the described pinhole model as follows.

$$s\tilde{m}=A[R\ \ t]\tilde{M} \tag{6}$$

wherein s is a scale factor and R and t are a rotation matrix and a displacement vector, respectively, which are extrinsic parameters. A is a matrix of the intrinsic parameter and is designated as a calibration matrix.

$$A=\begin{bmatrix}\alpha & \gamma & u_0\\ 0 & \beta & v_0\\ 0 & 0 & 1\end{bmatrix} \tag{7}$$

wherein α and β correspond to scale values to u and v axes, γ corresponds to a skewness of two image axes, and u_0 and v_0 are principal points. We use the abbreviation A^{-T} for $\left(A^{-1}\right)^T$ or $\left(A^T\right)^{-1}$.
Homography between the model plane and its image without loss of generality, we assume the model plane is on $Z=0$ of the world coordinate system. Let's denote the ith column of the rotation matrix R by r_i. From Eq. (6), we have

$$s\begin{bmatrix}u\\ v\\ 1\end{bmatrix}=A\begin{bmatrix}r_1 & r_2 & r_3 & t\end{bmatrix}\begin{bmatrix}X\\ Y\\ 0\\ 1\end{bmatrix}=A\begin{bmatrix}r_1 & r_2 & t\end{bmatrix}\begin{bmatrix}X\\ Y\\ 1\end{bmatrix} \tag{8}$$

By abuse of notation, we still use a point on the model plane, but $M=\begin{bmatrix}X & Y\end{bmatrix}^T$ since Z is always equal to 0. In turn, $\tilde{M}=\begin{bmatrix}X & Y & 1\end{bmatrix}^T$. Therefore, a model point M and its image m is related by a homography H :

$$s\tilde{m}=H\tilde{M}\ with\ H=\begin{bmatrix}r_1 & r_2 & t\end{bmatrix} \tag{9}$$

As is clear, the 3×3 matrix is defined up to a scale factor. Given an image of the model plane, an homography can be estimated. Let's denote it by $H=\begin{bmatrix}h_1 & h_2 & h_3\end{bmatrix}$. From Eq. (7), we have

$$\begin{bmatrix}h_1 & h_2 & h_3\end{bmatrix}=\lambda A\begin{bmatrix}r_1 & r_2 & t\end{bmatrix} \tag{10}$$

where λ is an arbitrary scalar. Using the knowledge that r_1 and r_2 are orthonormal, we have

$$h_1^T A^{-T} A^{-1} h_2 = 0 \tag{11}$$

$$h_1^T A^{-T} A^{-1} h_1 = h_2^T A^{-T} A^{-1} h_2 \tag{12}$$

These are the two basic constraints on the intrinsic parameters, given one homography. Because a homography has 8 degrees of freedom and there are 6 extrinsic parameters (3 for

rotation and 3 for translation), we can only obtain 2 constraints on the intrinsic parameters. Note that $A^{-T}A^{-1}$ actually describes the image of the absolute conic(Luong and Faugeras, 1997). We use the closed-form solution in order to solve camera calibration problem among nonlinear optimization methods.
Let

$$B = A^{-T}A^{-1} \equiv \begin{bmatrix} B_{11} & B_{12} & B_{13} \\ B_{21} & B_{22} & B_{23} \\ B_{31} & B_{32} & B_{33} \end{bmatrix}$$
$$= \begin{bmatrix} \frac{1}{\alpha^2} & -\frac{\gamma}{\alpha^2\beta} & \frac{v_0\gamma - u_0\beta}{\alpha^2\beta} \\ -\frac{\gamma}{\alpha^2\beta} & \frac{\gamma^2}{\alpha^2\beta^2}+\frac{1}{\beta^2} & -\frac{\gamma(v_0\gamma - u_0\beta)}{\alpha^2\beta^2}-\frac{v_0}{\beta^2} \\ \frac{v_0\gamma - u_0\beta}{\alpha^2\beta} & -\frac{\gamma(v_0\gamma - u_0\beta)}{\alpha^2\beta^2}-\frac{v_0}{\beta^2} & \frac{(v_0\gamma - u_0\beta)^2}{\alpha^2\beta^2}+\frac{v_0^2}{\beta^2}+1 \end{bmatrix} \tag{13}$$

Note that B is symmetric, defined by a 6D vector

$$b = \begin{bmatrix} B_{11} & B_{12} & B_{22} & B_{13} & B_{23} & B_{33} \end{bmatrix}^T \tag{14}$$

Let the ith column vector of H be $h_i = \begin{bmatrix} h_{i1} & h_{i2} & h_{i3} \end{bmatrix}^T$. Then, we have

$$h_i^T B h_j = v_{ij}^T b \tag{15}$$

with $v_{ij} = [h_{i1}h_{i1} \quad h_{i1}h_{i2}+h_{i2}h_{i1} \quad h_{i2}h_{i2} \quad h_{i3}h_{i1}+h_{i1}h_{i3} \quad h_{i3}h_{i2}+h_{i2}h_{i3} \quad h_{i3}h_{i3}]^T$.
Therefore, the two functional constraints (11) and (12), from a given homography, can be rewritten as 2 homogeneous equations in b:

$$\begin{bmatrix} v_{12}^T \\ (v_{11}-v_{22})^T \end{bmatrix} b = 0 \tag{16}$$

If n images of the model plane are observed, by stacking n such equations as (17) we have

$$Vb = 0 \tag{17}$$

where V is a $2n \times 6$ matrix. If $n \geq 3$, we will have in general a unique solution b defined up to a scale factor. If $n = 2$, we can impose the skewless constraint $\lambda = 0$. If b is solved, we can calculate A, $R = \begin{bmatrix} r_1 & r_2 & r_3 \end{bmatrix}$ and t as follows:

$$v_0 = \left(B_{12}B_{13} - B_{11}B_{23}\right) / \left(B_{11}B_{22} - B_{12}^2\right)$$
$$\lambda = B_{33} - \left[B_{13}^2 + v0\left(B12B13 - B11B23\right)\right] / B11$$
$$\alpha = \sqrt{\lambda / B_{11}}$$
$$\beta = \sqrt{\lambda B_{11} / \left(B_{11}B_{22} - B_{12}^2\right)}$$

$$\begin{aligned}
&\gamma = -B_{12}\alpha^2\beta / \lambda \\
&u0 = \gamma v_0 / \beta - B_{13}\alpha^2 / \lambda \\
&r_1 = sA^{-1}h_1 \\
&r_2 = sA^{-1}h_2 \\
&r_3 = r_1 \times r_2 \\
&t = sA^{-1}h_3 \\
&s = 1 / \left\|A^{-1}h_1\right\| = 1 / \left\|A^{-1}h_2\right\|
\end{aligned} \tag{18}$$

The above solution is obtained through minimizing an algebraic distance which is not physically meaningful. We can refine it through maximum likelihood inference. We are given n images of a model plane and there are m points on the model plane. Assume that the image points are corrupted by independent and identically distributed noise. The maximum likelihood estimation can be obtained by minimizing the following functional:

$$\sum_{i=1}^{n}\sum_{j=1}^{m}\left\|m_{ij} - \hat{m}(A,R_i,t_i,M_j)\right\|^2 \tag{19}$$

where $\hat{m}(A,R_i,t_i,M_j)$ is the projection of point M_j in image i, according to Eq. (7). A rotation R is parameterized by a vector of 3 parameters, denoted by r, which is parallel to the rotation axis and whose magnitude is equal to the rotation angle. R and r are related by the Rodrigues formula (Faugeras, 1993). Minimizing Eq. (20) is a nonlinear minimization problem, which is solved with the Levenberg-Marquardt algorithm (More, 1977). It requires an initial guess of A, $\{R_i \quad t_i | i = 1 \cdots n\}$ which can be obtained using the technique described in the previous subsection. Up to now, we have not considered lens distortion of a camera. However, a desktop camera usually exhibits significant lens distortion, especially radial distortion. Therefore, we only consider the first two terms of radial distortion.

Let $(u, \quad v)$ be the ideal (nonobservable distortion-free) pixel image coordinates, and $(\hat{u}, \quad \hat{v})$ the corresponding real observed image coordinates. The ideal points are the projection of the model points according to the pinhole model. Similarly, $(x, \quad y)$ and $(\hat{x}, \quad \hat{y})$ are the ideal(distortion-free) and real(distorted) normalized image coordinates. When we represent pixels of camera coordinates to $M_C = [XC \quad YC \quad ZC]^T$, normalized pixels is changed as follows:

$$M_n = \begin{bmatrix} X_C / Z_C \\ Y_C / Z_C \end{bmatrix} = \begin{bmatrix} x_n \\ y_n \end{bmatrix}, \tag{20}$$

where x_n, y_n is normalized coordinates vale.

Let $r = x_n^2 + y_n^2$. If M_d is assumed by distorted coordinates,

$$M_d = \begin{bmatrix} x_d \\ y_d \end{bmatrix} = M_n\left(1 + k_1 r + k_2 r^2 + \cdots\right) + t(M_n). \tag{21}$$

Where $M_n\left(1 + k_1 r + k_2 r^2 + \cdots\right)$ is the radial distortion and $t(M_n)$ is the tangential distortion. Finally, $t(M_n)$ can be solved as follows:

$$t(M_n) = \begin{bmatrix} 2p_1 x_n y_n + p_2(r + 2x_n^2) \\ p_1(r + 2y_n^2) + 2p_2 x_n y_n \end{bmatrix}. \tag{22}$$

Experimentally, we found the convergence of the above alternation technique is slow. A natural extension to Eq. (20) is then to estimate the complete set of parameters by minimizing the following functional:

$$\sum_{i=1}^{n} \sum_{j=1}^{m} \left\| m_{ij} - \hat{m}(A, k_1, k_2, p_1, p_2, R_i, t_i, M_j) \right\|^2 , \tag{23}$$

where $\hat{m}(A, k_1, k_2, R_i, t_i, M_j)$ is the projection of point M_j in image i according to Eq. (6), followed by distortion according to Eq. (11) and (12) (Zhang, 1997). This is a nonlinear minimization problem, which is solved with the Levenberg-Marquardt Algorithm (More, 1977). A rotation is again parameterized by a 3 vector r. An initial guess of A and $\{R_i \quad t_i | i = 1 \cdots n\}$ can be obtained using the prescribed technique. An initial guess of k_1 and k_2 can be obtained with the technique described in the last paragraph technique, or simply by setting them to 0.

A vector allowing the coordinates and the azimuth of the mobile robot to be known may be obtained by using a calibration equation, which is disclosed in detail in several theses as follows. A transform relationship of projecting a point on world coordinates to camera pixel coordinates will be described referring to Figure 9.

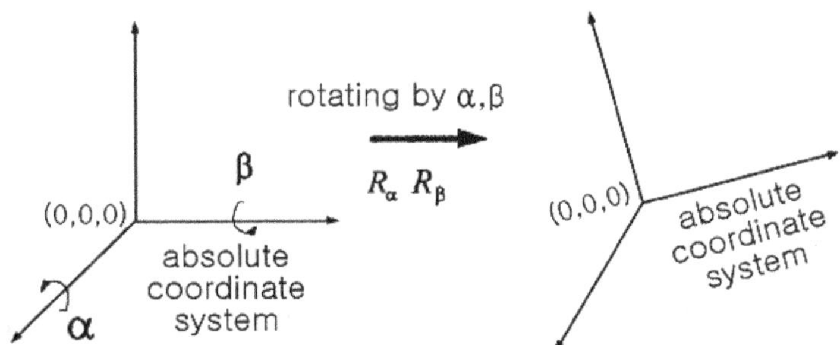

Fig. 9. Diagram illustrating a coordinate system when a camera rotates.

When a roll angle and a pitch angle corresponding to inclination of a camera are α and β, respectively, such a degree of inclination as α and β is expressed in a matrix as follows.

$$R_\alpha = \begin{bmatrix} 1 & 0 & 0 \\ 0 & \cos\alpha & \sin\alpha \\ 0 & -\sin\alpha & \cos\alpha \end{bmatrix} \tag{24}$$

$$R_\beta = \begin{bmatrix} \cos\beta & 0 & \sin\beta \\ 0 & 1 & 0 \\ -\sin\beta & 0 & \cos\beta \end{bmatrix} \tag{25}$$

wherein homogeneous pixel coordinates $\tilde{m}_{\alpha\beta}$ corresponding to a scale parameter are obtained as follows.

$$s\,\tilde{m}_{\alpha\beta} = R_\alpha R_\beta A[R \quad t]\tilde{M} \tag{26}$$

When assuming the displacement vector t to be known, the point M on the world coordinates may be obtained as follows.

$$M = sR^T\left(A^{-1}R_\beta^T R_\alpha^T \tilde{m}_{\alpha\beta} - t\right) \tag{27}$$

When the point M corresponding to a reference point is known, the displacement vector t that is finally to be calculated is obtained as follows, thereby calculating a self-localization of the mobile robot.

$$t = sA^{-1}R_\beta^T R_\alpha^T \tilde{m}_{\alpha\beta} - RM \tag{28}$$

As described above, the vector amount allowing the coordinates and azimuth of the mobile robot to be simultaneously known may be obtained by the vector operation using the three detected marks of the position recognition part and the calibration equation, thereby embodying a microprocessor at a low price.

4.4 Coordinates calculation for mobile robot

The proposed localization system includes a landmark indicating position information such as coordinates and an azimuth of a mobile robot, a position recognition apparatus, and method. A landmark indicating a position of a mobile robot will be described with reference to Figure 10.

According to an embodiment of the proposed localization system, the landmark is attached to a ceiling of a space, in which the mobile robot moves, and is photographed by a camera installed on the mobile robot or attached to a top of the mobile robot and photographed by a camera installed on the ceiling to be used for recognizing the position of the mobile robot. A landmark according to an exemplary embodiment of the present invention includes a position recognition part formed of three marks to recognize essential position information such as coordinates, an azimuth of a mobile robot, and an area recognition part formed of a plurality of marks to distinguish an individual landmark from others to recognize additional area information of the mobile robot. The position recognition part is formed of one mark B in any position and two marks C and A located on an X axis and Y axis, respectively, centered on the mark B. The three marks B, A, and C provide the landmark with a reference point and reference coordinates. Though there is mark shown in Figure 10, the number of the marks is not limited to this and more than two marks may be used. Though the area recognition part is formed of 4×4 marks inside the position recognition part as shown in Figure 10, a position and the number of the marks forming the position recognition part may be varied according to purpose. By giving an ID corresponding to the number and the position of each of the marks forming the area recognition part, each individual landmark may be distinguished from others. As shown in Figure 10, when the area recognition part is formed of the 3×3 marks, IDs of 512 is given. In this case, the position of the mark forming the area recognition part may be determined according to the reference coordinates provided by the position recognition part and each of the IDs may be

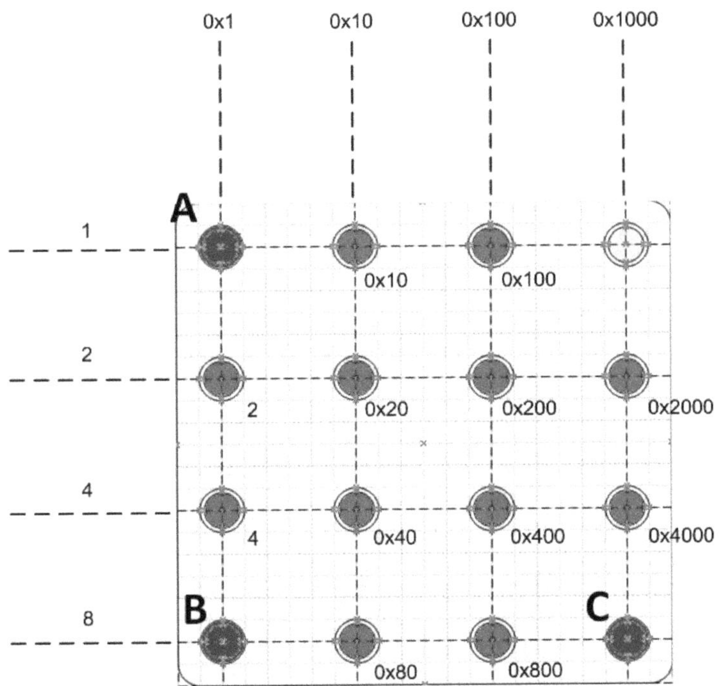

Fig. 10. Design of artificial landmark.

binary coded, thereby quickly recognizing area information of the mobile robot. On the other hand, an infrared reflection coating may be applied or a reflection sheet may be attached to the marks forming the landmark in order to diffusely reflect an infrared ray in a certain wavelength band, particularly, a wavelength band of 800 to 1200 nm. Accordingly, not only in the night but also when there exist reflected lights, only an infrared ray reflected by the mark is detected by an infrared camera, thereby quickly recognizing the position of the mobile robot without using other image processing methods. In this case, the mark may be formed in the shape of only a circle of a predetermined size or may be formed in the protruded shape of one of a circle on a plane and a hemisphere from a plane. The mark formed in the shape of one of the circle and the hemisphere may be used for easily obtaining a number, dispersion, and centric coordinates of pixels when detecting the mark. Though the marks may be formed identically with each other, the marks for the position recognition part are formed different from those for the area recognition part in size and/or color, thereby easily distinguishing the position recognition part from the area recognition part. The mark forming the landmark, described above, may be applied to a conventional mobile robot position recognition apparatus without using an infrared camera, and a use of the marks is not limited to a position recognition apparatus according to an exemplary embodiment of the proposed localization system.

Next, an apparatus and method for recognizing a position of a mobile robot, according to an exemplary embodiment of the proposed localization system, will be described in the order of operations. The embodiment may be applied to when a space in which the mobile robot moves or a space to which a landmark is attached has no bend and is flat. An infrared light emitting diode (LED) irradiates an infra ray to the landmark and an image reflected by the mark forming the position recognition part is photographed by a camera, thereby obtaining a binary image. Namely, the mark in the image obtained by the camera is set up as a bright light close to white and is converted into the binary image by selecting a predetermined threshold brightness value. Considering the camera in detail, the camera includes a plurality of infrared LEDs, an infrared light controller, a CMOS array, and a image processing controller around a wide angle lens. The camera is installed on one of the mobile robots and a ceiling of a space in which the mobile robot moves, to obtain an image of the landmark attached to one of the ceiling, a wall, and a top of the mobile robot. A partial image brightly displayed in the binary image is labeled, and the mark is detected from the number and/or dispersion of the labeled pixel. In this case, labeling indicates a procedure of recognizing an individual image, giving a reference number to the individual image, and making a label list to know a position and a size of the partial image brightly displayed in the binary image. After the labeling, centric coordinates are obtained for each label and the mark is detected from the number and/or dispersion of the labeled pixels. There may be various methods of detecting a mark from a label list. For example, one method may limit the number of pixels forming a label. Namely, since the mark is formed in the shape of a circle and has a uniform size, only a label having a certain number of pixels is selected as a mark candidate and labels having pixels more or less than the certain number are deleted from the label list.

Another method may determine a predetermined dispersion value corresponding to a dispersion index with respect to centric coordinates from the labels and delete labels in which pixels are not clustered from the label list, thereby determining a mark candidate, since the marks are clustered in the shape of a circle. The two methods of detecting a mark from labels, described above, may be used selectively or simultaneously if necessary.

On the other hand, when only the marks of the position recognition part exist in the landmark, three marks may be detected by using the above methods. However, when there are the marks of the area recognition part, whose size is identical with the mark of the position recognition part, only the marks corresponding to the position recognition part may be separately detected from the total marks by performing an additional process as follows. Namely, three labels whose distances from each other are similar and located in the shape of a right angle are detected from the determined mark candidates, thereby detecting the marks of the position recognition part. For example, an inner product of vectors connecting labels is obtained and a label whose inner product value is closest to a largest valid inner product value is detected, thereby detecting only the marks of the position recognition part from the total marks. When indexes of labels corresponding to A, B, and C of Figure 10 are designated by i, j, and k and a largest valid value of an inner product of vectors between the labels is the indexes whose difference of magnitudes is smallest among indexes whose inner product value corresponds to a range is obtained by using Eq. (6).

$$
\begin{aligned}
D(i, j, k) = \left\| \|\vec{ij}\| - \|\vec{kj}\| \right\|, \ \textit{where } \{i, j, k\} &= \arg_{i,j,k} \min D(i, j, k) \\
&\equiv \arg_{i,j,k} \left| \vec{ij} \bullet \vec{kj} \right| \prec \delta_{th}
\end{aligned}
\tag{29}
$$

When an existence and a position of the mark has been recognized by using Eq. (29), an identifier (ID) of the mark may be easily obtained by calculating the position by using a sum of position values and detecting whether the label exists in the position. Position information such as coordinates and an azimuth and area information of the mobile robot are detected by using the detected mark. The ID determined according to the number and position of the marks corresponding to the area recognition part from the detected marks may be quickly obtained, and the area information of the mobile robot may be obtained. In this case, the area information of the mobile robot is allocated to the ID and is an approximate position in which the mobile robot is located. Detailed position information of the mobile robot, such as the coordinates and the azimuth, may be obtained by using centric coordinates of the detected three marks A, B, and C forming the position recognition part. According to an exemplary embodiment of the present invention, coordinates of the mobile robot may be obtained by considering any point obtained from the centric coordinates of each of the three marks A, B, and C shown in Figure 10 as reference coordinates. In this case, the any point may be a center of gravity obtained by the centric coordinates of the three marks. In this case, since the center of gravity is an average of errors with respect to the centric coordinates of the three marks, an error with respect to the coordinates of the mobile robot obtained by using the center of gravity may be reduced. An azimuth of the mobile robot may be obtained based on one direction vector obtained by three centric coordinates, for example, a direction vector obtained by summation of a vector from B to A and a vector from B to C. A vector allowing both the coordinates and the azimuth of the mobile robot to be known may be obtained by using a calibration equation, which is disclosed in detail in several theses as follows (Hartley, 1994, Liebowitz and Zisserman, 1998).

The calibration equation is shown as follows.

$$\begin{aligned} s\tilde{m} &= A\begin{bmatrix} R & t \end{bmatrix}\tilde{M} \\ \Rightarrow \tilde{m} &= \frac{1}{s}A\left(RM + t\right) \end{aligned} \tag{30}$$

where $\tilde{m}_0$ is projected pixel coordinates corresponding to a reference position of the mobile robot, R_0 and t_0 are a rotation matrix and a displacement vector, respectively, s is a scale value, A is a calibration matrix, $\tilde{m}_1$ is pixel coordinates of a position to which the mobile robot rotates and moves, R_1 is a matrix corresponding to a rotation angle amount, and t_1 is a displacement vector, Eq. (31) and (32) may be obtained by using Eq. (30).

$$\tilde{m}_0 = \frac{1}{s}A\left(R_0 M + t_0\right) \tag{31}$$

$$\tilde{m}_1 = \frac{1}{s}A\left(R_0 R_1 M + R_1 t_1\right) \tag{32}$$

A coordinate value M is obtained by using Eq. (31), and the obtained value is assigned to Eq. (32). In Eq. (32), R_1 may be calculated by using a sum of vectors of the recognized marks. Since all values in Eq. (32) excluding t_1 are known, the displacement vector t_1 may be calculated. That is, the displacement vector of the mobile robot may be obtained by using Eq. (32).

$$t_1 = s\, R_1^{-1} A^{-1} \tilde{m}_1 - R_0 M \tag{33}$$

As described above, the vector allowing the coordinates and the azimuth of the mobile robot to be simultaneously known may be obtained by using the detected three marks of the position recognition part and a vector operation using the calibration equation, thereby embodying a microprocessor at a low price. Also, brief area information and detailed position information such as the coordinates and azimuth of the mobile robot may be converted into code information. The code information is transmitted to the mobile robot to perform necessary operations. An apparatus and method of recognizing a position of a mobile robot having an inclination correction function, according to another embodiment of the proposed localization system will be described referring to Figure 11.

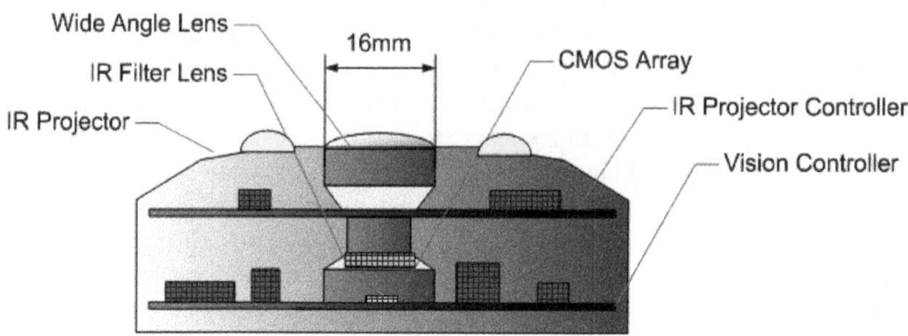

Fig. 11. The layout of proposed localization system.

The apparatus of Figure 11 further includes two axis inclinometers. An infrared light emitting diode (LED) irradiates an infra ray to the landmark and an image reflected by the mark forming the position recognition part is photographed by a camera, thereby obtaining a binary image. Namely, the mark in the image obtained by the camera is set up as a bright light close to white and is converted into the binary image by selecting a predetermined threshold brightness value. Considering the camera in detail, as shown in Figure 11, the camera includes a plurality of infrared LEDs, an infrared light controller, a CMOS array, a vision controller, and two axis inclinometers around a wide angle lens. The camera is installed on the mobile to obtain an image of the landmark attached to one of the ceiling and a wall.

Entire flowchart for localization algorithm is depicted in Figure 12. Position information such as coordinates and an azimuth of the mobile robot is detected by using the detected mark and the two-axis inclination information. The detailed position of the mobile robot, such as the coordinates and the azimuth of the mobile robot, may be obtained by using centric coordinates of the detected three marks A, B, and C forming the position recognition part.

4.5 Evaluation

A mobile robot can identify its own position relative to landmarks, the locations of which are known in advance. The main contribution of this research is that it gives various ways of making the self-localizing error smaller by referring to special landmarks which are developed as high gain reflection material and coded array associations. In order to prove the proposed localization system, we develop the embedded system using TMS320DM640 of Texas Instrument Co., Ltd. And then, the proposed localization system has been tested on mobile robot. The schematic diagram for embedded system is depicted in Figure 13. And embedded system on mobile robot is shown in Figure 14. This localization system is composed of a microprocessor and CMOS image sensor with a digital bus, a serial communication circuit, and infrared LED driver as shown in Figure 14. Calibration system for camera distortion has 3-axis motion controller in order to control gesture of reference

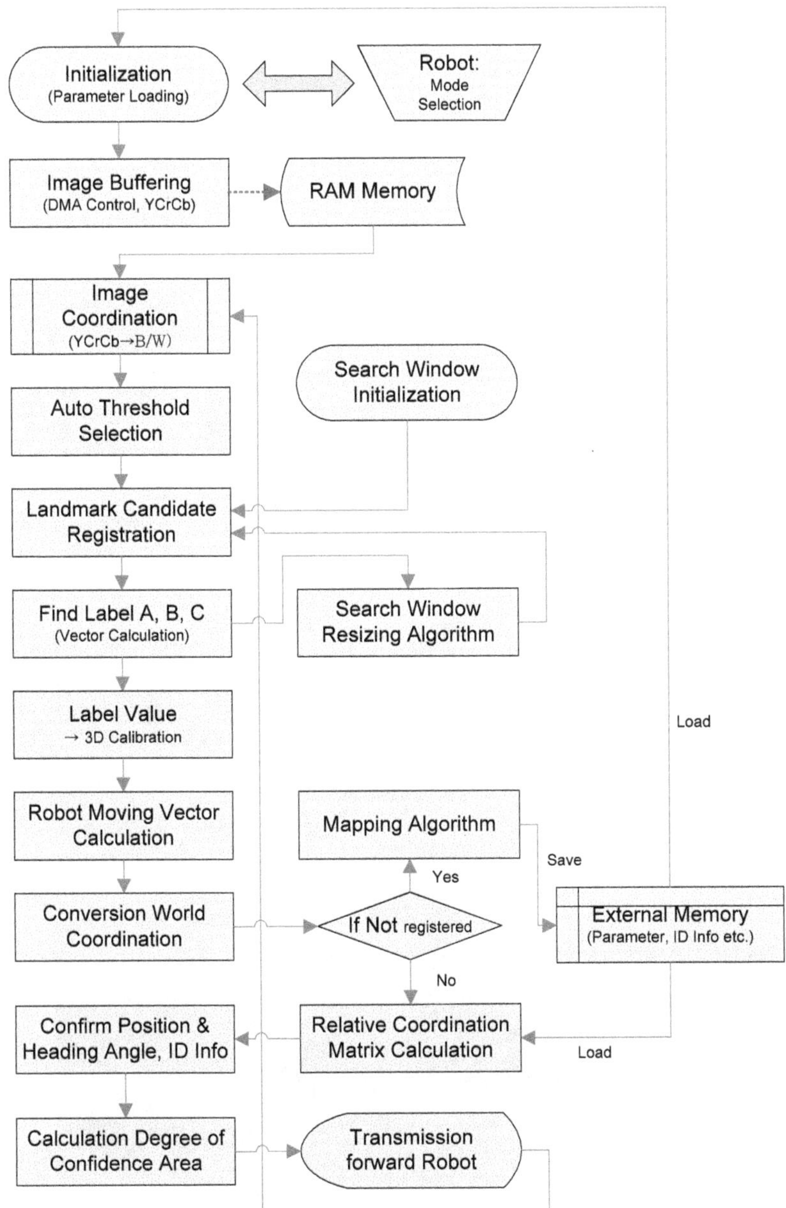

Fig. 12. Flowchart of localization algorithm.

plates. And reference images for camera calibration are show in Figure 15, which these can be acquired by calibration system as shown in Figure 26. In order to find label in reference images automatically, we use the Javis' March algorithm. This is perhaps the most simple-minded algorithm for the convex hull, and yet in some cases it can be very fast. The basic idea is as follows: Start at some extreme point, which is guaranteed to be on the hull. At each step, test each of the points, and find the one which makes the largest right-hand turn. That point has to be the next one on the hull. Because this process marches around the hull in counter-clockwise order, like a ribbon wrapping itself around the points, this algorithm also called the "gift-wrapping" algorithm.

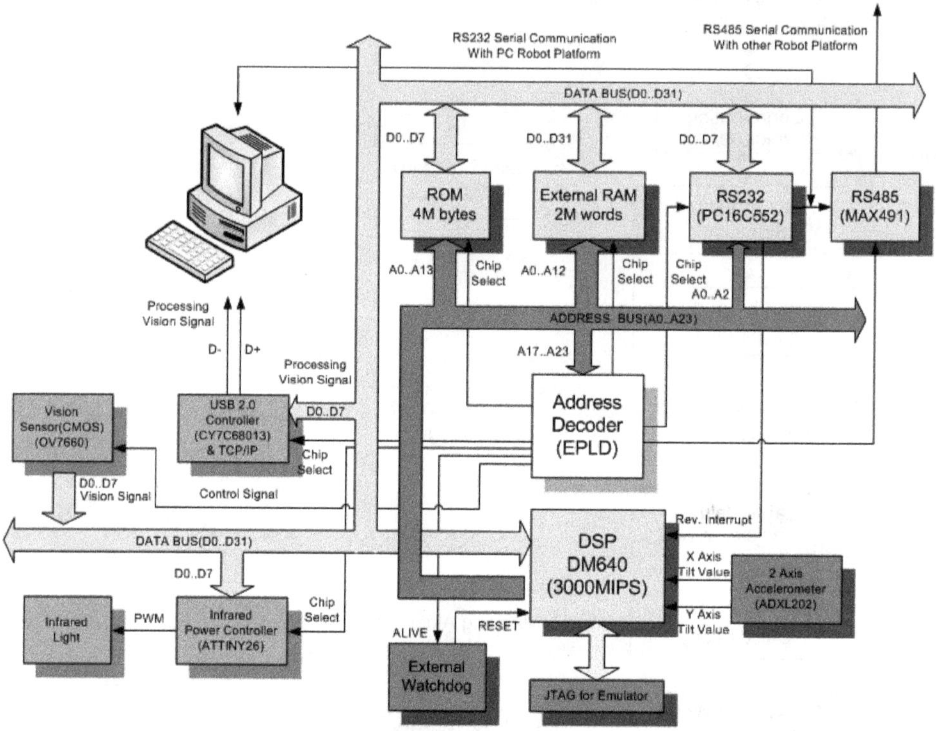

Fig. 13. Schematic diagram for embedded system.

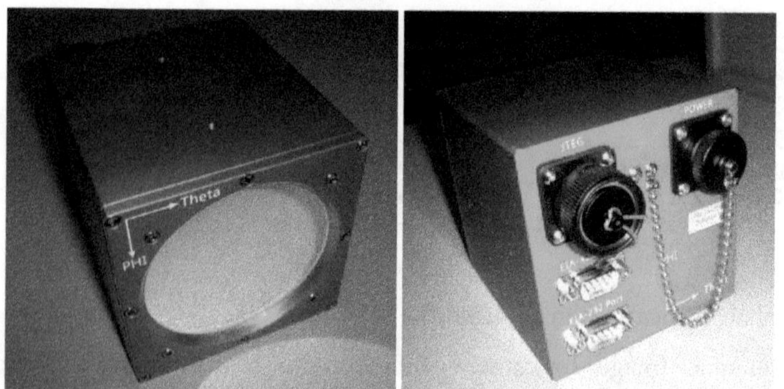

Fig. 14. Embedded system for localization.

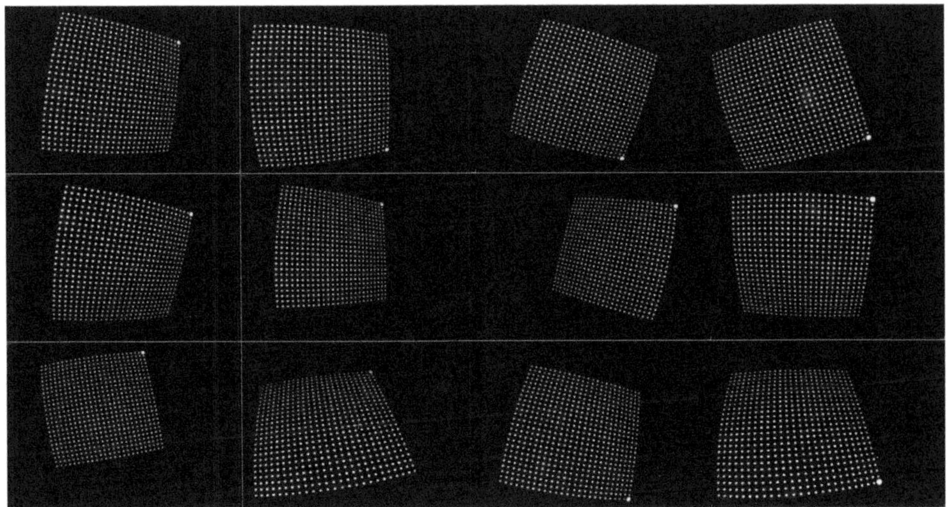

Fig. 15. Reference images for camera calibration.

The accuracy of self-localizing a mobile robot with landmarks based on the indices is evaluated as shown in Figure 17 and a rational way to minimize to reduce the computational cost of selecting the best self-localizing method. The simulation results show a high accuracy and a good performance as depicted in Figure 16. Also, the localization errors for the proposed algorithm are shown in Figures 17(a) and (b), respectively. In results of evaluation, the peak error is less than 3 cm as shown in Figure 17.

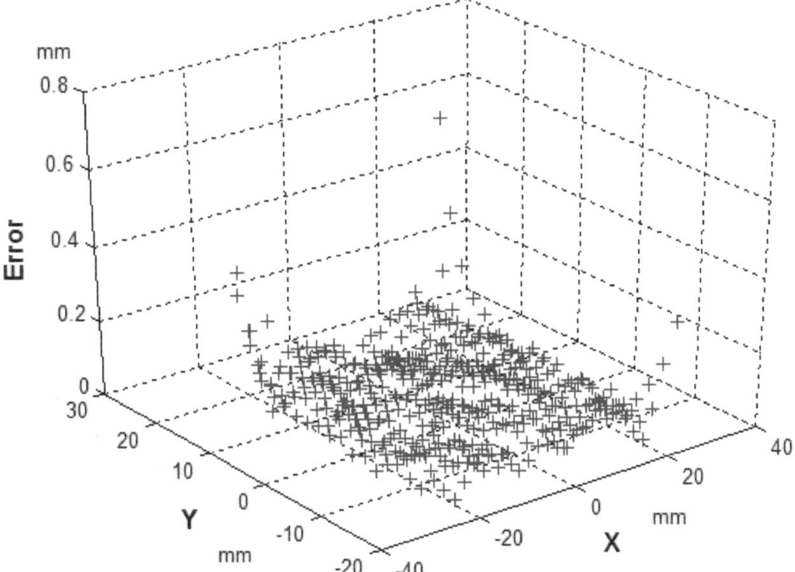

Fig. 16. Calibration errors for camera.

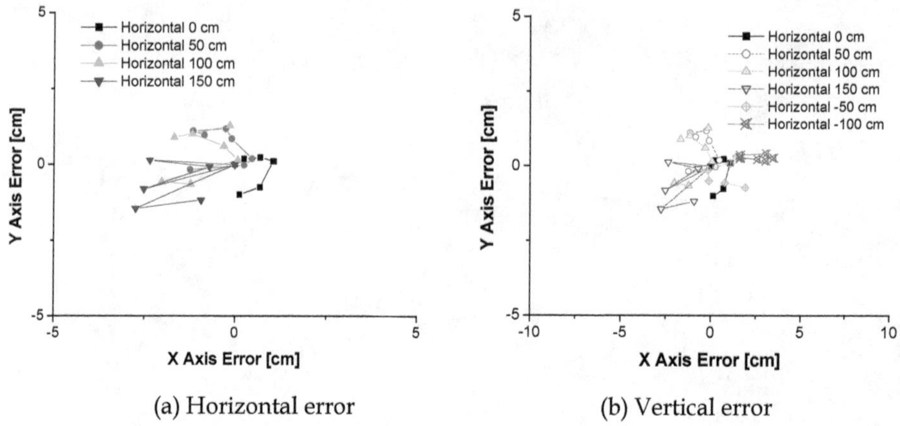

(a) Horizontal error (b) Vertical error

Fig. 17. Localization error.

5. Simulations and experiments

One of the most important cutting procedures taken into consideration is to properly use the cutting torch according the hot slab thickness ranged between 50 mm to 300 mm because choosing the right torch tip plays the major effects on the performance of cut quality and amount of usage for oxygen and gas. For automated oxygen cutting to blow the hot slab cobble into two pieces in the way of oxygen cutting equipped in the end of the robot arm, there are several things to consider as follows:

Step1: way of being lit with spark

Step2: control of flammable gas mixed with oxygen without human intervention

Step3: status of the metal hot enough to be melt down

Setp4: Status of end of blowing out of the melting

In order for the torch to be lit with spark, we installed the stepping motor with 300mm long guided arm equipped with the engine ignition plug used in the car engine. When signal is on, it is rolled out all the way to the bottom of torch tip and sparked until the gas is lit and then begin automatically adjusting the oxygen based on the oxygen insertion table [Table 1] that is previously decided.

Slab Thickness(mm)	Gas Pressure		Oxygen Flow rate
	Oxygen	LPG	
1~10	2.5	0.2~0.4	550
10~30	2.5	0.2~0.4	1,100
30~50	3.0	0.2~0.4	1,900
50~70	3.5	0.2~0.4	2,800
70~100	4.0	0.2~0.4	4,300

Table 1. Oxygen and LPG Insertion Table.

Before blowing out of the way, we must preheat starting area. In order to locate the starting area, we used limit switch mounted inside the torch holder wheel. The decision signal to determine the preheating status depends on the human decision. The most challenging task in the oxygen cutting is to hold the flame with blowing out of the focused area with the proper speed along the way to the end. To determine the proper speed according to the slab thickness, we designed the table that maps between the speed and slab thickness. Figures 18 are the pictures of gas cutting experiments. In summary, the first procedure in the developed oxygen cutting robot for the hot slab cobble is to choose the proper torch tip based on the slab thickness. Secondly, ignition procedure is operated based on the table explained in the Table 1. Thirdly, locating the starting area by using the limit switch inside the torch holder is automatically done. At last, the human inspects the right condition for preheating and send command for "Moving Forward" to the robot.

(a) Step 1: Start (b) Step 2: Ignition (c) Step 3: Cutting

(d) Step 4: Boom control (e) Step 5: Flame control (f) Step 6: Finish

Fig. 18. Experiment for cutting process.

6. Conclusion

In this chapter, we proposed the safety-oriented reactive controller implemented in an innovative and robust industrial mobile robot designed for cutting a poor strip in the steelworks. This mobile robot, called "APSCR", which provides the fully-autonomous operation and wireless controlled operation, is designed to take into consideration the functional safety because it controls the gas and oxygen without human intervention. To able to support the safety guarantee, we encapsulated the robot controller such as behavior and primitive action modules so that the behaviors do not care about how safety protection mechanism works behind the hardware part or software part. In this chapter, we propose a set of indices to evaluate the accuracy of self-localizing methods using the selective reflection landmark and infrared projector, and the indices are derived from the sensitivity enhancement using 3D distortion calibration of camera. And then, the accuracy of self-localizing a mobile robot with landmarks based on the indices is evaluated, and a rational way to reduce the computational cost of selecting the best self-localizing method is proposed. The simulation results show a high accuracy and a good performance. With the preliminary results, we have proved the robustness and reliability of the proposed control architecture. To prove that APSCR is able to use in the steelworks, we still have to perform further experiments with mechanical modification.

7. References

Faugeras, O. (1993). Three-Dimensional Computer Vision: a Geometric Viewpoint, *MIT Press*, 1993.

Faugeras, O., Luong, T., & Maybank, S. (1992). Camera self-calibration: theory and experiments, In G. Sandini, editor, *Proc 2nd ECCV, volume 588 of Lecture Notes in Computer Science,* pages 321-334, Santa Margherita Ligure, Italy, May 1992. Springer-Verlag.

Faugeras, O. , & Toscani, G. (1986). The calibration problem for stereo, *In Proceedings of the IEEE Conference on Computer Vision and Pattern Recognition,* pages 15-20, Miami Beach, FL, June 1986. IEEE.

Hartley, R. (1994). Self-calibration from multiple views with a rotating camera, In J.-O. Eklundh, editor, *Proceedings of the 3rd European Conference on Computer Vision,* volume 800-801 of Lecture Notes in Computer Science, pages 471-478, Stockholm, Sweden, May 1994. Springer-Verlag.

Hartley, R. (1995). In defence of the 8-point algorithm, *In Proceedings of the 5th International Conference on Computer Vision,* pages 1064-1070, Boston, MA, June 1995. IEEE Computer Society Press.

Hartley, R. I. (1994). An algorithm for self calibration from several views, *In Proceedings of the IEEE Conference on Computer Vision and Pattern Recognition,* pages 908-912, Seattle, WA, June 1994. IEEE.

Liebowitz, D. & Zisserman, A. (1998). Metric rectification for perspective images of planes, *In Proceedings of the IEEE Conference on Computer Vision and Pattern Recognition,* pages 482-488, Santa Barbara, California, June 1998. IEEE Computer Society.

Zhang, Z. (1997). Motion and structure from two perspective views: From essential parameters to euclidean motion via fundamental matrix, *Journal of the Optical Society of America A*, 14(11):2938-2950, 1997.

10

Feedback Equivalence and Control of Mobile Robots Through a Scalable FPGA Architecture

G.P. Moustris[1], K.M. Deliparaschos[2] and S.G. Tzafestas[1]

[1]*National Technical University of Athens*
[2]*Cyprus University of Technology*
[1]*Greece*
[2]*Cyprus*

1. Introduction

The control of mobile robots is an intense research field, having produced a substantial volume of research literature in the past three decades. Mobile robots present a challenge in both theoretical control design as well as hardware implementation. From a theoretical point of view, mobile robots exert highly non-linear kinematics and dynamics, non-holonomy and complex operational environment. On the other hand, hardware designs strive for scalability, downsizing, computational power and low cost.

The main control problems in robot motion control can be classified in three general cases; stabilization to a point, trajectory tracking and path following. Point stabilization refers to the stabilization of the robot to a specific configuration in its state space (pose). For example, if the robot at $t = t_0$ is at $p_0 = (x_0, y_0, \theta_0)$, then find a suitable control law that steers it to a goal point p_g=(x_g, y_g, θ_g). Apparently p_0 must be an equilibrium point of the closed loop system, exerting asymptotic stability (although practical stability can also be sought for). In the path following (or path tracking) problem, the robot is presented with a reference path, either feasible or infeasible, and has to follow it by issuing the appropriate control commands. A path is defined as a geometric curve in the robot's application space. The trajectory tracking problem is similar, although there is a notable difference; the trajectory is a path with an associated timing law i.e. the robot has to be on specific points at specific time instances.

These three problems present challenges and difficulties exacerbated by the fact that the robot models are highly non-linear and non-holonomic (although robots that lift the non-holonomic rolling constraint do exist and are called *omni-directional robots*. However the most interesting mobile robots present this constraint). The non-holonomy means that there are constraints in the robot velocities e.g. the non-slipping condition, which forces the robot to move tangentially to its path or equivalently, the robot's heading is always collinear to its velocity vector (this can readily be attested by every driver who expects his car to move at the direction it is heading and not sideways i.e. slip). For a more motivating example of holonomy, consider a prototypical and pedagogical kimenatic model for motion analysis and control; the unicycle robot, described by the equations,

$$\Sigma : \begin{bmatrix} \dot{x} \\ \dot{y} \\ \dot{\theta} \end{bmatrix} = \begin{bmatrix} \cos\theta \\ \sin\theta \\ 0 \end{bmatrix} v + \begin{bmatrix} 0 \\ 0 \\ 1 \end{bmatrix} \omega \tag{1}$$

This model is linear on the inputs and describes a linear combination of vector fields which evolve on a 3D configuration space $M = \mathbb{R}^3 \times S^1$ (a three dimensional manifold). The robot is controlled by two inputs v and ω, expressing the linear and angular velocities respectively. The generalized velocities live on the manifold's tangent space $T_q M$ at each point q, thus the system's equations express the available directions of movement. The non-holonomic no-slipping constraint is expressed by the Pfaffian equation,

$$0 = \dot{x}\sin(\theta) - \dot{y}\cos(\theta) \tag{2}$$

which can be put into the general form,

$$G(q)\dot{q} = 0 \tag{3}$$

Here $q \in M$ is the state vector and $G(q) \in \mathbb{R}^{1\times 3}$ is the constraint matrix (although in this case is just a vector). Each row vector (covector) of G lives in the manifold's cotangent space $T_q^* M$ at q, which is the dual space of $T_q M$. Equation 3 describes restrictions on the available directions of movement. Apparently, since the velocities must satisfy (3), it is evident that they live in the null space of G. One can move from Eq.(3) to Eq.(1) by solving (3) with respect to the velocities $\dot{q}$. Since the system is underdetermined (note that G is a one by three "matrix"), two generalized velocities can vary freely, which are precisely the two inputs of (1). The non-holonomy of the system derives from the fact that Eq.(2) is not integrable i.e. does not express the total derivative of some function. By the Frobenius theorem, if Δ is the distribution spanned by the two vector fields of Σ, the system is holonomic if Δ is involutive under Lie bracketing, a condition that is not satisfied by (1).

Due to these challenges, the path following problem has been attacked by several researchers from many angles, ranging from classical control approaches (Altafini, 1999; Kamga & Rachid, 1997; Kanayama & Fahroo, 1997), to nonlinear control methodologies (Altafini, 2002; Egerstedt et al., 1998; Koh & Cho, 1994; Samson, 1995; Wit et al., 2004) to intelligent control strategies (Abdessemed et al., 2004; Antonelli et al., 2007; Baltes & Otte, 1999; Cao & Hall, 1998; Deliparaschos et al., 2007; El Hajjaji & Bentalba, 2003; Lee et al., 2003; Liu & Lewis, 1994; Maalouf et al., 2006; Moustris & Tzafestas, 2005; Rodriguez-Castano et al., 2000; Sanchez et al., 1997; Yang et al., 1998). Of course, boundaries often blend since various approaches are used simultaneously. Fuzzy logic path trackers have been used by several researchers (Abdessemed et al., 2004; Antonelli et al., 2007; Baltes & Otte, 1999; Cao & Hall, 1998; Deliparaschos et al., 2007; El Hajjaji & Bentalba, 2003; Jiangzhou et al., 1999; Lee et al., 2003; Liu & Lewis, 1994; Moustris & Tzafestas, 2011; 2005; Ollero et al., 1997; Raimondi & Ciancimino, 2008; Rodriguez-Castano et al., 2000; Sanchez et al., 1997) since fuzzy logic provides a more intuitive way for analysing and formulating the control actions, which bypasses most of the mathematical load needed to tackle such a highly nonlinear control problem. Furthermore, the fuzzy controller, which can be less complex in its implementation, is inherently robust to noise and parameter uncertainties.

The implementation of Fuzzy Logic Controllers (FLC) in software suffers from speed limitations due to the sequential program execution and the fact that standard processors do not directly support many fuzzy operations (i.e., minimum or maximum). In an effort to reduce the lack of fuzzy operations several modified architectures of standard processors supporting fuzzy computation exist (Costa et al., 1997; Fortuna et al., 2003; Salapura, 2000). Software solutions running on these devices speed up fuzzy computations by at least one order of magnitude over standard processors, but are still not fast enough for some real-time applications. Thus, a dedicated hardware implementation must be used (Hung, 1995).

Due to the increased number of calculations necessary for the path tracking control, a high performance processing system to efficiently handle the task is required. By using a System-on-a-Chip (SoC) realised on an FPGA device, we utilize the hardware/software re-configurability of the FPGA to satisfy the needs of fuzzy logic path tracking for autonomous robots for high-performance onboard processing and flexible hardware for different tasks.

FPGAs provide several advantages over single processor hardware, on the one hand, and Application Specific Integrated Circuits (ASIC) on the other. FPGA chips are field-upgradable and do not require the time and expense involved with ASIC redesign. Being reconfigurable, FPGA chips are able to keep up with future modifications that might be necessary. They offer a simpler design cycle, re-programmability, and have a faster time-to-market since no fabrication (layout, masks, or other manufacturing steps) time is required, when compared to ASICs.

The use of FPGAs in robotic applications is noted in (Kongmunvattana & Chongstivatana, 1998; Leong & Tsoi, 2005; Li et al., 2003; Reynolds et al., 2001). A review of the application of FPGA's in robotic systems is provided be Leong and Tsoi in (Leong & Tsoi, 2005). A notable case study is the use of FPGA's in the Mars Pathfinder, Mars Surveyor '98, and Mars Surveyor '01 Lander crafts, analysed in (Reynolds et al., 2001).

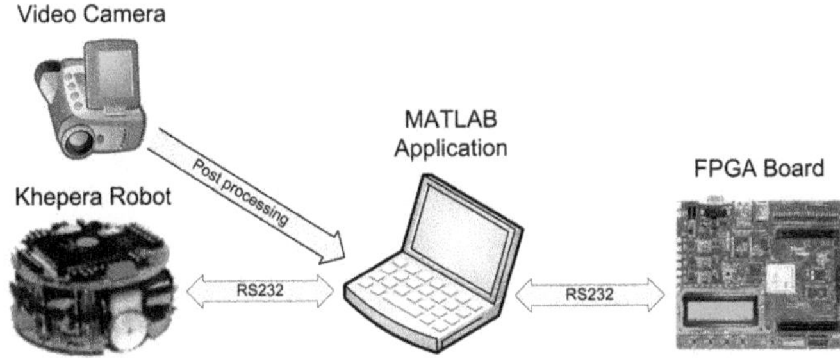

Fig. 1. Overview of the general system.

In this chapter we analyse a SoC implementation for the non-holonomic robot path tracking task using a fuzzy logic controller, along with a non-linear feedback-equivalence transformation which reduces path tracking to straight line tracking. The major components of the SoC are a parametrized Digital Fuzzy Logic Controller (DFLC) soft IP core Deliparaschos et al. (2006) Deliparaschos & Tzafestas (2006), implementing the fuzzy tracking algorithm, and Xilinx's Microblaze soft processor core as the top level flow controller. The system was tied to a differential drive robot and experiments were performed in order to asses the efficacy and performance of the overall control scheme. This was facilitated using an image analysis algorithm, presented in the following sections, which calculated the robot's position from a video stream captured using an overhead camera. The analysis was made off-line. The overall system setup can be seen in Fig 1.

2. Kinematics & odometry of the Khepera II robot

The mobile robot used in this work, is the Khepera II differential drive robot, described by the equations,

$$\begin{bmatrix} \dot{x} \\ \dot{y} \\ \dot{\theta} \end{bmatrix} = \begin{bmatrix} \frac{r}{2}\cos\theta \\ \frac{r}{2}\sin\theta \\ \frac{r}{L} \end{bmatrix} u_r + \begin{bmatrix} \frac{r}{2}\cos\theta \\ \frac{r}{2}\sin\theta \\ -\frac{r}{L} \end{bmatrix} u_l \tag{4}$$

Here, x, y are the coordinates of the middle point of the axis, L is the axis length (the distance between the two wheels), r is the wheel radius and u_l, u_r the angular wheel velocities. A diagram of the model is seen in Fig.(2). Equations 4 can be transformed to a model more akin to the unicycle by first noting that the linear velocity of a point on the wheel's circumference is $v = r\omega$ ($\omega_i \triangleq u_i$). It can be easily shown that the linear velocity of the wheel's center equals the velocity of its circumference. Thus, denoting the centres' velocities as v_r, v_l, then,

$$\begin{aligned} v_r &= r u_r \\ v_l &= r u_l \end{aligned} \tag{5}$$

and substituting them into Eq.4, the system takes the form,

$$\begin{bmatrix} \dot{x} \\ \dot{y} \\ \dot{\theta} \end{bmatrix} = \begin{bmatrix} \cos\theta/2 \\ \sin\theta/2 \\ 1/L \end{bmatrix} v_l + \begin{bmatrix} \cos\theta/2 \\ \sin\theta/2 \\ -1/L \end{bmatrix} v_r \tag{6}$$

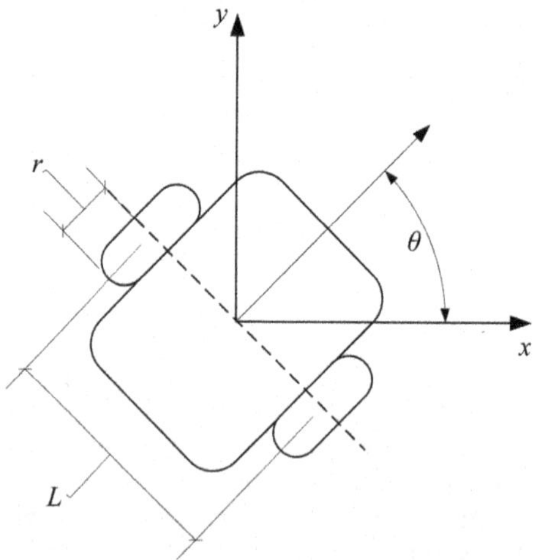

Fig. 2. Depiction of the generalized coordinated for the Differential Drive model

If we further apply a new input transformation,

$$\begin{aligned} u_s &= \frac{v_r + v_l}{2} \\ u_\theta &= \frac{v_r - v_l}{2} \end{aligned} \tag{7}$$

we get the familiar unicycle model, i.e.

$$\begin{bmatrix} \dot{x} \\ \dot{y} \\ \dot{\theta} \end{bmatrix} = \begin{bmatrix} \cos\theta \\ \sin\theta \\ 0 \end{bmatrix} u_s + \begin{bmatrix} 0 \\ 0 \\ 1 \end{bmatrix} u_\theta \tag{8}$$

For the physical interpretation of the inputs u_s, u_θ, consider that the robot performs a turn of radius R with respect to the axis' middle point (point K), centred at the point O (Fig 3). The point O is called the *Instantaneous Centre of Rotation* or ICR. If ω is the robot's angular velocity (actually the angular velocity of K), then its linear velocity is $v_s = \omega R$. It also holds that,

$$\begin{aligned} v_r &= \omega(R + L/2) \\ v_l &= \omega(R - L/2) \end{aligned} \tag{9}$$

By adding (9) together and solving for the linear velocity v_s, we get,

$$v_s = \frac{v_r + v_l}{2} \tag{10}$$

Subtracting (9) we come up with the robot's angular velocity ω,

$$v_\theta = \frac{v_r - v_l}{L} = \omega \tag{11}$$

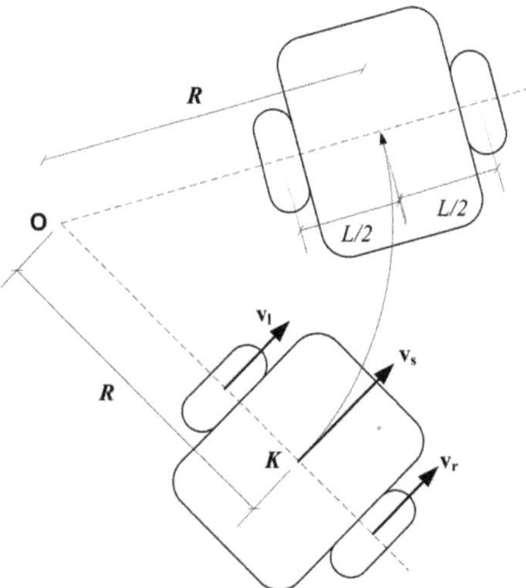

Fig. 3. Depiction of the velocities of the Differential Drive model

Observe that v_s, v_θ are actually u_s, u_θ, thus the new input variables in Eq.7 are actually the robot's linear and angular velocities. What we have proved so far is that the unicycle (or the Dubins Car, which is the unicycle with a constant speed) is related to the Differential Drive by an input transformation, hence they are equivalent. This means that the Differential Drive can *emulate* these models. Consequently, we can develop a controller for either system and apply it to the others by using this transformation (this is feasible if the input transformation is actually *bijective*. If it is not, then the controller can by ported to one direction i.e. from model A to model B, and not the other way around). As mentioned earlier, the Dubins Car is a simplified model where the velocity is considered constant. Furthermore, the path following problem involves the tracking of a purely geometric curve, where indeed the velocity is irrelevant. Hence, the model (8) can be transformed to a more suitable control form by using the curvature $\kappa \triangleq u_\kappa$. The curvature is related to the angular and linear velocities by the well known formula $\omega = \kappa v$, or, using our nomenclature,

$$u_\theta = u_\kappa u_s \tag{12}$$

Since the linear velocity $u_s = v$ is constant, by applying this transformation to the system (8), we get,

$$\begin{bmatrix} \dot{x} \\ \dot{y} \\ \dot{\theta} \end{bmatrix} = \begin{bmatrix} v\cos\theta \\ v\sin\theta \\ 0 \end{bmatrix} + \begin{bmatrix} 0 \\ 0 \\ v \end{bmatrix} u_\kappa \tag{13}$$

This model is of control-affine form, with a non-vanishing drift term, where the only input is the curvature u_κ. By controlling the curvature in this model, we expect to control the actual system i.e. the Khepera robot, which is a differential drive and has two inputs. Thus, starting with the curvature, in order to calculate the true input vector to the robot, we need a second equation. This is of course the velocity equation $u_s = v$, which is considered known. By combining (10), (11), (12), the wheel velocities are calculated as,

$$\begin{aligned} v_r &= v(1 + u_\kappa L/2) \\ v_l &= v(1 - u_\kappa L/2) \end{aligned} \tag{14}$$

Equation 14 produces the linear wheel velocities of the robot, given its linear velocity and curvature. Since the linear velocity is constant, the only input is the curvature which is output by the fuzzy controller implemented on the FPGA.

In order to calculate its position, MATLAB queries the robot about its wheel encoder readings every 12.5 msec. The robot returns the 32bit encoder position, expressed in pulse units, with each unit corresponding to 0.08 mm. Consequently, by multiplying the units with 0.08 we get the *total length* each wheel has travelled since the beginning of the experiment. Now let $S_R(t), S_L(t)$ be the travel length of the right and left wheels at time t, and $S_R(t-1), S_L(t-1)$ be the length at $t-1$. We assume that in the interval Δt the robot moves with a constant curvature, and thus traces an arc. This translates to constant wheel velocities (Eq. 14). Then, using (11) we have,

$$\omega = \frac{\Delta\theta}{\Delta t} = \frac{v_r - v_l}{L} = \frac{\Delta S_R - \Delta S_L}{\Delta t L} \Leftrightarrow \Delta\theta = \frac{\Delta S_R - \Delta S_L}{L} \tag{15}$$

If the robot's initial heading θ_0, with respect to the world frame, is known, then at time t it holds that,

$$\theta(t) = \theta_0 + \sum_{\tau=0}^{t} \Delta\theta(t) = \theta_0 + \sum_{\tau=0}^{t} \frac{\Delta S_R(\tau) - \Delta S_L(\tau)}{L} \tag{16}$$

Using 10, the length travelled by the point K in Δt is found to be,

$$v_s = \frac{\Delta S}{\Delta t} = \frac{v_r + v_l}{2} \Leftrightarrow \Delta S = \frac{\Delta S_R + \Delta S_L}{2} \tag{17}$$

To calculate the robot's x, y position, we must solve the kinematics for the interval Δt, setting the input constant. The solution can be easily shown to be,

$$\begin{aligned} \Delta x(t) &= 2\frac{\Delta S}{\Delta\theta}\sin(\frac{\Delta\theta}{2})\cos(\theta_{t-1}+\frac{\Delta\theta}{2}) \\ \Delta y(t) &= 2\frac{\Delta S}{\Delta\theta}\sin(\frac{\Delta\theta}{2})\sin(\theta_{t-1}+\frac{\Delta\theta}{2}) \end{aligned} \tag{18}$$

In the case that the robot moves in a straight line, hence $\Delta\theta = 0$, taking the limit of 18 gives the equations,

$$\begin{aligned} \Delta x(t) &= \Delta S\cos(\theta_{t-1}+\frac{\Delta\theta}{2}) \\ \Delta y(t) &= \Delta S\sin(\theta_{t-1}+\frac{\Delta\theta}{2}) \end{aligned} \tag{19}$$

To get the absolute coordinates of the robot at t, Eq.19 must be integrated, leading to the odometric equations,

$$\begin{aligned} x(t) &= x_0 + \sum_{\tau=0}^{t}\Delta x(\tau) \\ y(t) &= y_0 + \sum_{\tau=0}^{t}\Delta y(\tau) \end{aligned} \tag{20}$$

Using the previous formulas (Eq.16 and Eq.20) we have succeeded in reconstructing the robot's state vector, i.e. the states (x, y, θ). Note that *time* is absent from the odometry equations. This has been chosen deliberately since it reduces the estimation error significantly. To consider this further, suppose that we get the wheel velocities from the robot and use the odometry equations involving the time Δt. The use of the velocities in the formulas inserts two types of errors; the first is the estimation of the velocities themselves from the robot. In the time between two queries to the robot, which is 12.5 msec, the velocity cannot be computed with adequate precision; the second error derives from the calculation of the interval Δt, which is inserted into the equations. This interval is not constant since there is always a small computational overhead in the software in order to setup and issue the command, communication delays etc. Furthermore, the queries to the robot are implemented in MATLAB using a *timer object*. The timer period however, is not guaranteed and is affected by the processes running on the computer at each instant. Thus, Δt can vary from its nominal value, something which was also seen in actual experiments and must be minimized.

3. Strip-Wise Affine Map

The Strip-Wise Affine Map (SWAM) is the first step towards constructing a feedback equivalence relation which transforms the robot's model to a suitable form, under specific requirements. The equivalence relation however, exerts the useful property of *form invariance* on the mobile robot equations. The SWAM is defined for a robot model *and* a reference path, being applied to tracking control. To begin with, consider a reference polygonal path in the original physical domain, i.e. the actual domain where the robot dwells. Denote this physical domain as D_p (w-plane) and the transformed canonical domain as D_c (z-plane). Then, the strip-wise affine map is a homeomorphism $\Phi : D_c \to D_p$ that sends the real line of the canonical domain to the reference polygonal path in the physical domain. The SWAM is a piecewise linear homeomorphism between the two spaces (Groff, 2003; Gupta & Wenger,

1997) . It acts by inducing a strip decomposition on the planes and then applying an affine map between them. Note that the map acts on the entire domains, not just on a bounded region.

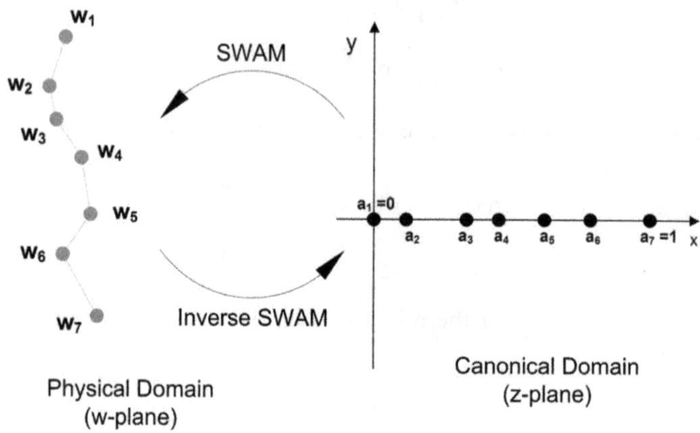

Fig. 4. Illustration of the Strip-Wise Affine Map

In more rigorous terms, let $A = \{w_1, w_2, \ldots, w_n\}, w_i \in C$ be a polygonal chain on the *complex* physical domain. This represents the original reference path. Each vertex w_i of the chain is projected to a point a_i on the real axis in the canonical domain according to its normalized length,

$$a_i = \sum_{k=1}^{i} \frac{S_k}{S}, \text{i} = 1,2,3, \ldots, \text{n} \tag{21}$$

where $S_k = |w_k - w_{k-1}|, S = \sum_{k=1}^{n} S_k$ and $S_1 = 0$. The polygon edge from w_{k-1} to w_k is linearly projected onto $[a_{k-1}, a_k]$. The transformation of $[a_{k-1}, a_k]$ onto its respective polygon edge is done using the function

$$f_{k-1}(x) = w_{k-1} + S \cdot (x - a_{k-1}) \cdot e^{j \cdot \arg(w_k - w_{k-1})} \tag{22}$$

Each interval $[a_{k-1}, a_k]$ is transformed by a respective transformation $f_1, f_2,, f_{n-1}$. Now, consider the following rectangle pulse function on the canonical real axis,

$$\psi_k = \begin{cases} 1, x \in [a_k, a_{k+1}) \\ 0, \text{elsewhere} \end{cases} \tag{23}$$

The pulse is a complex function of $z = x + jy$ in the canonical domain. Each function f_{k-1} is multiplied by the corresponding pulse and the products are summed to account for the general transformation that projects the interval [0,1) onto the polygonal chain,

$$f(x) = \sum_{k=1}^{n-1} f_k(x)\psi_k \tag{24}$$

Extension to the intervals (-∞,0) and [1,+ ∞) can be performed by appending an edge that begins from infinity ending at w_1, for the first case, and another edge starting from w_n escaping to infinity, on the second case. The edges have a direction of $\theta_{-\infty}$ and $\theta_{+\infty}$ respectively. The formulas that account for these branches are given by:

$$f_{-\infty}(x) = (w_1 + S \cdot (x - a_1) \cdot e^{j \cdot \theta_{-\infty}})\psi_0, \\ f_{+\infty}(x) = (w_n + S \cdot (x - a_n) \cdot e^{j \cdot \theta_{+\infty}})\psi_n \tag{25}$$

Here ψ_0 is an open-left pulse with a single falling edge at x=a_1, and ψ_n is the open-right pulse with a single rising edge at x=a_n. Combining the above and using the conventions $\theta_k = arg(w_{k+1} - w_k)$, $a_0 = -\infty$, $a_{n+1} = +\infty$, $w_0 \triangleq$ the point at infinity corresponding to a_0, $w_{n+1} \triangleq$ the point at infinity corresponding to a_{n+1}, $\theta_0 = arg(w_1 - w_0) = \theta_{-\infty}$, $\theta_n = arg(w_{n+1} - w_n) = \theta_{+\infty}$, the extended transformation takes the following form,

$$f(x) = \sum_{k=0}^{n} (w_k + S \cdot (x - a_k) \cdot e^{j \cdot \theta_k})\psi_k \tag{26}$$

where the functions $f_{-\infty}, f_{+\infty}$, correspond to k=0 and k=n respectively. In order to extend this transformation to the entire complex plane, let $z = x + jy$ be a complex variable in the canonical domain and consider the mapping,

$$\Phi(z) = y \cdot S \cdot e^{j \cdot \theta_s} + f(x) \tag{27}$$

where θ_s is the *shifting angle* in [$-\pi/2, \pi/2$].The complex variable $w = u + jv$ in the physical domain, is identified with the transformation $\Phi(z)$, i.e. $w = u + jv = \Phi(z)$. This transformation is the *direct strip-wise affine map* and produces a linear displacement of the polygon along the direction θ_s. Each edge of the polygon produces an affine transformation that applies only in the "strip" that the edge sweeps as it is being translated. Thus, the transformation $\Phi(z)$ can be described as a "strip-wise affine" transformation. The invertibility of the map depends firstly on the geometry of the chain and secondly on the shifting angle. It can be shown (Moustris & Tzafestas, 2008) that necessary and sufficient conditions for the mapping to be invertible are that the chain must be a strictly monotone polygonal chain (Preparata & Supowit, 1981) and the shifting angle must not coincide with the angle of any of the chain's edges i.e. the chain must not be shifted along one of its edges. The inverse strip-wise affine map can be expressed in matrix form by treating $\Phi(z)$ as an $\mathbb{R}^2$ to $\mathbb{R}^2$ mapping since it cannot be solved analytically with respect to z. The inverse mapping equations can be calculated as,

$$\begin{bmatrix} x \\ y \end{bmatrix} = A^{-1} \begin{bmatrix} u \\ v \end{bmatrix} - \begin{bmatrix} C/J - a_k \\ -D/J \end{bmatrix} \tag{28}$$

where $C = S\sum_{k=0}^{n} (w_k^R \sin\theta_s - w_k^I \cos\theta_s)\psi_k$, $D = S\sum_{k=0}^{n} (w_k^R \sin\theta_k - w_k^I \cos\theta_k)\psi_k$ and J is the map Jacobian given by $J = S^2 \sum_{k=0}^{n} \sin(\theta_s - \theta_k)\psi_k$. A^{-1} is the inverse system matrix, i.e.

$$A^{-1} = \begin{bmatrix} \sin\theta_s & -\cos\theta_s \\ -\sum\limits_{k=0}^{n} \psi_k \sin\theta_k & \sum\limits_{k=0}^{n} \psi_k \cos\theta_k \end{bmatrix} / S \sum_{k=0}^{n} \sin(\theta_s - \theta_k)\psi_k \tag{29}$$

Besides Eq.28, one also needs to know the activated pulse since the rectangle pulses are functions of the variable x, and thus (28) does not provide a complete solution to the inversion problem. If this information is provided, the sums in Eq.28 degenerate and the equation provides the inverse system. The activated pulse can be calculated algorithmically by doing

a point-in-strip test. Consider an axis orthogonal to the direction θ_s such that the projection of w_1 corresponds to 0. Furthermore let the projections of each w_i onto that axis be denoted as b_i, and the projection of the current mapping point denoted as b_c. The projections of w_i apparently partition the axis into consecutive line segments $[b_i b_{i+1}]$ which are into one-to-one correspondence with the edges of the polygonal chain. Then, in order to find the current pulse one needs to find the segment into which the point b_c resides. This can be performed optimally by a binary search algorithm in $O(logn)$.
Since the SWAM transforms the robot's application space, it also affects its model's equations. Denoting all the state variables in the physical and canonical domains with a subscript of p and c respectively, then the u-v plane (physical domain) is mapped to the x-y plane (canonical domain), i.e. the state vector $q_p = [x_p, y_p, \theta_p]^T$ is transformed to $q'_p = [x_c, y_c, \theta_p]^T$. The homeomorphism Φ defines an equivalence relation between the two states. Notice that the state θ_p remains unaffected. By introducing a new extended homeomorphism Ψ that also maps the heading angle θ_p, one can send the *canonical state-space* to the *physical state-space*, i.e. $q_p = \Psi(q_c)$. This transformation acts on all state variables and the new system state is $q_c = [x_c, y_c, \theta_c]^T$. The map is then defined by,

$$\begin{bmatrix} x_p \\ y_p \\ \theta_p \end{bmatrix} = \begin{bmatrix} y_c S \cos\theta_s + \mathrm{Re}(f(x_c)) \\ y_c S \sin\theta_s + \mathrm{Im}(f(x_c)) \\ \tan^{-1}\left(\frac{\sum_{\kappa=0}^{n} \sin\theta_\kappa \psi_\kappa + \sin\theta_s \tan\theta_c}{\sum_{\kappa=0}^{n} \cos\theta_\kappa \psi_\kappa + \cos\theta_s \tan\theta_c}\right) \end{bmatrix} = \Psi(q_c) \tag{30}$$

and the new system is,

$$\tilde{\Sigma}: \begin{bmatrix} \dot{x}_c \\ \dot{y}_c \\ \dot{\theta}_c \end{bmatrix} = \begin{bmatrix} v_c \cos\theta_c \\ v_c \sin\theta_c \\ 0 \end{bmatrix} + \begin{bmatrix} 0 \\ 0 \\ S^3 \gamma^3 J^{-1} v_c \end{bmatrix} \kappa_p \tag{31}$$

J is the Jacobian of Φ and $\gamma = \sqrt{1 + \sin 2\theta_c \sum_{\kappa=0}^{n} \cos(\theta_s - \theta_\kappa)\psi_\kappa}$. The input κ_p of the system remains unaffected. However, since it expresses the curvature of the physical system, it can also be transformed under Ψ. Thus by including the transformation of the input and extending the homeomorphism Ψ to $\hat{\Psi} = (\Psi, \Omega)$, where

$$\kappa_p = \Omega(\kappa_c, q_c) = S^{-3} J \gamma^{-3} \kappa_c$$

is the input map that sends the controls from the canonical input space to the controls in the physical input space, one gets the new extended system,

$$\Sigma_c: \begin{bmatrix} \dot{x}_c \\ \dot{y}_c \\ \dot{\theta}_c \end{bmatrix} = \begin{bmatrix} v_c \cos\theta_c \\ v_c \sin\theta_c \\ 0 \end{bmatrix} + \begin{bmatrix} 0 \\ 0 \\ v_c \end{bmatrix} \kappa_c \tag{32}$$

The systems Σ_p and Σ_c are feedback-equivalent (Gardner & Shadwick, 1987; 1990) since they are related by a state and input transformation. The input transformation is actually a feedback transformation of Σ_c that feeds the states q_c back to the input. A more careful look at Eq.(32) shows that it expresses the Dubins Car in the canonical domain, thus $\hat{\Psi}$ presents a kind of *form invariance* on the model.

Now, let p_{ref} be a reference path in the physical domain and let $u_c(q_c, I_c, t)$ be a straight line tracking controller for Σ_c, where I_c denotes the line segment $y_c = \{0/x_c \in [0,1]\}$ of the C-plane, i.e. the reference path in the canonical domain. This controller is transformed to a path tracking controller for Σ_p under the equation,

$$u_p(q_p, p_{ref}, t) = \Omega(u_c(q_c, I_c, t), q_c) = \Omega(u_c(\Psi^{-1}(q_p), \Phi^{-1}(p_{ref}), t), \Psi^{-1}(q_p)) \tag{33}$$

However, since Σ_c is the Dubins Car in the canonical domain, the straight line tracking controller $u_c(q_c, I_c, t)$ for Σ_c is actually a straight line tracker for the Dubins Car, and in order to build a path tracker for the Dubins Car, one has but to build a straight line tracker and use Eq.(33) to promote it to a path tracker for strictly monotone polygonal chains. Furthermore, one could also use existing path trackers for the Dubins Car to track the straight line in the canonical domain. In this case, these controllers can be simplified since in general, straight line tracking is simpler than path tracking.

4. Fuzzy Logic Controller

The controller used in this work for the path following task, is based on a Fuzzy Logic Controller (FLC) developed by the authors, which has been deployed in previous experiments. The original Fuzzy Logic tracker is described in (Moustris & Tzafestas, 2005), and further modified in (Deliparaschos et al., 2007) in order to be implemented on a FPGA. Specifically the tracker is a zero-order Takagi-Sugeno FLC with the two inputs partitioned in nine triangular membership functions each, while the output is partitioned in five singletons (Fig. 5).

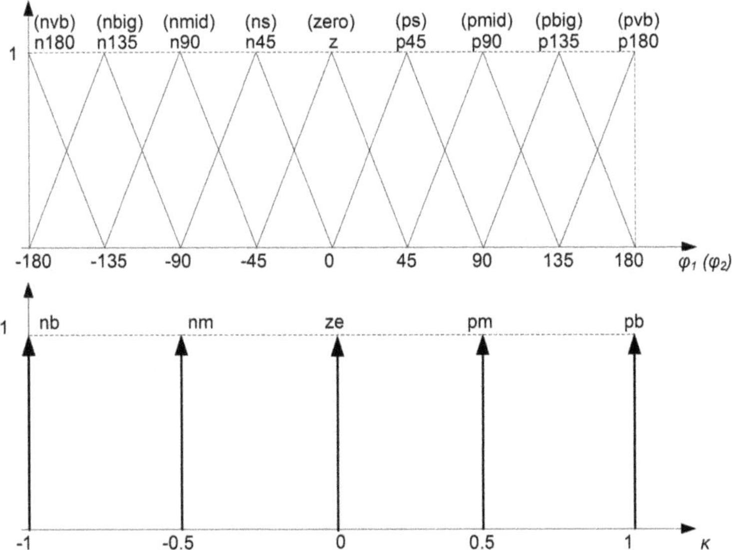

Fig. 5. Input and output membership functions of the fuzzy controller

The FL tracker uses two angles as inputs, and outputs the control input u_K. It is assumed that the path is provided as a sequence of points on $\mathbb{R}^2$. The fuzzy rule base consists of 81 rules, which are presented in Table 1. The implication operator is the min operator.

	pvb	pbig	pmid	ps	zero	ns	nmid	nbig	nvb
p180	pb	pb	pb	pm	ze	nm	nb	nb	pb
p135	pb	pb	pb	pb	pm	nm	nb	Pb	pb
p90	pb	pb	pb	pm	pm	pm	pb	pb	Pb
p45	pb	pb	pb	pm	pm	ze	nb	pb	pb
z	pb	pb	Pb	pm	ze	nm	nb	nb	nb
n45	nb	nb	pb	ze	nm	nb	nb	nb	nb
n90	nb	nb	nb	nm	nm	nb	nb	nb	nb
n135	nb	nb	pb	pm	nm	nb	nb	nb	nb
n180	pb	pb	pb	pm	zero	nm	nb	nb	pb

Table 1. FLC Rule Base

In each control loop the closest path point is picked up and the two input angles are calculated. These angles are the angle ϕ_1 of the closest point with respect to the current robot heading and the direction ϕ_2 of the tangent of the path at that point, as depicted in Fig. 6a. Using the SWAM, we can move the tracking task to the canonical domain where the path is a straight line. In this case, the oriented straight line splits the plane into two half-planes, which present two general cases for the placement of the robot. Due to the symmetry of the cases only one will be analyzed. Consider that the robot resides in the positive half-plane (Fig. 6b) and that the distance from the closest path point P is D.

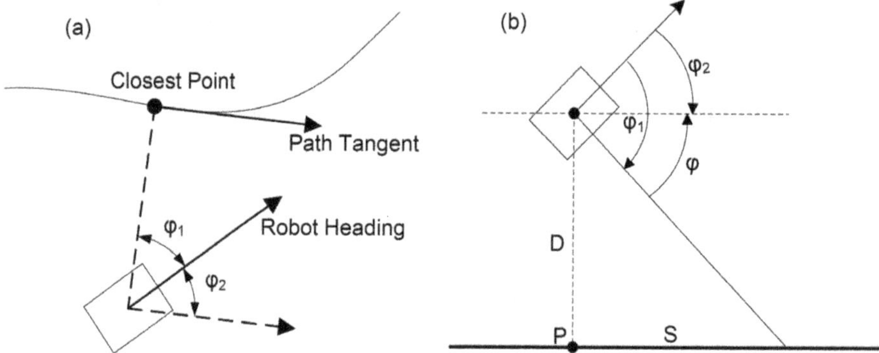

Fig. 6. Illustration of the controller inputs for path tracking in the general case (a) and in the straight line case (b).

Furthermore, one can consider that the robot tracks not the closest point P but the one found distance S ahead. The "sightline" of the robot to that point forms an angle ϕ with the path. In this situation, by varying the angle ϕ_2 one can discern four cases for the relation between the angles ϕ, ϕ_1 and ϕ_2 with the three of them being the same, namely,

$$\begin{aligned} \varphi_1 - \varphi_2 &= -\varphi \,, \varphi_2 \in [-\pi + \varphi, 0] \cup [0, \varphi] \cup [\varphi, \pi] \\ \varphi_1 - \varphi_2 &= 2\pi - \varphi \,, \varphi_2 \in [-\pi, -\pi + \varphi] \end{aligned} \tag{34}$$

When the robot resides in the positive half-plane, the angle ϕ is also positive. On the contrary, when it is in the negative half-plane, the angle changes sign although Eqs.(34) remain the same. With respect to the point being tracked, we discern two cases; either fixing the sightline, i.e. fixing the angle ϕ, or fixing the look-ahead distance S, i.e. tracking the point that is distance

S ahead of the one nearest to the robot (point P). Of course, the nearest point P can be easily found since its coordinates are $(x_c, 0)$, where x_c is the x-axis coordinate of the robot in the canonical space. In the case of a constant S, the angle ϕ varies from $[-\pi/2, \pi/2]$ and the tuple (ϕ_1, ϕ_2) is constrained in a strip. This can also lead to a rule reduction in the FLC rule base since some rules are never activated ((Moustris & Tzafestas, 2011)). For the control of the robot, the FLC and the SWAM were calculated on-line using a dedicated FPGA SoC, as described in the next section.

5. DFLC & SoC architecture

This section discusses the System on Chip (SoC) implemented on an FPGA chip for the robot path tracking task using fuzzy logic. The SoC design was implemented on the Spartan-3 MB development kit (DS-KIT-3SMB1500) by Digilent Inc. The Spartan-3 MB system board utilizes the 1.5 million-gate Xilinx Spartan-3 device (XC3S1500-4FG676) in the 676-pin fine-grid array package. A high level and a detailed architecture view of the SoC is shown in Fig.7 and 8 respectively.

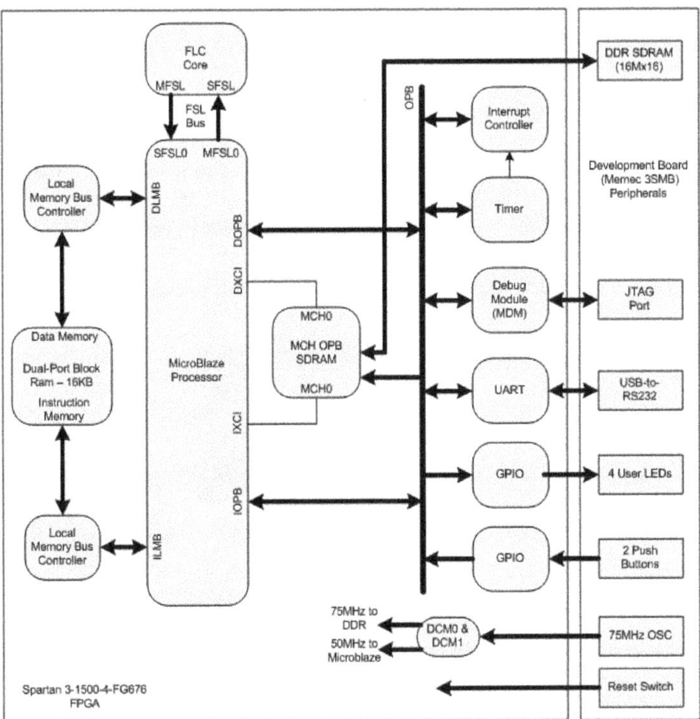

Fig. 7. Overview of the SoC's hardware architecture

A design on an FPGA could be thought as a "hard" implementation of program execution. The Processor based systems often involve several layers of abstraction to help schedule tasks and share resources among multiple processes. The driver layer controls hardware resources and the operating system manages memory and processor bandwidth. Any given processor core can execute only one instruction at a time, and processor based systems are

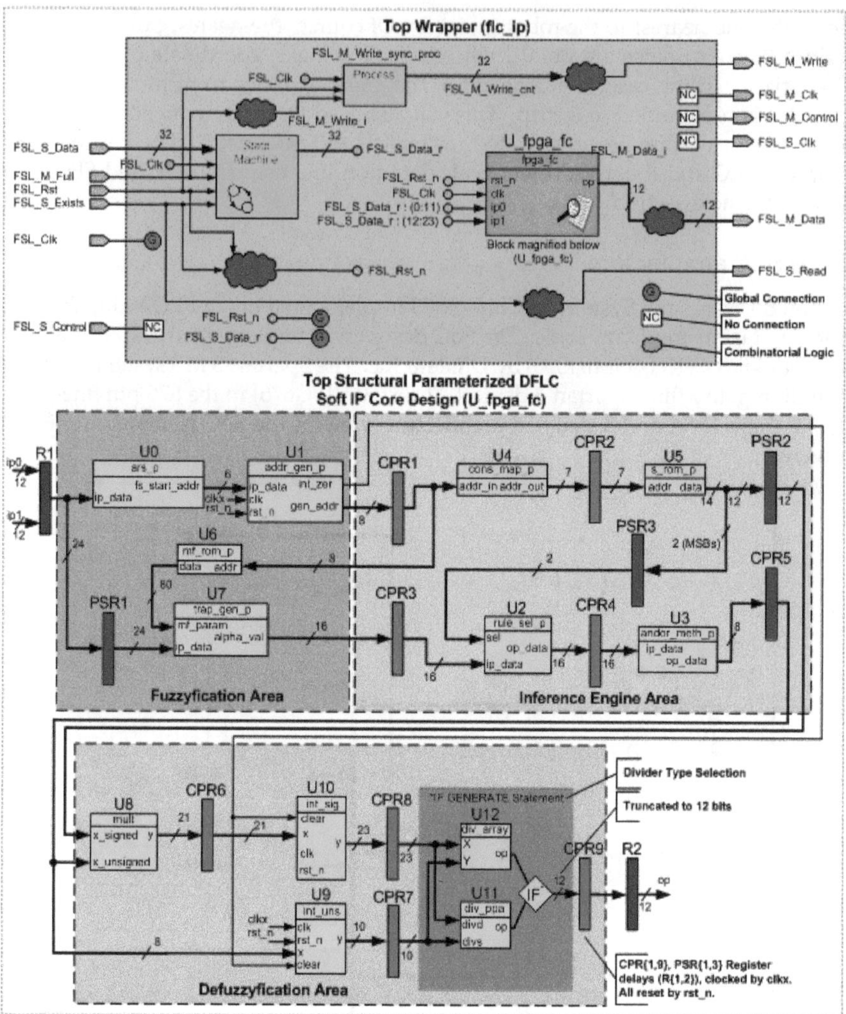

Fig. 8. Architecture of the DFLC IP core

continually at risk of time critical tasks preempting one another. FPGAs on the other hand do not use operating systems and minimize reliability concerns with true parallel execution and deterministic hardware dedicated to every task (see Fig.9).

Today's FPGAs contain hundreds of powerful DSP slices with up to 4.7 Tera-MACS throughput; 2 million logic cells with clock speeds of up to 600MHz, and up to 2.4 Tera-bps high-speed on-chip bandwidth capable to outperform DSP and RISC processors by a factor of 100 to 1,000. Taking advantage of hardware parallelism, FPGAs exceed the computing power of digital signal processors (DSPs) by breaking the paradigm of sequential execution and accomplishing more per clock cycle.

The main unit of the SoC is a parametrized Digital Fuzzy Logic Controller (DFLC) soft IP core Deliparaschos et al. (2006)Deliparaschos & Tzafestas (2006) that implements the fuzzy

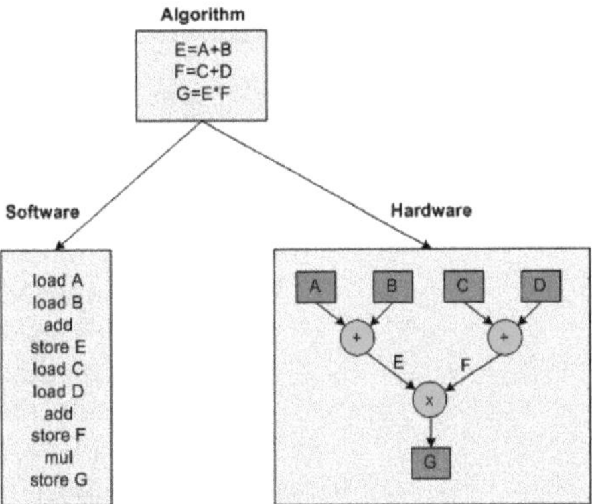

Fig. 9. Illustration of HW vs SW computation process

tracking algorithm, and a Xilinx's Microblaze soft processor core which acts as the top level flow controller. The FPGA board hosting the SoC controls the Kephera robot, used in the experiments of the tracking scheme.

In our application the DFLC facilitates scaling and can be configured for different number of inputs and outputs, number of triangular or trapezoidal fuzzy sets per input, number of singletons per output, antecedent method (t-norm, s-norm), divider type, and number of pipeline registers for the various components in the model. This parametrization enabled the creation of a generic DFLC soft IP core that was used to produce a fuzzy controller of different specifications without the need of redesigning the IP from the beginning. The fuzzy logic controller architecture assumes overlap of two fuzzy sets among adjoining fuzzy sets, and requires 2^n (n is the number of inputs) clock cycles at the core frequency speed in order to sample the input data (input sample rate of 78.2960ns), since it processes one active rule per clock cycle. In its present form the SoC design achieves a core frequency speed of 51.1 MHz. To achieve this timing result, the latency of the chip architecture involves 9 pipeline stages each one requiring 19.574ns. The featured DFLC IP is based on a simple algorithm similar to the zero-order Takagi-Sugeno inference scheme and the weighted average defuzzification method. By using the chosen parameters of Table 2, it employs two 12-bit inputs and one 12-bit output, 9 triangular membership functions (MFs) per input and 5 singleton MFs at the output with 8-bit and 12-bit degree of truth resolution respectively.

The FPGA SoC implements the autonomous control logic of the Kephera II robot. It receives odometry information from the robot and issues steering commands output by the FL tracker. The encoding and decoding of the information packets (i.e., encoding of steering control into data frames) is handled by the MATLAB application. Therefore the MATLAB application implements the actual framer/deframer for the I/O communication with the Kephera robot and downloads the tracking path to the SoC. The top-level program that supervises these tasks, treats synchronization and timing requirements, is written in C and executes in the Microblaze soft processor core. The SWAM algorithm is also implemented on the FPGA, in order to reduce the computation time.

The MATLAB application displays information about the robot's pose and speed, as well as some other data used for the path tracking control. It also calculates the robot's position relative to the world and the local coordinate frames. Another important function of the MATLAB application is to provide a path for the robot to track. The current work deals only

Parameters (VHDL generics)	Value	Generic Description
ip_no	2	Number of inputs
ip_sz	12	Input bus resolution (bits)
op_no	1	Number of outputs
op_sz	12	Output bus resolution (bits)
MF_ip_no	9	Number of input MFs (same for all inputs)
dy_ip	8	Input MFs degree of truth resolution (bits)
MF_op_no	5	Number of output MFs (singletons)
MF_op_sz	12	Output MFs resolution (bits)
sel_op	0	Antecedent method connection: 0: min, 1: prod, 2: max, 3: probor
div_type	1	Divider Model: 0: restoring array, 1: LUT reciprocal approx.
PSR		**Signal Path Route**
psr1_no	1	ip_set→psr1_no→trap_gen_p
psr2_no	4	s_rom→psr2_no→mult
psr3_no	1	s_rom→psr3_no→rul_sel_p
psr4_no	1	cpr5→psr→int_uns
CPR		**Component (Entity) Name**
cpr1_no	1	addr_gen_p
cpr2_no	1	cons_map_p
cpr3_no	3	trap_gen_p
cpr4_no	0	rule_sel_p
cpr5_no	2	minmax_p
cpr6_no	1	mult
cpr7_no	0	int_uns
cpr8_no	0	int_sig
cpr9_no	2	div_array

Table 2. DFLC soft IP core chosen parameters

with the path tracking task and not path planning. To compensate for this, the path is drawn in MATLAB, encoded properly and downloaded to the SoC. Then, the SoC begins the tracking control.

The Microblaze soft processor core is licensed as part of the Xilinx Embedded Development Kit (EDK) and is a soft core, meaning that it is implemented using general logic primitives rather than a hard dedicated block in the FPGA. The Microblaze is based on a RISC architecture which is very similar to the DLX architecture described in (Patterson & Hennessy, 1997)(Sailer et al., 1996). It features a 3-stage pipeline with most instruction completing in a single cycle. Both the instruction and data words are 32 bits. The core alone can obtain a speed of up to 100MHz on the Spartan 3 FPGA family. The Microblaze processor can connect to the OPB bus for access to a wide range of different modules, it can communicate via the LMB bus for a fast access to local memory, normally block RAM (BRAM) inside the FPGA.

Moreover, the Fast Simplex Link (FSL) offers the ability to connect user soft core IP's acting as co-processors to accelerate time critical algorithms. The FSL channels are dedicated unidirectional point-to-point data streaming interfaces. Each FSL channel provides a low latency interface to the processor pipeline allowing extending the processor's execution unit with custom soft core co-processors. In this work the DFLP IP core is playing the role of such

a co-processor and is connected to the Microblaze via the FSL bus. The architecture of the present SoC consists mainly of the DFLP that communicates with the Microblaze Processor through the Fast Simplex Bus (FSL), the utilized block RAMs (BRAM) through the LMB bus, and other peripherals such as the general purpose input/output ports (GPIO), and UART modules via the OPB bus. Here, the DFLP incorporates the fuzzy tracking algorithm, whereas the Microblaze processor mainly executes the C code for the flow control.
The parametrized zero-order TSK type Fuzzy Logic Controller (FLC) core exchanges data with the MicroBlaze processor via the FSL bus. The scope of the FLC core is to serve as high-speed fuzzy inference co-processor to the Microblaze. The DFLC core was implemented with the following parameters (see Table3).

Property	Value
Inputs	2
Input resolution	12
Outputs	1
Output resolution	12 bit
Antecedent Membership Functions (MF's)	9 Triangular MF's
Degree of Truth resolution	8 bit
Consequent MF's	5 Singletons
MF resolution	8 bit
Number of fuzzy inference rules	81
Rule activation method	MIN
Aggregation method	SUM
Implication method	PROD
MF overlapping degree	2
Defuzzification method	Weighted average

Table 3. DFLC core characteristics

Besides these two main cores and buses, the design consists of 16 KB local memory, 32 MB DDR, timer, interrupt controller, a UART, a debug peripheral (MDM) and a couple of General Purpose Inputs/Outputs (GPIOs). A Multi-CHannel (MCH) On-chip Peripheral Bus (OPB) Double Data Rate (DDR) Synchronous DRAM (SDRAM) memory controller (MCH OPB DDR with support for asynchronous DDR clock) is used in this design. This allows the MicroBlaze system to run at a lower speed of 51 MHz, which is more reasonable for Spartan-3, while the DDR is running at 75 MHz, which is the minimum required frequency for the Micron DDR chip. The on-chip Digital Clock Manager (DCM) is used to create the various clock frequencies and phases required to make this system work, all based on the 75 MHz oscillator on the 3SMB board. The FLC core runs at the same speed as the OPB and MicroBlaze, which is 51 MHz. Based on the place and route report, the design occupies 4174 out of 13312 slices of the Xilinx Spartan 3 FPGA (XC3S1500-4FG676).

6. Hardware/software co-design

On the beginning of the co-design process one starts with an architecture independent description of the system functionality. Since the description of the system functionality is independent of the HW and SW, several system modelling representations may be used, such as finite state machines (FSMs) for example. The modelled system can then be described by means of a high level language, which is next compiled into an internal representation

such as a data control flow description. This description which serves as a unified system representation allows to perform HW/SW functional partitioning. After the completion of the partitioning, the HW and SW blocks are synthesized and evaluation is then performed. If the evaluation does not meet the required objectives, another HW/SW partition is generated and evaluated (Rozenblit & Buchenrieder, 1996)(Kumar, 1995).
A general HW/SW co-design schema followed in this SoC implementation is illustrated in Fig.10).

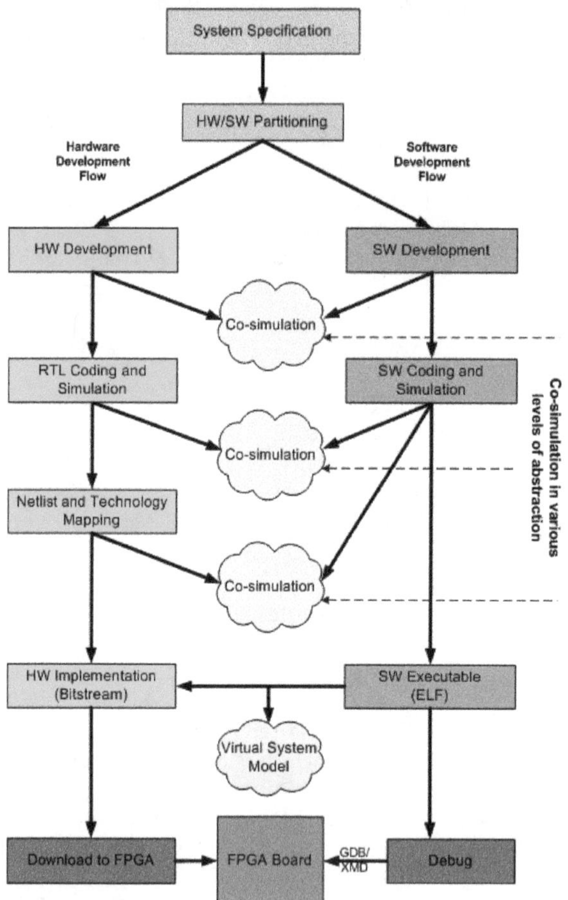

Fig. 10. HW/SW Co-design Flow

The DFLC core implementation follows a sequential design manner (see Fig.11) (Navabi, 1998). The starting point of the design process was the functional modelling of the fuzzy controller in a high level description (i.e., MATLAB/Simulink). This serves a two purpose role, first to evaluate the model and second to generate a set of test vectors for RTL and timing verification. The model was coded in register transfer level (RTL) with the use of hardware description language VHDL. Extensive use of VHDL generic and generate statements was used through out the coding of the different blocks, in order to achieve a parameterized DFLC core. The DFLC core is scalable in terms of the number of inputs/bus resolution, number of input/output fuzzy sets per input and membership resolution. More specifically A VHDL package stores the above generic parameters together with the number of necessary pipeline stages for each block. An RTL simulation was performed to ensure the correct functionality of the fuzzy controller. The DFLC core was independently synthesized with Synopsys Synplify logic synthesis tool (as it produced better synthesis results and meet timing constraints), whereas the rest of the SoC cores were synthesised with Xilinx synthesis tool XST. The Xilinx EDK studio was used for the integration flow of different SoC cores (i.e., DFLC, Microblaze, etc) and Xilinx ISE tool for the placement and routing of the SoC on the FPGA. More analytically, the place and route tool accepts the input netlist file (.edf), previously created by the synthesis tool and goes through the following steps. First, the translation program translates the input netlist together with the design constraints to a database file. After the translation program has run successfully, the logical design is mapped to the Xilinx FPGA device. Lastly, the the mapped design is placed and routed onto the chosen FPGA family and a device configuration file (bitstream) is created. Xilinx's SDK used for C programming and debugging the SoC's Microblaze soft processor. RTL and timing simulation to verify the correct functionality was handled with the use of Mentor's Modelsim simulator.

7. FPGA design and performance evaluation

The Component Pipeline Registers (CPR) blocks in Fig.8 indicate the number of pipeline stages for each component; the Path Synchronization Registers (PSR) blocks point to registers used for synchronizing the data paths, while the "U" blocks represent the different components of the DFLC Deliparaschos & Tzafestas (2006).

The U_fpga_fc component is embedded in the flc_ip top structural entity wrapper which is compliant with the FSL standard and provides all the necessary peripheral logic to the DFLC soft IP core in order to send/receive data to/from the FSL bus. The flc_ip wrapper architecture is shown in Fig.8 while the chosen (generic) parameters (VHDL package definition file) for the parameterized DFLC IP (U_fpga_fc) and its characteristics are summarized in Table 2 and Table 3 respectively.

The U_fpga_fc alone was synthesized using Synplify Pro synthesizer tool, while the rest of the design components were synthesized with Xilinx Synthesis Tool (XST) through the EDK Platform Studio. The produced .edf file for the U_fpga_fc is been seeing by the flc_ip wrapper as a blackbox during the XST flow. The placement and routing of the SoC design into the FPGA was done through the EDK by calling the Xilinx ISE tool.

According to the device utilization report from the place and route tool (see Table 4), the SoC design (including the DFLC) occupies 4,494 (16%) LUTs, 15 Block Multipliers (MULT18X18s), and 18 Block RAMs. The implemented design uses two Digital Clock Manager (DCM) Modules (DCM_0 for the system clock and DCM_1 for clocking the external DDR RAM) that produce the different clocks in the FPGA. The DFLC core itself occupies 1303 or 4% LUTs, 8 Block Multipliers, 12 64x1 ROMs (ROM64X1) and 54 256x1 ROMs (ROM256X1). The SoC achieves a minimum clock operating period of 19.574ns or a maximum frequency of $\sim$51.1 MHz respectively (the DFLC with the chosen parameters reports a frequency of 85MHz when implemented alone).

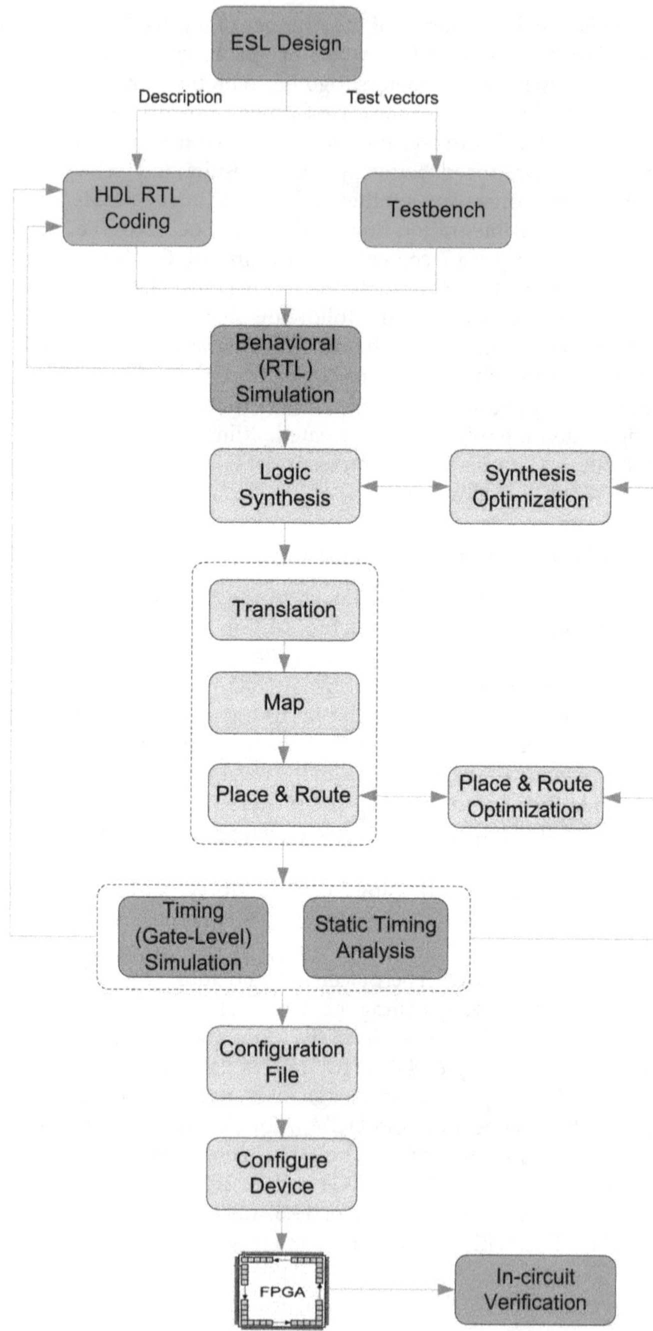

Fig. 11. HW/SW Hardware design flow

Logic Utilization		
Number of Slice Flip Flops	3,288 out of 26,624	12%
Number of 4 input LUTs	4,494 out of 26,624	16%
Logic Distribution		
Number of occupied Slices	4,174 out of 13,312	31%
Number of Slices containing only related logic	4,174 out of 4,174	100%
Number of Slices containing unrelated logic	0 out of 4,174	0%
Total Number of 4 input LUTs	5,893 out of 26,624	22%
Number used as logic	4,494	
Number used as a route-thru	165	
Number used for Dual Port RAMs	432	
(Two LUTs used per Dual Port RAM)		
Number used as 16x1 ROMs	432	
Number used as Shift registers	370	
Number of bonded IOBs	62 out of 487	12%
IOB Flip Flops	94	
IOB Dual-Data Rate Flops	23	
Number of Block RAMs	18 out of 32	56%
Number of MULT18X18s	15 out of 32	46%
Number of GCLKs	6 out of 8	75%
Number of DCMs	2 out of 4	50%
Number of BSCANs	1 out of 1	100%
Total equivalent gate count for design	1,394,323	
Additional JTAG gate count for IOBs	2,976	

Table 4. SoC design summary

8. Experimental results

The experiments consist of tracking predefined paths and analysing the displacement error. The paths are drawn by hand in the MATLAB application and downloaded to the FPGA. Then the control algorithm running on the board, calculates the steering command (curvature), relays it to MATLAB, which in turn passes it to the robot. Conversely, the MATLAB application receives odometric data from the robot which are then relayed to the FPGA. Essentially, the MATLAB application acts as an intermediary between the board and the robot, transforming commands and data to a suitable form for each party. Note also that the actual odometry is being performed by MATLAB (estimation of the robot's pose (x, y, θ) using the data from the robot's encoders). A key detail in the above process is that odometry provides an estimation of the actual pose. Thus in order to analyse the efficacy of the tracking scheme, we need to know the actual pose of the robot. Position detection of the robot is achieved using a camera hanging above the robot's activity terrain and utilizing a video tracking algorithm which extracts the robot's trajectory in post-processing. This algorithm tracks a red LED placed at the center of the robot.

The video tracking algorithm uses the high contrast of the LED with its surrounding space. Specifically, each video frame is transformed from the *RGB* color space to the generalized *rgb* space. This space expresses the percentage of each color at each pixel, i.e.,

$$
\begin{aligned}
r &= \frac{R+G+B}{R} \\
g &= \frac{R+G+B}{G} \\
b &= \frac{R+G+B}{B}
\end{aligned}
\tag{35}
$$

It is the hyper-space of the so-called *rg* chromaticity space, which consists of the first two equations. Following the transformation, in order to enhance the contrast, each image pixel is raised to the 3rd power,

$$(r', g', b')_{(u,v)} = (r^3, g^3, b^3)_{(u,v)} \tag{36}$$

The new image is then re-transformed according to the *rgb* transform, essentially computing the color percentage of the percentage. Then, we apply a thresholding on the r channel, producing a binary image. The threshold was empirically set to "0.6". This procedure produces the "patch" of pixels corresponding to the red LED. The next step is, of course, to calculate a single pixel value from this patch, and thus get the robot's position. To this end, we calculate the median (row, col) value of the patch's pixels. This algorithm is applied to the first video frame, and is rather slow since the image dimensions are large (1280×720 pixels). In order to speed up the process, the algorithm processes an image region of interest in each consecutive frame. This ROI is a 17×17 pixel square, centred at the point extracted from the previous frame. The square dimensions are appropriate since the robot is not expected to have moved far between frames. The precision of the algorithm is about ± 2 pixels, translating to 2.4 mm.

Previous to all runs, the camera was calibrated using the Camera Calibration Toolbox by J.Y. Bouguet, extracting the camera's intrinsic and extrinsic parameters. Two runs were performed; one tracking a straight line and one tracking a curved path (snapshots of the two videos are seen in Fig. 12). For the first run, the reference, odometry and camera paths are presented in Fig. 13(UP). The minimum distance versus the path length of the odometry and the camera paths are shown in Fig.13(DOWN). Likewise, for the second experiment the results are presented in Fig14.

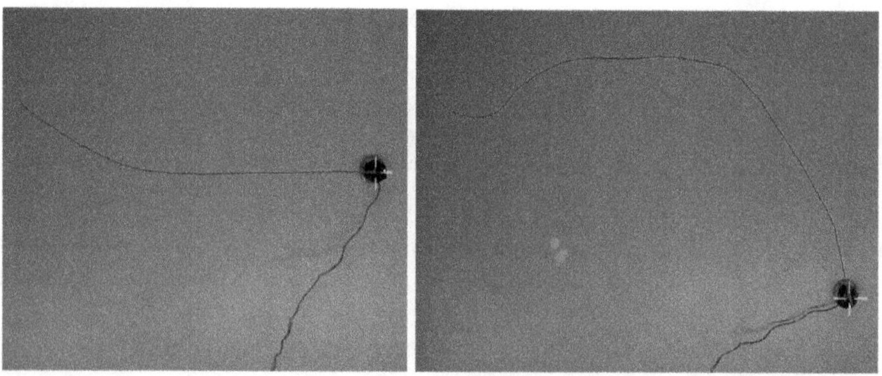

Fig. 12. Snapshots of the first (LEFT) and the second (RIGHT) experiments. The red line is the robot's path calculated off-line from the video camera.

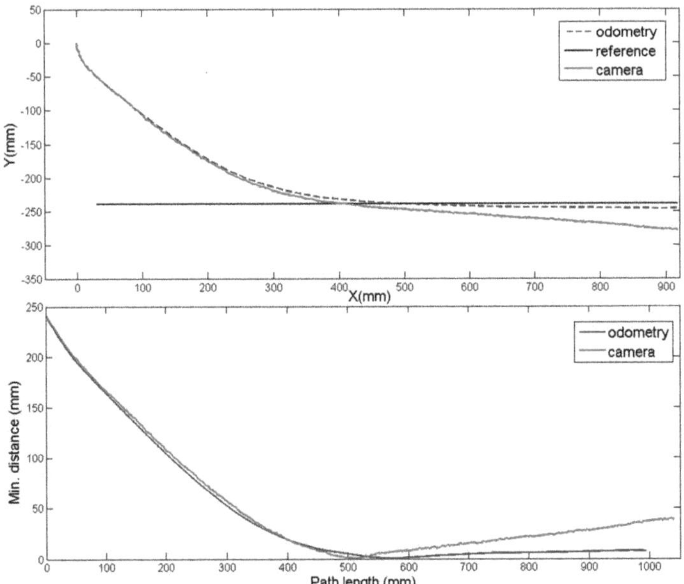

Fig. 13. (UP) The odometry (blue), camera (red) and reference (dashed black) paths for the first experiment. (DOWN) Minimum distance of the odometry and camera paths to the reference path versus path length experiment.

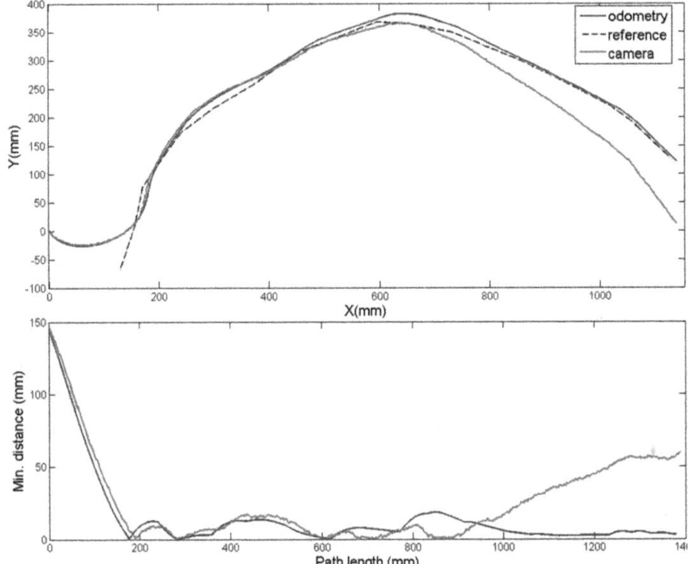

Fig. 14. (UP) The odometry (blue), camera (red) and reference (dashed black) paths for the second experiment. (DOWN) Minimum distance of the odometry and camera paths to the reference path versus path length.

As one can see, the performance of the tracking scheme is satisfactory maintaining the minimum distance to the reference path at about 50mm in the worst case. However, by taking a closer look at Figures 13 and 14, it is clear that the performance degradation is attributed not to the algorithm *per se* but to the odometry. The error accumulation of odometric data forces the robot to diverge from the actual path. But the actual odometry solution is very close to the reference path, meaning that based solely on odometry (as is the case in these experiments), the tracker maintains the robot very close to the reference path (the minimum distance is below 10mm in both cases). This implies that if a better localization technique is used, our tracking scheme would perform with more accuracy.

9. Conclusions

In this chapter we have analysed and demonstrated the applicability of the strip-wise affine transform in the path tracking task for mobile robots. The transformation was translated to hardware and implemented into an FPGA chip with the use of VHDL and advanced EDA software. The scalability of the fuzzy controller core allowed easy parameter adaptation of the theoretic fuzzy tracker model. The experiments show that the tracking scheme performs satisfactory but is degraded by the accumulation of errors of the odometry used in estimating the robots position.

10. References

Abdessemed, F., Benmahammed, K. & Monacelli, E. (2004). A fuzzy-based reactive controller for a non-holonomic mobile robot, *Robotics and Autonomous Systems* 47(1): 31–46.

Altafini, C. (1999). A Path-Tracking criterion for an LHD articulated vehicle, *The International Journal of Robotics Research* 18(5): 435–441.

Altafini, C. (2002). Following a path of varying curvature as an output regulation problem, *Automatic Control, IEEE Transactions on* 47(9): 1551–1556.

Antonelli, G., Chiaverini, S. & Fusco, G. (2007). A Fuzzy-Logic-Based approach for mobile robot path tracking, *Fuzzy Systems, IEEE Transactions on* 15(2): 211–221.

Baltes, J. & Otte, R. (1999). A fuzzy logic controller for car-like mobile robots, *Computational Intelligence in Robotics and Automation, 1999. CIRA '99. Proceedings. 1999 IEEE International Symposium on*, Monterey, CA , USA, pp. 89–94.

Cao, M. & Hall, E. L. (1998). Fuzzy logic control for an automated guided vehicle, *Intelligent Robots and Computer Vision XVII: Algorithms, Techniques, and Active Vision* 3522(1): 303–312.

Costa, A., Gloria, A. D., Giudici, F. & Olivieri, M. (1997). Fuzzy logic microcontroller, *IEEE Micro* 17(1): 66–74.

Deliparaschos, K. M., Nenedakis, F. I. & Tzafestas, S. G. (2006). Design and implementation of a fast digital fuzzy logic controller using FPGA technology, *Journal of Intelligent and Robotics Systems* 45(1): 77–96.

Deliparaschos, K. M. & Tzafestas, S. G. (2006). A parameterized T-S digital fuzzy logic processor: soft core VLSI design and FPGA implementation, *International Journal of Factory Automation, Robotics and Soft Computing* 3: 7–15.

Deliparaschos, K., Moustris, G. & Tzafestas, S. (2007). Autonomous SoC for fuzzy robot path tracking, *Proceedings of the European Control Conference 2007*, Kos, Greece.

Egerstedt, M., Hu, X. & Stotsky, A. (1998). Control of a car-like robot using a dynamic model, *Robotics and Automation, 1998. Proceedings. 1998 IEEE International Conference on*, Vol. 4, pp. 3273–3278 vol.4.

El Hajjaji, A. & Bentalba, S. (2003). Fuzzy path tracking control for automatic steering of vehicles, *Robotics and Autonomous Systems* 43(4): 203–213.

Fortuna, L., Presti, M. L., Vinci, C. & Cucuccio, A. (2003). Recent trends in fuzzy control of electrical drives: an industry point of view, *Proceedings of the 2003 International Symposium on Circuits and Systems*, Vol. 3, pp. 459–461.

Gardner, R. B. & Shadwick, W. F. (1987). Feedback equivalence of control systems, *Systems & Control Letters* 8(5): 463–465.

Gardner, R. B. & Shadwick, W. F. (1990). Feedback equivalence for general control systems, *Systems & Control Letters* 15(1): 15–23.

Groff, R. E. (2003). *Piecewise Linear Homeomorphisms for Approximation of Invertible Maps*, PhD thesis, The University of Michigan.

Gupta, H. & Wenger, R. (1997). Constructing piecewise linear homeomorphisms of simple polygons, *J. Algorithms* 22(1): 142–157.

Hung, D. L. (1995). Dedicated digital fuzzy hardware, *IEEE Micro* 15(4): 31–39.

Jiangzhou, L., Sekhavat, S. & Laugier, C. (1999). Fuzzy variable-structure control for nonholonomic vehicle path tracking, *Intelligent Transportation Systems, 1999. Proceedings. 1999 IEEE/IEEJ/JSAI International Conference on*, pp. 465–470.

Kamga, A. & Rachid, A. (1997). A simple path tracking controller for car-like mobile robots, *ECC97 Proc.*

Kanayama, Y. & Fahroo, F. (1997). A new line tracking method for nonholonomic vehicles, *Robotics and Automation, 1997. Proceedings., 1997 IEEE International Conference on*, Vol. 4, pp. 2908–2913 vol.4.

Koh, K. & Cho, H. (1994). A path tracking control system for autonomous mobile robots: an experimental investigation, *Mechatronics* 4(8): 799–820.

Kongmunvattana, A. & Chongstivatana, P. (1998). A FPGA-based behavioral control system for a mobile robot, *Circuits and Systems, 1998. IEEE APCCAS 1998. The 1998 IEEE Asia-Pacific Conference on*, pp. 759–762.

Kumar, S. (1995). *A unified representation for hardware/software codesign*, PhD thesis, University of Virginia. UMI Order No. GAX96-00485.

Lee, T., Lam, H., Leung, F. & Tam, P. (2003). A practical fuzzy logic controller for the path tracking of wheeled mobile robots, *Control Systems Magazine, IEEE* 23(2): 60– 65.

Leong, P. & Tsoi, K. (2005). Field programmable gate array technology for robotics applications, *Robotics and Biomimetics (ROBIO). 2005 IEEE International Conference on*, pp. 295–298.

Li, T., Chang, S. & Chen, Y. (2003). Implementation of human-like driving skills by autonomous fuzzy behavior control on an FPGA-based car-like mobile robot, *Industrial Electronics, IEEE Transactions on* 50(5): 867– 880.

Liu, K. & Lewis, F. (1994). Fuzzy logic-based navigation controller for an autonomous mobile robot, *Systems, Man, and Cybernetics, 1994. 'Humans, Information and Technology'., 1994 IEEE International Conference on*, Vol. 2, pp. 1782–1789 vol.2.

Maalouf, E., Saad, M. & Saliah, H. (2006). A higher level path tracking controller for a four-wheel differentially steered mobile robot, *Robotics and Autonomous Systems* 54(1): 23–33.

Moustris, G. P. & Tzafestas, S. G. (2011). Switching fuzzy tracking control for mobile robots under curvature constraints, *Control Engineering Practice* 19(1): 45–53.

Moustris, G. & Tzafestas, S. (2005). A robust fuzzy logic path tracker for non-holonomic mobile robots., *International Journal on Artificial Intelligence Tools* 14(6): 935–966.

Moustris, G. & Tzafestas, S. (2008). Reducing a class of polygonal path tracking to straight line tracking via nonlinear strip-wise affine transformation, *Mathematics and Computers in Simulation* 79(2): 133–148.

Navabi, Z. (1998). *VHDL: analysis and modeling of digital systems*, McGraw-Hill Professional.

Ollero, A., Garcia-Cerezo, A., Martinez, J. L. & Mandow, A. (1997). Fuzzy tracking methods for mobile robots, *in* M. Jamshidi, A. Titli, L. Zadeh & S. Boverie (eds), *Applications of fuzzy logic: Towards high machine intelligence quotient systems*, Prentice-Hall, New Jersey.

Patterson, D. A. & Hennessy, J. L. (1997). *Computer Organization and Design: The Hardware/Software Interface*, 2 edn, Morgan Kaufmann.

Preparata, F. P. & Supowit, K. J. (1981). Testing a simple polygon for monotonicity, *Info. Proc. Lett.* 12(4): 161–164.

Raimondi, F. & Ciancimino, L. (2008). Intelligent neuro-fuzzy dynamic path following for car-like vehicle, *Advanced Motion Control, 2008. AMC '08. 10th IEEE International Workshop on*, pp. 744–750.

Reynolds, R., Smith, P., Bell, L. & Keller, H. (2001). The design of mars lander cameras for mars pathfinder, mars surveyor '98 and mars surveyor '01, *Instrumentation and Measurement, IEEE Transactions on* 50(1): 63–71.

Rodriguez-Castano, A., Heredia, G. & Ollero, A. (2000). Fuzzy path tracking and position estimation of autonomous vehicles using differential GPS, *Mathware Soft Comput* 7(3): 257–264.

Rozenblit, J. & Buchenrieder, K. (1996). *Codesign: Computer-aided Software/Hardware Engineering*, I.E.E.E.Press.

Sailer, P. M., Sailer, P. M. & Kaeli, D. R. (1996). *The DLX Instruction Set Architecture Handbook*, 1st edn, Morgan Kaufmann Publishers Inc.

Salapura, V. (2000). A fuzzy RISC processor, *IEEE Transactions on Fuzzy Systems* 8(6): 781–790.

Samson, C. (1995). Control of chained systems application to path following and time-varying point-stabilization of mobile robots, *Automatic Control, IEEE Transactions on* 40(1): 64–77.

Sanchez, O., Ollero, A. & Heredia, G. (1997). Adaptive fuzzy control for automatic path tracking of outdoor mobile robots. application to romeo 3R, *Fuzzy Systems, 1997., Proceedings of the Sixth IEEE International Conference on*, Vol. 1, pp. 593–599 vol.1.

Wit, J., Crane, C. D. & Armstrong, D. (2004). Autonomous ground vehicle path tracking, *J. Robot. Syst.* 21(8): 439–449.

Yang, X., He, K., Guo, M. & Zhang, B. (1998). An intelligent predictive control approach to path tracking problem of autonomous mobile robot, *Systems, Man, and Cybernetics, 1998. 1998 IEEE International Conference on*, Vol. 4, pp. 3301–3306 vol.4.

Permissions

The contributors of this book come from diverse backgrounds, making this book a truly international effort. This book will bring forth new frontiers with its revolutionizing research information and detailed analysis of the nascent developments around the world.

We would like to thank Andon V. Topalov, for lending his expertise to make the book truly unique. He has played a crucial role in the development of this book. Without his invaluable contribution this book wouldn't have been possible. He has made vital efforts to compile up to date information on the varied aspects of this subject to make this book a valuable addition to the collection of many professionals and students.

This book was conceptualized with the vision of imparting up-to-date information and advanced data in this field. To ensure the same, a matchless editorial board was set up. Every individual on the board went through rigorous rounds of assessment to prove their worth. After which they invested a large part of their time researching and compiling the most relevant data for our readers. Conferences and sessions were held from time to time between the editorial board and the contributing authors to present the data in the most comprehensible form. The editorial team has worked tirelessly to provide valuable and valid information to help people across the globe.

Every chapter published in this book has been scrutinized by our experts. Their significance has been extensively debated. The topics covered herein carry significant findings which will fuel the growth of the discipline. They may even be implemented as practical applications or may be referred to as a beginning point for another development.

The editorial board has been involved in producing this book since its inception. They have spent rigorous hours researching and exploring the diverse topics which have resulted in the successful publishing of this book. They have passed on their knowledge of decades through this book. To expedite this challenging task, the publisher supported the team at every step. A small team of assistant editors was also appointed to further simplify the editing procedure and attain best results for the readers.

Our editorial team has been hand-picked from every corner of the world. Their multi-ethnicity adds dynamic inputs to the discussions which result in innovative outcomes. These outcomes are then further discussed with the researchers and contributors who give their valuable feedback and opinion regarding the same. The feedback is then collaborated with the researches and they are edited in a comprehensive manner to aid the understanding of the subject.

Apart from the editorial board, the designing team has also invested a significant amount of their time in understanding the subject and creating the most relevant covers. They scrutinized every image to scout for the most suitable representation of the subject and create an appropriate cover for the book.

The publishing team has been involved in this book since its early stages. They were actively engaged in every process, be it collecting the data, connecting with the contributors or procuring relevant information. The team has been an ardent support to the editorial, designing and production team. Their endless efforts to recruit the best for this project, has resulted in the accomplishment of this book. They are a veteran in the field of academics and their pool of knowledge is as vast as their experience in printing. Their expertise and guidance has proved useful at every step. Their uncompromising quality standards have made this book an exceptional effort. Their encouragement from time to time has been an inspiration for everyone.

The publisher and the editorial board hope that this book will prove to be a valuable piece of knowledge for researchers, students, practitioners and scholars across the globe.

List of Contributors

Amur S. Al Yahmedi and Muhammed A. Fatmi
Sultan Qaboos University, Oman

Srđan T. Mitrovic´
Defense University, Military Academy, Serbia

Željko M. Đurovic´
University of Belgrade, School of Electrical Engineering, Serbia

Mattias Wahde, David Sandberg and Krister Wolff
Department of Applied Mechanics, Chalmers University of Technology, Göteborg, Sweden

Eugene Kagan and Irad Ben-Gal
Tel-Aviv University, Israel

Seung-Hun Kim
Korea Electronics Technology Institute, Korea

Leonardo Leottau
Advanced Mining Technology Center (AMTC), Department of Electrical Engineering, University of Chile, Santiago, Chile

Miguel Melgarejo
Laboratory for Automation, Microelectronics and Computational Intelligence (LAMIC), Faculty of Engineering, Universidad Distrital Francisco José de Caldas, Bogotá D.C., Colombia

Krzysztof Okarma and Piotr Lech
West Pomeranian University of Technology, Szczecin, Poland

Luis F. Lupián and Josué R. Rabadán-Martin
Mobile Robotics & Automated Systems Lab, Universidad La Salle, Mexico

KiSung You, HwangRyol Ryu and Chintae Choi
Research Institute of Industrial Science and Technology, Korea

G.P. Moustris and S.G. Tzafestas
National Technical University of Athens, Greece

K.M. Deliparaschos
Cyprus University of Technology, Cyprus

Printed in the USA
CPSIA information can be obtained
at www.ICGtesting.com
JSHW011811301024
72690JS00002B/45

9 781632 403520